Shallow Subterranean Habitats

T0177892

Shallow Subterranean Habitats

Ecology, Evolution, and Conservation

David C. Culver

Department of Environmental Science, American University, Washington, DC, USA

Tanja Pipan

Karst Research Institute, Research Centre of the Slovenian Academy of Sciences and Arts, Postojna, Slovenia

OXFORD
UNIVERSITY PRESS

Shallow Subterranean Habitats. David C. Culver and Tanja Pipan.
© David C. Culver and Tanja Pipan 2014. Published 2014 by Oxford University Press.

OXFORD
UNIVERSITY PRESS

Great Clarendon Street, Oxford, OX2 6DP,
United Kingdom

Oxford University Press is a department of the University of Oxford.
It furthers the University's objective of excellence in research, scholarship,
and education by publishing worldwide. Oxford is a registered trade mark of
Oxford University Press in the UK and in certain other countries

© David C. Culver and Tanja Pipan 2014

The moral rights of the authors have been asserted

First Edition published in 2014

Impression: 1

All rights reserved. No part of this publication may be reproduced, stored in
a retrieval system, or transmitted, in any form or by any means, without the
prior permission in writing of Oxford University Press, or as expressly permitted
by law, by licence or under terms agreed with the appropriate reprographics
rights organization. Enquiries concerning reproduction outside the scope of the
above should be sent to the Rights Department, Oxford University Press, at the
address above

You must not circulate this work in any other form
and you must impose this same condition on any acquirer

Published in the United States of America by Oxford University Press
198 Madison Avenue, New York, NY 10016, United States of America

British Library Cataloguing in Publication Data
Data available

Library of Congress Control Number: 2013956986

ISBN 978–0–19–964617–3

Printed in Great Britain by
Clays Ltd, St Ives plc

Links to third party websites are provided by Oxford in good faith and
for information only. Oxford disclaims any responsibility for the materials
contained in any third party website referenced in this work.

Preface

In 2009, we wrote an introductory text, *The Biology of Caves and Other Subterranean Habitats*, designed to be a 'first book' on subterranean biology for students and others interested in the topic. As we were writing the book, we became aware that the 'other subterranean habitats' part of the topic was relatively undeveloped and poorly understood. Much of what we know about subterranean biology comes from the study of caves, in part because of the adventure and excitement of visiting and exploring caves, certainly more exciting than visiting, for example, talus slopes. And part of the reason that caves have played an important role in understanding subterranean life is that they are more accessible than other subterranean habitats, which we cannot directly explore and must sample indirectly. Nevertheless, as other speleobiologists before us, we became aware that caves were only a small part of the picture. One of us (TP) began her study of subterranean biology with a non-cave habitat—epikarst—albeit one that was sampled by collecting dripping water from the ceilings of caves. Epikarst water in the Slovenian caves that she studied contained a treasure trove of eyeless, depigmented species, especially copepods. On the other side of the Atlantic, one of us (DC) was beginning to study the eyeless and depigmented 'cave' animals that had been reported from minute seepage springs in Washington, DC, nearly a hundred kilometres from any known cave. As we worked together, first on epikarst and then on seepage springs (which have the wonderful technical name of the hypotelminorheic), we began to see important connections between these very different habitats in terms of some basic physical and biological features.

This led us to a more general exploration of shallow subterranean habitats, ones that were intermediate in many parameters between surface and deeper subsurface habitats, save that of light. There is a rich tradition of study, particularly in Europe, of another shallow subterranean aquatic habitat—the underflow of rivers and streams. These hyporheic habitats also harbour eyeless and depigmented species, albeit typically smaller in size than those found in caves. We also became aware of the remarkable discoveries our Australian colleagues were making in the shallow calcrete aquifers of Western Australia, habitats that may rival caves in terms of species richness. From our study of these aquatic habitats, we became convinced that these less extreme yet aphotic aquatic habitats were key to understanding aquatic subterranean life.

Although our research experience was largely with aquatic subterranean systems, it became clear to us that parallel habitats and processes occurred in terrestrial shallow subterranean habitats, and that these shallow habitats included lava tubes and the soil, as well as intermediate-sized habitats, including talus slopes and lava clinker. We were very fortunate in being able to visit a variety of these sites, visits that were crucial to the development of our ideas about shallow subterranean habitats, which are the focus of this book.

Unlike our first book, we see this book as directed towards researchers and students interested in research in subterranean biology. Our first book was in a way a consensus view of subterranean biology, and we did not emphasize the controversies of the field. This book is different, and we have put forward our views, even when it is perhaps a minority view. The book has two main parts—the first is a detailed description of shallow subterranean habitats, and the second is an exploration of the biological consequences of the existence of these habitats.

It is designed to stimulate research and discussion, and we hope it serves that purpose. It is designed to be a starting point for research, not a 'bible' of information.

A number of colleagues read chapters and offered useful criticisms: Florian Malard, Université Lyon I, France (chapter 1.); Daniel Fong, American University, USA (chapter 2.); Maria Cristina Bruno, Fondazione E. Mach Research and Innovation Centre, Trento, Italy (chapter 3.); Pedro Oromí, Universidad de La Laguna, Tenerife, Spain (chapters 4 and 8); William F. Humphreys, Western Australia Museum (chapter 5.); Remko Leijs, South Australia Museum (chapter 5.); Michelle Guzik, University of Adelaide, Australia (chapter 5.); Kym Abrams, University of Adelaide, Australia (chapter 5.); Pierre Marmonier, Université Lyon I, France (chapter 6.); Louis Deharveng, Museum National d'Histoire Naturelle, Paris, France (chapter 7.); Anne Bedos, Museum National d'Histoire Naturelle, Paris, France (chapter 7.); Cene Fišer, Univerza v Ljubljani, Slovenia (chapters 9 and 12); Vlastimil Růžička, Czech Academy of Sciences, České Budějovice, Czech Republic (chapter 10.); Sanja Gottstein, University of Zagreb, Croatia (chapter 10.); Kevin S. Simon, University of Auckland, New Zealand (chapter 11.); Peter Trontelj, Univerza v Ljubljani, Slovenia (chapters 13 and 14); Maja Zagmajster, Univerza v Ljubljani, Slovenia (chapter 15.); and William Jeffery, University of Maryland, USA (chapter 16.).

This book would not have been possible without a number of researchers taking time to show us their research sites and help us understand their systems: Marie-José Dole-Olivier (France), Charles Gers (France), Heriberto López (Canary Islands), Pedro Oromí (Canary Islands), Pierre Marmonier (France), Slavko Polak (Slovenia), Ana Sofia Reboleira (Portugal), Vlastimil Růžička (Czech Republic), Fred Stone (Hawaii), and Miloslav Zacharda (Czech Republic).

A number of colleagues gave us comments, answers to questions, preprints, reprints etc.: Anne Bedos, Gloria J. Chepko, Louis Deharveng, Leon Drame, David Eme, Cene Fišer, Daniel W. Fong, William F. Humphreys, William R. Jeffery, William K. Jones, Leonardo Latella, Florian Malard, Pierre Marmonier, Janez Mulec, Diana Northup, Tone Novak, Pedro Oromí, Slavko Polak, Megan Porter, Ana Sofia Reboleira, Vlastimil Růžička, Trevor Shaw, Kevin S. Simon, Tadej Slabe, Stanka Šebela, Peter Trontelj, Maja Zagmajster, and Nadja Zupan Hajna.

Jure Hajna of the Karst Research Institute at ZRC SAZU provided professional help with the illustrations. Florita Gunasekara and Kristina Hsu of American University helped draft the maps. All reasonable effort has been made to contact the holders of copyright in materials reproduced in this book. Any omissions will be rectified in future printings if notice is given to the publishers.

Financial support came from the College of Arts and Sciences of American University to DCC, Karst Research Institute at ZRC SAZU (Slovenia) to DCC and TP, and Slovenian Research Agency to DCC and TP.

A project of this magnitude was a burden on both of our families, and we are especially grateful to our spouses, Gloria Chepko and Miran Pipan, for providing both understanding and support.

Postojna, Slovenia
May, 2014

Contents

Glossary

α-diversity Local within-site diversity (species richness); in contrast to β-diversity (among sites diversity).

A'a lava Jagged, irregular volcanic rock formed from the solidification of highly viscous lavas.

Adaptation A process of genetic change resulting in improvement of a character with reference to a specific function, or feature that has become prevalent in a population because of a selective advantage.

Adaptive Shift Hypothesis (ASH) Hypothesis, originally due to Howarth, that species actively colonized subterranean habitats to exploit new resources, and that subsequent speciation was sympatric.

Adit A roughly horizontal passage introduced into a mine for the purpose of access or drainage.

Aeolian Windblown or wind-driven.

Albinism The complete loss of melanin pigment.

Allochthonous Originating elsewhere; in this context it relates to organic matter that has been carried into a habitat from another one.

Allopatric speciation Speciation after geographical isolation of a population.

Allozyme Different forms of an enzyme, differing slightly in amino acid sequence, which are the products of alternative alleles at a given locus.

Alluvial Sediment deposited by a flowing river.

Anchialine (or anchihaline) Subterranean habitats, with more or less extensive connections to the sea, and showing noticeable marine as well as terrestrial influences.

Anophthalmic Lacking eyes.

ANOVA (Analysis of Variance) A procedure for analysing the statistical significance of differences between the mean scores of two or more groups on one or more independent variables, involving resolution of the total variance of the set of observations into components associated with different sources of variation.

Aphotic The absence of light.

Aquiclude Body of relatively impermeable rock or sediment acting as a boundary to an aquifer.

Aquifer A groundwater reservoir, usually a rock of high permeability capable of delivering water to a well.

Aquitard Geological formation of a rather impervious and semi-confining nature that transmits water at a very slow rate compared with an aquifer.

Archaea One of the three domains of life. They are similar to bacteria in lacking a cell nucleus and similar to Eukaryota in the method of genetic transcription and translation.

Autotrophy The production of organic substances (food) using light or inorganic substances as an energy source.

Baermann funnel A modification of the Berlese funnel, used to force nematodes out of soil by filling the funnel with warm water, driving the nematodes into a vessel below.

Berlese funnel A device for extracting invertebrates from soil or leaf litter. A funnel contains the soil or litter, and a heat source such as an electric lamp heats the litter. Animals escaping from the desiccation of the litter descend through a filter into a preservative liquid in a receptacle.

Biocentric Regarding life in general (rather than just human life) as a central fact of the universe.

Bioclimatic model A model developed by How-arth including the 'tropical winter' effect to explain the restriction of troglobionts to deep cave zones and small subsurface voids (mesocaverns) with constant high humidity.

Biofilm A coating on rocks and other surfaces composed of microorganisms, extracellular polysaccharides, other materials that the organisms produce, and particles trapped or precipitated within the biofilm.

Bou-Rouch pump A special hand pump designed to collect water samples from shallow interstitial aquifers.

Buccal Of or relating to the mouth cavity.

Calcrete aquifer Carbonate deposits that form in the vicinity of the water table as a result of the evaporation of groundwater. Known from western Australia.

Caliche A hardened deposit of calcium carbonate. Caliche occurs worldwide, generally in arid or semiarid regions.

Canonical Correspondence Analysis (CCA) A multivariate method to elucidate the relationship between biological assemblages of species and their environment by extracting synthetic environmental gradients from datasets, and visualizing species preferences in an ordination diagram.

Canonical Variate Analysis (CVA) A multivariate method that examines the interrelationships between a number of groups, similar to Principal Components Analysis, but the axes are chosen to maximize the separation between the groups.

Chao estimates A formula to estimate total species richness based on the abundance or number of species occurring in one and two samples.

Chelicerae Mouthparts of the Chelicerata, including the arachnids. Some are hollow and contain venom glands.

Chemoautotrophy Ability of an organism to obtain nourishment from chemical reactions involving inorganic substances, such as the oxidation of sulphides to sulphates, as contrasted to using the sun's energy as with green plant photoautotrophs.

Circadian An endogenously driven, roughly 24-hour cycle in biochemical, physiological, or behavioural processes.

Clade A species or population and all its descendants.

Cladistic Pertaining to branching patterns; a cladistic classification classifies organisms on the basis of the historical sequences by which they diverged from common ancestors.

Clay A stiff, viscous earth found, in many varieties, in beds or other deposits near the surface of the ground and at various depths below it: with water it forms a tenacious paste.

Climate Relict Hypothesis (CRH) Hypothesis that species became isolated in subterranean habitats because of unfavourable surface conditions which caused the extirpation of the surface population. Speciation is allopatric.

Clinker The surface of a'a lava is covered with a layer of partly loose, very irregular fragments commonly called clinker, formed as the lava cools.

Clypeus In insects, the clypeus delimits the lower margin of the face, with the labrum articulated along the ventral margin.

Coastal plain An area of flat, low-lying land adjacent to a seacoast and separated from the interior by other features.

Coefficient of variation (CV) The ratio of the standard deviation to the mean, usually multiplied by 100 to give a percentage. It is a measure of variation independent of the size of the mean.

Colluvial Loose bodies of sediment that have been deposited or built up at the bottom of a low-grade slope, transported by gravity.

Conductivity Measure of the electrical conductance per unit distance in an aqueous solution, and a measurement of the ionic content.

Conservation easement A power invested in a qualified private land conservation organization (often called a 'land trust') or government (municipal, county, state or federal) to constrain, as to a specified land area, the exercise of rights otherwise held by a landowner so as to achieve certain conservation purposes. The conservation easement 'runs with the land,' meaning it is applicable to both present and future owners of the land.

Convergence The independent evolution of similar traits in two or more lineages.

Cryptic species Genetically distinct species with no discernable morphological differences.

Cryptozoa All eyeless and depigmented species fall into the cryptozoa, which occur in aphotic and in some cases dimly lit habitats, as defined by Peck.

Cyclicity The quality of recurring at regular intervals.

Delmarva Peninsula The peninsula between Chesapeake Bay and the Atlantic Ocean, in the eastern United States, including parts of the states of Delaware, Maryland, and Virginia, hence the name.

Denitrification A process by which oxidized forms of nitrogen, e.g. nitrate (NO_3^-), are transformed by denitrifying bacteria to form nitrites (NO_2^-), nitrogen oxides (NO_x), ammonia (NH_3), or nitrogen (N_2).

Detritivore Species that feed on dead plants and animals and their waste products.

Disjunct distribution A species range with one or more gaps.

Dissolution The process by which a rock or mineral dissolves (usually in water).

Distributary A stream that branches off and flows away from a main stream channel.

DOC Dissolved organic carbon.

Dolina Simple closed circular depression with subterranean drainage, and commonly funnel-shaped.

Downwelling Vertical movement of stream water into the hyporheic and phreatic zones.

Dry-channel hyporheic Hyporheic zone beneath a dry stream channel.

Ecosystem services Goods and services to human populations provided by ecosystems that would otherwise have to be accomplished in some other way.

Ecotone A zone of transition between adjacent ecological systems, having a set of characteristics uniquely defined by space and time scales, and by the strength of the interactions between adjacent ecological systems.

Endemism The ecological state of being unique to a defined geographical location, such as an island, nation, or other defined zone, or habitat type.

Edaphic Living in the soil.

Edaphobiont An obligate soil-dwelling species.

Edaphomorphy (euedaphomorphy) Morphological characteristics associated with life in the soil, especially miniaturization, appendage reduction, and body thinning.

Edaphophile A species that completes its entire life cycle in the soil but can also complete its entire life cycle elsewhere.

Edaphoxene A species that is regularly found in soil but does not complete its life cycle there.

Endogean Beneath the surface of the ground, usually taken to mean the soil.

Endopod (endopodite) The inner ramus of a biramous limb of a crustacean.

Epigean Living on or near the surface of the ground, as opposed to the subsurface (subterranean, hypogean) environment.

Epikarst The boundary region between soil and rock in karst, usually honeycombed with small fractures, solution pockets, and solutionally widened trenches.

Epikarst endemic Species limited to epikarst.

Estuary A partly enclosed coastal body of brackish water with one or more rivers or streams flowing into it, and with a free connection to the open sea.

Eutrophic cave Cave with a large amount of available organic matter/energy, especially that provided by guano or debris taken in by water (especially during flooding).

Exaptation Adaptation for one function serving for another function.

Exopod (exopodite) The outer ramus of a biramous limb of a crustacean.

Extinction vortex A combination of factors (genetic inbreeding, environmental stochasticity, and small population size) that lead to population or species extinction.

Flank-margin cave A cave formed in young sea-coast limestone in the freshwater–seawater mixing zone, along the flanks of a freshwater lens.

Fluvial Of or pertaining to rivers.

Flysch Sequence of interbedded shales and sandstones deposited contemporaneously with mountain building.

Freezing core technique A technique for obtaining *in situ* samples of hyporheic, using liquid nitrogen to freeze the substrate around a standpipe driven into the hyporheic.

F_{ST} A measure of population differentiation due to genetic structure. It is frequently estimated from genetic polymorphism data.

Furcula The 'spring-tail' of Collembola that allows them to jump.

Gondwana The southern supercontinent formed following the fragmentation of Pangaea in the Mesozoic era, about 150 million years ago, largely comprising the present South America, Africa, India, Madagascar, Australia, New Zealand, and Antarctica.

Granulometry The measurement of the size distribution in a collection of sediments, especially from the soil.

Groundwater That portion of the Earth's water that is held within the pore spaces (openings) in geologic formations including water in the saturated (the pores are completely filled with water) and in the unsaturated (the pores are only partially filled with water) zones.

Guano A mixture of faeces and urine deposited by bats that serves as an important food resource for other cave organisms, or accumulation of faeces of particular animals in time and space, especially of bats, birds (oilbirds and swiftlets), and crickets.

Guanobionts (guanobites) Species that exclusively inhabit guano deposits in caves, and whose entire biological cycle takes place in this substrate. Some authors include species found outside of caves but only on guano in caves (e.g. Gnaspini).

Guanophiles Species that may live and reproduce both in guano piles and in other substrates in the cave environment.

Guanoxenes Species that, when in caves, may be found feeding and/or reproducing on guano deposits but depend on other substrate(s) in the caves to complete their biological cycle.

Haplotype A mitochondrial genotype, which is haploid.

Hard pan See caliche.

Hardness (**Of water**) the concentration of multivalent cations.

Helocrene spring Spring where water seeps out of the ground slowly and is usually temporarily confined to small holes or ditches.

Heterozygosity In genetics the frequency of individuals that carry two different genes at the same locus.

Hjulstrom curve A graph used to determine whether a river will erode, transport, or deposit sediment.

Holocene A geological epoch that began 11,700 years ago and continues to the present.

Holotype A holotype is a single physical example (or illustration) of an organism, known to have been used when the species (or lower-ranked taxon) was formally described.

Homoplasy In cladistics, a homoplasy or a homoplastic character state is a trait (genetic, morphological, etc.) that is shared by two or more taxa because of convergence, parallelism, or reversal.

Horse latitudes Subtropical latitudes between 30 and 35 degrees both north and south. This region, under a ridge of high pressure called the subtropical high, is an area which receives little precipitation and has variable winds mixed with calm.

Hotspot An area of relatively high number of species or high number of endemics.

Humus Any organic matter that has reached a point of stability where it will break down no further and might, if conditions do not change, last for centuries.

Hydraulic conductivity The ease with which a fluid (usually water) can move through pore spaces or fractures. It depends on the intrinsic permeability of the material, the degree of saturation, and on the density and viscosity of the fluid.

Hydraulic head The sum of energy that a given volume of fluid has resulting from various combinations of pressure, motion, and elevation, which provides the basis for understanding fluid movement.

Hydraulic gradient The drop in hydraulic head divided by the distance of water flow. It is equivalent to the rate of energy loss within a hydrological system.

Hydraulic potential The pressure of groundwater flow, usually directly related to the gradient of the water table.

Hypogean The subsurface or subterranean environment as opposed to the surface (epigean) environment.

Hypogenic (cave) An origin deep within the earth's surface.

Hyporheic (hyporheos) Interstitial spaces within the sediments of a stream bed; a transition zone between surface water and groundwater.

Hyporheic refuge hypothesis (HRH) Hypothesis that the hyporheic could serve as a refuge for the invertebrate inhabitants of surface streams from environmental stresses, including flooding and drying.

Hypotelminorheic A persistent wet spot, a kind of perched aquifer; fed by subsurface water in a slight depression in an area of low to moderate slope; rich in organic matter; underlain by a clay layer typically 5–50 cm beneath the surface; with a drainage area typically of less than 10,000 m^2; and with a characteristic dark colour derived from decaying leaves which are usually not skeletonized.

Infiltration Flow of water through the soil surface into a porous medium.

Interstitial Subsurface habitats with small spaces, such as gravel aquifers, compared to the large spaces of caves.

Interstitial highway The hypothesis that interstitial habitats, especially along rivers and streams, are a corridor for subterranean dispersal.

Isotropy Uniformity in all orientations.

Jaccard index A measure of overlap of two communities $[= c/(a + b - c)]$ where a and b are the number of species in the two communities, and c is the number of shared species.

Jack-knife estimates Systematically recomputing the statistical estimate leaving out one or more observations at a time from the sample set. From this new set of replicates of the statistic, an estimate for the bias and an estimate for the variance of the statistic can be calculated.

Karst Landscape in soluble rock where solution rather than erosion is the primary geomorphic agent, typically with caves, sinkholes, and springs.

K selection Selection for life-history characteristics that increase fitness in stable environments where populations reach densities near environmental carrying capacity.

Lampenflora Proliferation of phototrophic organisms near artificial light sources, especially in caves.

Lapidocolous Living under a stone.

Laterite Soil types rich in iron and aluminium, formed in hot and wet tropical areas. Nearly all laterites are rusty-red because of iron oxides.

Limnocrene spring Spring where water comes out of the ground and creates a pond at the source, before flowing out slowly.

Macaronesia Collective name for several groups of islands in the North Atlantic Ocean off the coast of Europe and Africa, including the Azores, Canary, Madeira, and Cape Verde Islands.

Macrocavern Underground voids >20cm, especially caves and lava tubes.

Macropores A pore in soil of such size that water drains from it by gravity and is not held by capillary action. They are typically created by burrowing activities of animals and plant roots.

Mao Tau estimates An analytical estimate of the accumulation of species numbers with increasing samples, based on expectations for an infinite number of randomizations.

Meiofauna Assemblage of animals that pass through a 500 μm sieve but are retained by a 40 μm sieve.

Mesocavern Cavities smaller than caves, between 0.1 and 20 cm in diameter.

Mesonotum The dorsal portion of the mesothorax of insects, the middle of three divisions of the thorax.

Methanogenesis Process where Archaea oxidize hydrogen and reduce CO_2 to methane in chemoautotrophy.

Microbial mats A multi-layered sheet of micro-organisms, mainly bacteria and Archaea. Microbial mats grow at interfaces between different types of material, mostly on submerged or moist surfaces.

Microcavern Cavities smaller than mesocaverns, less than 1 mm in diameter.

Microphthalmic Possessing tiny, but complete eyes.

Milieu souterrain superficiel (MSS) Interconnected cracks and crevices in scree slopes and similar habitats, including those covered with soil or moss according to some authors.

Monophyletic Having arisen from one ancestral form; in the strictest sense, from one initial population.

Moonmilk A white, plastic calcareous cave deposit composed of calcite, huntite, or magnesite.

MSS See **milieu souterrain superficiel**.

Mycorhizae (also miccorhizae) The symbiotic association between fungi and the roots of vascular plants.

NAPL Non-aqueous phase liquid.

Neutral mutation A genetic mutation that has no advantage or disadvantage to the organism.

Niche The total requirements of a population or species for resources and physical conditions.

Nitrification The production of nitrates.

Nutrient spiralling The cycling between inorganic and organic phases of a nutrient, and the distance along a stream reach that this requires.

Occasional hyporheic Species that could spend part of their life cycle (typically nymphal stages) in the hyporheic.

Organicism According to Vandel, lineages undergo three stages: birth, specialization, and senescence. He hypothesized that all subterranean species were in the senescent phase.

Pahoehoe lava Volcanic rock with a smooth, ropy surface formed from the solidification of fluid lavas.

Parafluvial The area of the bankfull channel that is to some extent annually scoured by flooding, and is thus lateral to the normal stream channel.

Parapatric Pertaining to species or populations that have contiguous but non-overlapping geographical distributions.

Paratype Each specimen of the type series other than the holotype.

Parsimony A particular non-parametric statistical method for constructing phylogenies. In this application, the preferred phylogenetic tree is the tree that supposes the least evolutionary change to explain observed data (maximum parsimony).

Peat bog A mire that accumulates peat, a deposit of dead plant material—often mosses, and in a majority of cases, *Sphagnum* moss.

Pedogenic The processes of soil development and evolution.

Pegmatite A very crystalline, intrusive igneous rock composed of interlocking crystals usually larger than 2.5 cm in size.

Percolating water Water moving vertically from epikarst through the unsaturated zone.

Peripheral isolates Populations on the periphery of the range of a species that may differentiate as a result of selection and genetic drift.

Phreatic Groundwater, usually implying permanent groundwater.

Phreatobiology Coined by Orghidan, it is the study of biology of groundwater.

Phylogeography The study of the historical processes that may be responsible for the contemporary geographical distributions of individuals.

Piedmont A region of Italy, literally meaning the foot of the mountain. Also, a region in the United States immediately east of the Blue Ridge Mountains.

Pit A near-vertical cave passage. Pits may open to the surface or may be located completely inside the cave.

Playa A desert basin with no outlet which periodically fills with water to form a temporary lake.

POC Particulate organic carbon.

Porosity Ratio of the volume of the interstices in a given sample of a porous medium, e.g. soil, to the gross volume of the porous medium, inclusive of voids.

Pre-adaptation The possession of a morphological, behavioural, or physiological character that enhances an organism's ability to survive or exploit a novel situation.

Principal Components Analysis (PCA) A mathematical procedure that uses an orthogonal transformation to convert a set of observations of possibly correlated variables into a set of values of linearly uncorrelated variables called principal components. The number of principal components is less than or equal to the number of original variables.

Progenesis Acceleration of sexual maturation relative to the rest of development.

Quantile Each of a set of values of a variate which divide a frequency distribution into a certain number of equal groups, each group containing the same fraction of the total population.

Rarity Being uncommon geographically, numerically, or because the habitat is uncommon.

Recharge That part of precipitation or surface water that penetrates the Earth's surface and eventually reaches the water table.

Redox A reversible reaction in which one compound is oxidized and another reduced.

Refugium (plural **refugia**) A region or habitat in which certain organisms are able to persist during a period in which most of their original geographical range becomes uninhabitable because of climatic change.

Regolith Unconsolidated solid material covering bedrock.

Regressive evolution The loss of morphological and behavioural characters that accompanies isolation in caves.

Resurgence Spring where a stream, which has a course on the surface higher up, reappears at the surface.

Riffle In a stream course, areas of shallower, faster-moving water often associated with whitewater. Alternates with pools.

Riparian Pertaining to the banks of a river or stream.

Schiner-Racoviţă system Ecological classification of subterranean animals into troglobionts (troglobites), troglophiles, trogloxenes, and accidentals, based on ecological criteria.

Schist A crystalline rock whose component minerals are arranged in a more or less parallel manner.

Scree A collection of broken rock fragments at the base of crags, mountain cliffs, volcanoes, or valley shoulders that has accumulated through periodic rockfall from adjacent cliff faces.

Seep (seepage spring) A small spring where water oozes out of the ground. Often associated with hypotelminorheic habitats.

Shallow subterranean habitat (SSH) Subterranean habitats within 10 m of the surface, characterized by spaces larger than the organisms, and organisms adapted to darkness.

Source populations Populations in a metapopulation with positive growth rate in the absence of immigrants.

Spatial subsidy When resources from one system (e.g. surface) are transferred to another adjoining system (e.g. SSHs).

Speleobiology The branch of biology dealing with subterranean organisms and their habitats.

Speleothanatic zone The uppermost zone of karst, one of rapid destruction, according to Šušteršič.

Spring Water emerging from the earth.

SSH See **shallow subterranean habitat**.

Standardized coefficients Variable converted to one with a mean of zero and a standard deviation of one.

Stoichiometry The relative quantities of reactants and products in chemical reactions.

Straminicolous Living in leaf litter.

Stridulum Structure on the forewing of some Orthoptera, which allows production of sound.

Stygobiont Obligate, permanent resident of aquatic subterranean habitats.

Stygophile A species that completes its entire life cycle in subterranean waters but can also complete its entire life cycle in surface waters.

Stygoxene Organisms appearing sporadically in groundwater systems, called accidentals by some authors.

Superficial subterranean habitat See shallow subterranean habitat.

Sympatric speciation Speciation without geographical isolation, as a result of genetic differentiation within a population, and resulting in species with overlapping distributions.

Synapomorphy A trait that is shared ('symmorphy') by two or more taxa and inferred to have been present in their most recent common ancestor, whose own ancestor in turn is inferred not to possess the trait.

Talus Piles of loose boulders that accumulate at the bottoms of cliffs and other steep slopes. See also **scree**.

TOC Total organic carbon.

Troglobiont Obligate, permanent resident of terrestrial subterranean habitats; used by some authors for aquatic species as well (see **stygobiont**).

Troglobionts in training The hypothesis that troglophiles and non-troglomorphic troglobionts are phylogenetically young inhabitants of subterranean habitats, and given time, will evolve into troglobionts.

Troglomorphic Pertaining to morphological and behavioural characters that are convergent in subterranean populations.

Troglophile Species able to live and reproduce in subterranean habitats as well as in the epigean domain.

Trogloxene Species appearing sporadically in subterranean habitats; called accidentals by some authors.

Tympanum Hearing organ of insects.

Upwelling Vertical movement of hyporheic and phreatic water up into the stream.

Vadose The zone above the water table in which water moves by gravity and capillarity. Water does not fill all the openings and does not build up pressures greater than atmospheric.

Vernal pool Temporary pools of water that provide habitat for distinctive plants and animals. They are considered to be a distinctive type of wetland usually devoid of fish, and thus allow the safe development of natal amphibian and insect species unable to withstand competition or predation by fish.

Vicariance Isolation (often followed by speciation) as a result of range disruption, typically the result of some non-biological process.

Wet campo Marsh type in the cerrado region of Brazil.

Wetland A land area that is saturated with water, either permanently or seasonally, such that it takes on the characteristics of a distinct ecosystem. Primarily, the factor that distinguishes wetlands from other land forms or water bodies is the characteristic vegetation that is adapted to its unique soil conditions.

Würm The last glacial period in Europe, approximately 110,000–10,000 years ago in the late Pleistocene.

Zeitgeber An external clock setter for circadian clocks.

Gazetteer

Major locations for SSHs, by habitat. Because of scaling, some sites are lumped together. Not shown are Shihua Cave in China (epikarst), lava tubes in Hawaii, lava tubes in the Galapagos Islands, and calcrete aquifers in western Australia.

Hypotelminorheic habitats

H1—Mid-Atlantic, USA, including Rock Creek, George Washington Memorial Parkway, and Prince William Forest Park
H2—Medvenica Mountain, Croatia
H3—Nanos Mountain, Slovenia

Epikarst habitats

E4—Natural Bridge Cavern, Texas, USA
E5—Slovenian caves, including Huda luknja, Postojnska jama, Pivka jama, Črna jama, Škocjanske jame, Snežna jama na planini Arto, Županova jama, Dimnice, and Jama v Kovačiji (terrestrial)
E6—Organ Cave, West Virginia, USA
E7—Romanian caves, including Peştera Ciur Izbuc, Peştera cu Apă din Valea Leşului, Peştera Doboş, Peştera Ungurului, and Peştera Vadu Crişului
E8—Grotte du Cormoran, France
E9—White Scar Cave, England

Intermediate-sized terrestrial habitats (MSS)

M10—Slovenia, including Mašun and Jama v Kovačiji

M11—Canary Islands, including Teno (Tenerife), La Guancha (Tenerife), Sima de la Perdiz (Tenerife), and La Gomera
M12—Czech Republic, including Plešivec u Jincŭ, Kamenec Hill, Lovoš Hill, Kammená Hŭra, and Klíč
M13—Pyrenees Mountains, France, including Massif de l'Arize, Tour Laffont, and La Ballongue

Interstitial habitats along rivers and streams (hyporheic)

I14—Roseg River, Switzerland
I15—Flathead River, Montana, USA
I16—Sycamore Creek, Arizona, USA
I17—Lachein Creek and Baget basin, Midi-Pyrenees, France
I18—Rhône Alpes, France, including Morcilles, Méant, and Rhône River

Soil

S19—Pyrenees Mountains, France, Spain, and Andorra, including Sainte-Cécile, Spain
S20—Hranice na Moravě, Czech Republic
S21—Belgium, including Epraves, Han, and Nou Maulin
S22—Wales

Lava tubes

L23—Tenerife, Canary Islands, including Cueva del Mulo, Cueva del Viento, and Cueva del Felipe Reventón
L24—Azores Islands, including Furna dos Montanheiros (Pico)

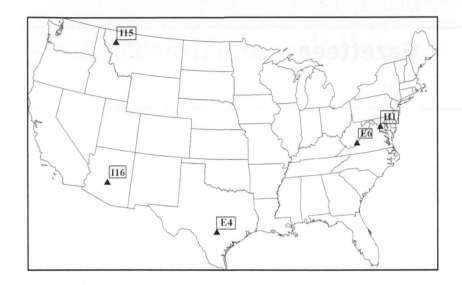

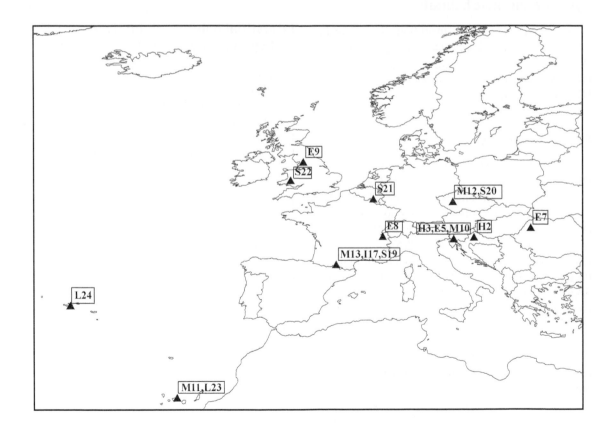

CHAPTER 1

The shallow subterranean domain

1.1 Introduction

The subterranean, or subsurface, realm has long been distinguished from the surface, or epigean realm. A remarkably large number of words have been employed to identify the subterranean realm, including hypogean, endogean, stygian, as well as the more colloquial subterranean. While these words have somewhat different meanings, e.g. endogean often refers to the soil, they are all still in use and often used interchangeably. Furthermore, the subterranean realm is often taken to mean simply caves, i.e. cavities enterable by humans and at least in part in total darkness. For us, the key distinguishing feature of the two realms is that of light. The subterranean realm is aphotic, and the surface realm is photic.

Authors such as Botosaneanu (1986) distinguished two main types of subterranean habitat, in this case limiting attention to aquatic habitats—large cavity habitats, especially caves, and small cavity habitats, especially the habitats between gravels or even grains of sand, collectively referred to as interstitial habitats. Other authors, such as Juberthie (2000) emphasize the dichotomy between karst and non-karst habitats. Karst refers both to a region of southwest Slovenia and northeast Italy and to a landscape where the geomorphological processes of solution dominate over those of erosion. *De facto*, this comes to a very similar distinction to that of Botosaneanu (1986) because large-scale cavities are typically limited to karst, with the notable exception of some volcanic terrains (Palmer 2007). There are yet other meanings of the word subterranean, including cavities created by organisms such as moles and mole rats.

Many, but not all of the organisms that are found only in small cavity or large cavity subterranean habitats are without eyes or pigment (Pipan and Culver 2012a). Organisms found in large cavities also tend to have elongated appendages, large size, and elaborated extra-optic sensory structures. Christansen (1962, 2012) used the term troglomorphy to describe this set of convergent traits of organisms living in large cavities. Organisms found in small cavities tend to have reduced appendages and reduced size, an adaptation for movement through tight spaces. Aquatic species limited to subterranean habitats, whether or not they show morphological modifications, are called stygobionts (sometimes aquatic troglobionts), and terrestrial species, with the exception of soil dwellers, limited to subterranean habitats, whether or not they show morphological modifications, are called troglobionts. Facultative subterranean dwellers, species which have populations both on the surface and in subterranean habitats, are given the names stygophiles and troglophiles. Species limited to the soil are given the name edaphobionts.

The utility of these terms has been questioned (Romero 2009) even in the face of a continuing refinement and elaboration of these and related terms (Sket 2008). For the time being we use these terms because they are a convenient shorthand to describe both the morphological features (troglomorphy) and distributional features (troglobiotic, etc.) of subterranean species. We return to the question of the usefulness of these terms in chapters 9 and 12.

The dichotomy of subterranean habitats, whether it is based on habitat size or the presence or absence of solutional processes, is one that is in many ways misleading. The reason that biologists found and continue to find speleobiology interesting is almost

Shallow Subterranean Habitats. David C. Culver and Tanja Pipan.
© David C. Culver and Tanja Pipan 2014. Published 2014 by Oxford University Press.

always related to the morphological distinctiveness of subterranean animals, especially the loss of eyes and pigment and the hypertrophy of extra-optic sensory structures. This interest includes the evolutionary and developmental questions, such as eye loss in cavefish (e.g. Jeffery 2009, Yamamoto et al. 2009), and ecological and biogeographical questions, such as the comparison of numbers and patterns of such eyeless and blind species (e.g. Culver et al. 2006a, Pipan and Culver 2012a). We believe that it is the absence of light that is the defining characteristic of subterranean environments, especially if we are interested in all those environments, large or small, karst and non-karst, where eye and pigment loss occurs.

The characteristic morphology of subterranean-dwelling animals is not limited to eye and pigment loss but also involves a suite of other changes, such as appendage elongation, elaboration of extra-optic sensory structures, and decreased metabolic rate in cave animals (Christiansen 2012), and miniaturization in interstitial animals (Coineau 2000). Many authors, dating back until at least the 1960s (Christiansen 1961, Poulson 1963), have invoked the morphology of subterranean animals as strong examples of convergent evolution, driven by the selective forces of darkness, resource scarcity, and environmental uniformity.

The large-small dichotomy of Botosaneanu (1986) is inadequate for several reasons. First and foremost it fails to take into account the range of subterranean habitats, especially those with eyeless and pigmentless species. There are a number of aphotic habitats that have features and commonalities not captured by the large–small dichotomy— ones that we collectively call shallow (superficial) subterranean habitats, or SSHs (Culver and Pipan 2008, 2009, 2011, Pipan and Culver 2012b, Pipan et al. 2011a). These are subterranean habitats very close to the surface (we use the arbitrary cutoff of 10 m), but are aphotic. Some of these habitats are large cavities close to the surface, especially lava tubes, while others are small cavity habitats, especially the underflow of streams and rivers, and the soil.

But, there is an especially interesting set of SSHs that do not fit into either category. These habitats include talus and scree slopes, in both carbonate

(soluble) and non-carbonate rocks, including volcanic rocks. The name *milieu souterrain superficiel* (MSS), originally used by Juberthie et al. (1980b) for erosional features in non-calcareous rock has been extended to include the habitats listed above. Whatever name we give it, it is an SSH with intermediate-sized space with many close connections with the surface. *Epikarst*, the uppermost layer of karst formed largely by solutional processes (Gabrovšek 2004) that may at any given time be air- or water-filled, occupies a similar vertical position to that of the MSS, but perhaps with smaller spaces. The most superficial of SSHs are the miniature perched aquifers (isolated wetlands) given the name *hypotelminorheic* by Meštrov (1962) that exit through seepage springs, diffuse discharges when the flow cannot be immediately observed but the land surface is wet compared to the surrounding area (Kresic 2010). Finally, *calcrete aquifers* are rather bizarre shallow aquifers formed under arid conditions by evaporation along a length of river draining into a salt lake. In some areas of Western Australia, the only region where they have been studied, they are always less than 10 m deep (Yilgarn) but in others they are typically deeper (Pilbara).

These four SSHs (MSS, epikarst, hypotelminorheic, and calcrete aquifers), which do not extend beyond a few metres in depth and which have intermediate-sized habitats we call strict sense shallow subterranean habitats. Soil and shallow interstitial aquifers have small habitat spaces, while lava tubes, although rarely extending below 10 m, are much larger habitats. These three habitats plus the four strict sense SSHs, collectively constitute broad sense shallow subterranean habitats.

There are other subterranean habitats that fall within the zone of SSHs. All caves, except those with artificial entrances, have at least a section near the entrance that is a shallow subterranean habitat. However, all karst areas have caves that extend well below the 10 m cutoff, and have close connections with underlying groundwater (phreatic habitats). Thus, karst caves are not distinct from deeper subterranean habitats in the same way broad sense SSHs are not distinct from strict sense SSHs, and we will not consider karst caves further. In the case of lava tubes, the entire cave is typically within 10 m or less of the surface (Palmer 2007).

What habitats lie below SSHs? In many areas there are aquifers which typically are an important source of water for human activity. Nearly all aquifers, even ones more than 1 km deep have at least microbial activity (Frederickson et al. 1989), and nematodes have been found in a 3-km-deep fracture aquifer intersected by gold mines in South Africa (Borgonie et al. 2011). In karst regions, which cover about 15 per cent of earth's surface (Jones and White 2012), caves extend well below 10 m. In regions of extensive cave development, such as the Dinaric karst in Slovenia, densities of cave entrances average one per km^2 but reach more than 14/km^2 in a 25 km^2 area north of Planinsko polje (J. Hajna, pers. comm.). During construction of a motorway between Klanec and Črni Kal, 67 caves were intersected in a 6.5-km stretch (Knez and Slabe 2007). The lateral extent of caves varies from a few metres to nearly 600 km in the case of Mammoth Cave, Kentucky, USA. Caves extend vertically to a depth of more than 2 km, in Krubera (Voronja) Cave in Abkhazia (Klimchouk 2012). Average densities, lengths, and depths vary greatly from region to region. For example, Slovenia and Abkhazia have many deep caves while England and eastern USA have many long caves but few deep caves. Finally, there are regions, especially non-karst regions along ocean margins with few or no deep subterranean habitats. The eastern USA Coastal Plain is an example of an area with SSHs but no deep subterranean habitats (Culver et al. 2012a).

1.2 Shallow subterranean habitats

1.2.1 Hypotelminorheic habitats

Meštrov (1962) gave the name hypotelminorheos to shallow groundwater habitats, miniature drainage basins that were vertically isolated from the water table and were constituted of humid soils in the mountains, rich in organic matter and traversed by moving water. Based on Meštrov's definition and his sketch of the habitat, Culver et al. (2006b) proposed that the term hypotelminorheic be used to describe habitats with the following major features:

1. A perched aquifer fed by subsurface water that creates a persistent wet spot
2. Underlain by a clay or other impermeable layer typically 5–50 cm below the surface
3. Rich in organic matter compared with other aquatic subterranean habitats.

The water exits at a seepage spring (Kresic 2010).

The densest known concentration of hypotelminorheic habitats is along the lower Potomac River drainage in the vicinity of Washington, DC, USA. There are several reasons for this. One is that this area has been intensively sampled, and the initial work on seepage springs in the 1970s was the result of collecting efforts by Roman Kenk (1977), a taxonomist at the US National Museum. Another is that conditions appear to be very favourable for the existence of hypotelminorheic habitats and seepage springs, including the existence of a layer of clay at a depth of 1–2 m in much of the area. A seepage spring is shown in Fig. 1.1 and is in a shallow depression in a forested area along Scott's Run, a tributary of the Potomac River. As with most seepage springs in the region, it is in a US parkland, and is a typical example of the topographic setting for these habitats. Unlike the site Meštrov described in Croatia, this is not in an area of high elevational relief, and the total relief from the Potomac River to the highest point is no more than 50 m. No direct measurements have been made of the discharge but it is unlikely it ever reaches even 10 cm^3 per second (Culver et al. 2012a). For most of the year, stygobionts can be found in the leaf litter layer, which appears as a darkened area in Fig. 1.1. During the summer months when temperatures are higher and evapotranspiration is high, few if any denizens of the hypotelminorheic can be found in seeps. In typical years flow from the seepage spring drops to zero during the summer months, but the site remains wet down to the clay layer, which is less than 1 m from the surface in the seep in Fig. 1.1. Beneath the clay layer, the substrate is dry. In this particular seepage spring, stygobiotic (and troglomorphic) snails (*Fontigens bottimeri*), isopods (*Caecidotea kenki*), and amphipods (*Stygobromus tenuis potomacus*) are found, along with a few individuals of non-specialist species of amphipods and isopods.

Figure 1.1 The authors at a seepage spring at Scott's Run Park, near Washington, DC, USA. Photo by W.K. Jones, with permission. (see Plate 1)

1.2.2 Epikarst

Epikarst is the uppermost layer of karst, the skin of karst (Bakalowicz 2004). It is a more or less permanently saturated zone with a considerable volume of water close under the surface, and is the reason that cave streams rarely dry up. Epikarst aquifers are even important as a source of water for trees growing over epikarst (Huang et al. 2011). A principal characteristic of epikarst is its heterogeneity with many semi-isolated solution pockets whose water chemistry is also variable (Pipan 2005a, Kogovšek 2010).

Epikarst biology has been most thoroughly studied in central Slovenia, especially the Postojna Planina Cave system (Pipan 2005a, Pipan and Culver 2007a, b, Kogovšek 2010). A photograph of a drip from the system is shown in Fig. 1.2. The diversity of organisms in epikarst is remarkable. Up to seven species of stygobiotic copepods have been found in a drip in the Pivka jama section of the Postojna Planina Cave system. Overall, in Pivka jama, eight stygobiotic and three generalist copepod species have been found. In addition, stygobiotic Bathynella and Amphipoda have been found, as well as a number of generalists, including Oligochaeta, Nematoda, Acari, and Ostracoda

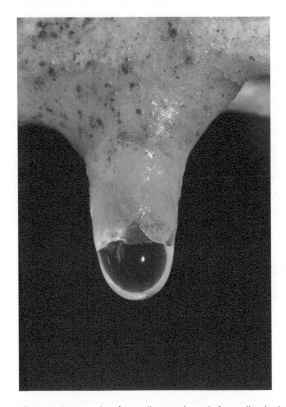

Figure 1.2 Water drop from epikarst at the end of a small stalactite in Postojna Planina cave system, Slovenia. Photo by J. Hajna, with permission. (see Plate 2)

(Pipan 2005a). Temperature of the water in the drips is relatively constant, varying only one or two degrees (Kogovšek 2010), but discharge rates can vary more than one hundred fold. The epikarst habitat itself is highly heterogeneous with respect to water chemistry and residence time of the water, as well as physical structure.

1.2.3 Milieu souterrain superficiel

The study of terrestrial SSHs—the milieu souterrain superficiel (MSS) of Juberthie et al. (1980b)—began with non-calcareous moss-covered talus and scree slopes in the Pyrenees (e.g. Crouau-Roy 1987, Gers 1992) and was expanded to similar habitats in Europe, Japan, and China (Juberthie and Decu 1994). Medina and Oromí (1990) extended the habitat to include volcanic terrains, especially in the Canary Islands. Pipan et al. (2011) included karst regions in their study. All of these regions were mountainous or at least in areas of considerable topographic relief. In non-volcanic terrains, MSS habitats result from erosion and debris flow. The origin of such habitats is more complex in volcanic terrains, and can be the result not only of debris flows and erosion, but may be depositional, that is, cracks and crevices formed as the lava cooled. The nature of MSS habitats varied considerably from region to region (see chapter 4) and a general operational definition was the presence of troglobiotic species.

The Mašun region, a flank of Snežnik Mountain, has extensive MSS developed from erosion of exposed limestone (Fig. 1.3), in a mature *Fagus–Acer* forest (S. Polak, pers. comm.). As is typical of MSS sites, there is a clear annual temperature cycle, with a range of 17 °C at a depth of 10 cm to 12 °C at a depth of 50 cm (Pipan et al. 2011a). Winter temperatures reach below freezing in the shallow depths. In the shallower depths, there is even a daily temperature cycle. The site is covered mostly by moss. Two troglobiotic species of beetles (*Bathysciotes khevenhuelleri* and *Anophthalmus schmidti*) and one millipede (genus undetermined and probably new) are known from the Mašun MSS, as well as several generalist species.

1.2.4 Calcrete aquifers

Calcrete is a rock type created by the cementing of sand or gravel by calcium carbonate. Typically associated with arid climates, it can form in a variety of ways and geological contexts (Reeves 1976), and can extend to varying depths. In the Yilgarn region of Western Australia, it forms shallow aquifers, less than 10 m in depth (Fig. 1.4). Their formation in the Yilgarn is directly associated with concentration processes by groundwater evaporation, rather than with soil development (Humphreys 2001). They expand upwards being older near the surface and consequentially they are a superficial deposit overlying clays in the underlying palaeovalley. It is

Figure 1.3 Photo of exposed MSS in a *Fagus* forest at Mašun, Slovenia. Photo by T. Pipan. (see Plate 3)

Figure 1.4 Calcrete aquifers in southwestern Australia. **A**. Disused and uncapped water extraction bores (wells) in Lake Austen calcrete aquifer formerly supplying water to Big Bell gold mine near Cue, Western Australia. **B**. Deliberate exposure of groundwater in calcrete aquifer to provide watering point for stock. Ngalia Basin, Northern Territory. Photos by W. Humphreys, with permission of the photographer and the Western Australian Museum. (see Plate 4)

notable that many Yilgarn calcrete aquifers have highly salinity-stratified water columns being hypersaline within several metres of the water surface (Humphreys et al. 2009). Water in Yilgarn calcrete aquifers is well oxygenated and exposed to open spaces, which may help to explain the very high diversity of aquatic beetles in the family Dytiscidae. The Yilgarn extends for many hundreds of kilometres, and individual calcrete aquifers can be 100 km long and 5 km wide (Humphreys 2001). Overall there are over 210 stygobiotic species in the Yilgarn, many of them Dytiscidae (Humphreys 2008). Many chemical and physical aspects of the calcrete aquifers are surprising and poorly understood: hydraulic conductivity is low, and salinity

is highly variable. Genetic analysis (mtDNA) indicates little movement among individual animals even over distances of less than 1 km (Bradford et al. 2010). Sources of organic carbon are unclear but are likely from percolating water, roots, or possibly chemoautotrophy, although there is no evidence for this.

1.2.5 Aquatic interstitial—hyporheic

Shallow aquatic interstitial habitats are comprised of water-filled spaces between grains of unconsolidated sediments (Fig. 1.5). These habitats occur in littoral sea bottoms and beaches, freshwater lake bottoms, river beds, and the hyporheic zone (the porous aquifer beneath and lateral to streams). The hyporheic zone, the best-studied of all interstitial habitats, is the surface–subsurface hydrological exchange zone beneath and alongside the channels of rivers and streams. The connection between the hyporheic and permanent groundwater (phreatic water) can be very direct or without any direct connection at all. In the case of direct connections between the hyporheic and permanent groundwater, fauna showing adaptation to subterranean life is often found. When there are unconsolidated sediments along the stream bank, the hyporheic can extend tens of metres from the stream bank.

Formed by a meander arm, the Lobau wetlands, an alluvial aquifer, are part of the floodplain of the Danube River near Vienna, Austria, and comprise the Danube Flood Plain National Park. This UNESCO Biosphere Reserve, with an area of 0.8 km^2, has been extensively sampled for decades (Pospisil 1994, Danielopol et al. 2000, Danielopol et al. 2001, Danielopol and Pospisil 2001). A small 900 m^2 area of this flood plain, called 'Lobau C' (Pospisil 1994), was monitored and sampled intensively. Loosely packed gravel, alternating with a thin layer of finer sediments, extends from 4 to 8 m beneath a thin soil cover. Both oxygen and dissolved organic matter concentrations are very heterogenous, even at scales of 1 m or less. Animals were found throughout the depth of gravel, but were most common 0.5 m beneath the surface, and rare below 2 m. In Lobau C, 27 species were found, 11 of them stygobionts.

Figure 1.5 Photo of Lachein Creek, France, one of the first sites where the hyporheic was studied (Rouch 1991). Pipes into the stream are sites for sampling chemical and physical parameters as well as pump sites for the fauna. In the photo are Charles Gers, Raymond Rouch, and Thomas C. Kane. Photo by D. Culver. (see Plate 5)

1.2.6 Soil

Soil forms a thin mantle over the earth's surface and acts as the interface between the atmosphere and the lithosphere. It is a multiphase system consisting of mineral material, plant roots, water and gases, and organic matter at various stages of decay (Fig. 1.6). The soil also provides a medium in which an outstanding variety of organisms live (Bardgett 2005). Typically soils reach a depth of between 20 and 75 cm. The terrestrial subterranean equivalent of shallow aquatic interstitial habitats is the soil.

For more than 100 years, soil has been a curious exclusion from the list of subterranean habitats (e.g. Racoviţă 1907, Sket 2004). The exclusion is partly the result of different sets of biologists being interested in caves and the soil (though see Coiffait 1958), but also because the morphology of cave and soil organisms is so different, at least on a superficial level. Soil animals are miniaturized with short appendages, and cave animals are often, relatively speaking, phyletic giants with long appendages. Nonetheless, they typically share a lack of eyes and pigment.

Figure 1.6 Photo of a matrix of soil, rocks, and roots in the Pyrenees in Ariège, France, one of the sites studied by Gers (1992). The predominant habitat is soil, but the MSS and MSS fauna are present. Photo by T. Pipan. (see Plate 6)

A detailed study of biodiversity in the soil was that of Coiffait (1958) for a series of 100 sites mostly in southern France. Among these were 20 sites in the central Pyrenees, in the same region as the pioneering studies on MSS by Juberthie (1983), Gers (1992), and Crouau-Roy (1987, see chapter 4). At a site at 200 m above sea level (asl) in Andorra, Coiffait (1958) found that the daily temperature range in July was 5–40 °C at the surface of the soil and only 10–13 °C at a depth of 10 cm. He also indicated that no light penetrates to this depth. Although he did not record annual temperature changes at this site, he did report a range of 7–16 °C at a depth of 30 cm. Coiffait's (1958) study emphasized beetles and he found a total of 52 species of beetles in 12 families. He used an ecological classification scheme parallel to that used by speleobiologists—edaphobionts (obligate soil dwellers), edaphophiles (facultative soil dwellers), and edaphoxenes (accidental soil dwellers). Edaphobionts comprised 40 per cent, edaphophiles 53 per cent, and edaphoxenes four per cent. The remainder were species associated with ant colonies. Interestingly, he found no troglobionts. Coleoptera themselves were a minor component of the soil fauna, representing only 3.2 per cent of the individuals extracted by Berlese funnels, and Collembola and Acarina predominated.

1.2.7 Lava tubes

Caves in volcanic rock are often called lava tubes. Their origin is entirely different from those in soluble rock. Most lava tubes are byproducts of volcanic processes and caves are the same age as the rock, unlike karst caves, where caves form well after rock deposition (Palmer 2007). Lava tubes are most common in pahoehoe basalt flows, which has a smooth surface with ropy wrinkles. Lava tubes are formed by the outflow of fluid lava beneath a hardened and cooler crust. Unlike solution caves which may take millions of years to form, lava tubes form almost instantaneously. Furthermore, lava tubes are quite transient, rapidly being eroded or covered by new lava flows. Even Kazumura Cave in Hawaii[1], with over 65 km of passages, a vertical extent of more than 1100 m, and 101 entrances, is only 350–500 years old (Allred 2012). Typically lava tubes are quite shallow, often less than 5 m below the surface. The roots of plants on the surface often penetrate the ceiling of the lava tube. The root fauna is both unique and diverse (Stone et al. 2012).

La Cueva del Viento (Fig. 1.7) on the island of Tenerife in the Canary Islands is more than 17 km long (Martín Esquivel and Izquierdo Zamora 1999), most

Figure 1.7 Photo of typical passage in La Cueva del Viento on the island of Tenerife, Canary Islands, Spain. Photo by J. S. Socorro Hernández, with permission. (see Plate 7)

---

[1] We follow convention and use Hawaii for the state and Hawai'i for the big island.

of which is between 2.5 and 7 m from the surface (P. Oromí pers. comm). Hourly temperatures, taken over the course of a year, ranged between 7.5 °C and 18.6 °C at the entrance but only between 13.0 °C and 15.0 °C at a site 100 m in the cave. Small piles of soil, occasional dead small mammals, and tree roots are the main sources of organic matter. Tree roots typically occur along the walls of the lava tube, where plant hoppers of the genus *Tachycixius* are the primary consumers. Oromí (1995) reports a total of 80 species in the cave, with 24 trogloxenes, 24 troglophiles, and 24 troglobionts, as well as six accidentals and two of indeterminate association. This makes La Cueva del Viento one of the most diverse subterranean sites in the world with respect to number of troglobionts (Culver and Pipan 2013). Only four other caves, including one other in the Canary Islands, are known to harbour more troglobionts. The cave is especially rich in spiders, with nine troglobionts, four troglophiles, and three trogloxenes.

1.3 General features of shallow subterranean habitats

1.3.1 Absence of light

These oddball and widely disparate habitats are united by several features. First and foremost, they are aphotic. For many of these habitats the absence of light is obvious. This is the case for epikarst, calcrete aquifers, and lava tubes. In other habitats, such as MSS habitats, hyporheic, hypotelminorheic, and even the soil, there is not always a clear line defining the absence of light. However, habitats with stygobionts and troglobionts appear to be aphotic. A light-recording datalogger in a seepage spring at a depth of less than 5 cm in the substrate on Nanos Mountain, Slovenia, where stygobionts were present (see Culver et al. 2006b), indicated that no light penetrated the habitat for a period of 173 days, from 5 June 2010 to 24 November 2010. Another light-recording datalogger placed at a depth of 10 cm in a MSS site on Mašun, Slovenia where troglobionts are present (see Pipan et al. 2011a) indicated no light penetrated for a period of 238 days, from 31 October 2010 to 15 June 2011. These data suggest that light rarely penetrates more than 10 cm below the surface.

1.3.2 Surface–subsurface connections

A second feature that unites SSHs is that they have relatively close connections to surface environments. These connections have three main impacts. The first is that of the influence on chemical and physical factors. This is easiest to understand in the case of temperature. In the absence of close connections with the surface, a subterranean habitat will have a nearly invariant annual temperature that is approximately the mean annual surface temperature (Palmer 2007). This damping of fluctuations occurs even in shallow lava tubes. An example is provided in Fig. 1.8 and Table 1.1 for hourly temperature records over the course of a year for a surface site 10 m outside the entrance to Pahoa Cave in Hawaii and a site 200 m inside the cave. The coefficient of variation[2] for the surface site was 14.66 per cent while the coefficient of variation for the in-cave site was only 1.38 per cent, more than a tenfold reduction. At another site in Pahoa Cave 300 m inside, a datalogger was placed near an area with extensive roots of *Metrosideros polymorpha*; temperature was still relatively invariant, but less so than at the cave site 200 m in the cave but not near roots. The coefficient of variation at the tree root site was 1.50 per cent (Table 1.1). Another feature that distinguishes the lava tube temperature pattern from the surface temperature pattern is that the cave temperatures showed a platykurtic distribution, a 'fat' distribution with thin tails. That is, temperature extremes in the cave were less common than on the surface. We return to lava tubes in chapter 8 and the general topic of environmental variability in chapter 10, but for now we just note that the tree root site was where animals were found, even though variability was higher.

In SSHs where there are closer surface connections, as occurs with strict sense SSHs, environmental conditions are more variable, and in some cases come to approximate surface conditions. An example of intermediate level of variability is provided by another volcanic site, in this case erosional MSS in the Teno area, northwest Tenerife, Canary Islands (Pipan et al. 2011a). The coefficient of variation of hourly temperature 70 cm below the surface in MSS

---

[2] Standard deviation × 100/mean

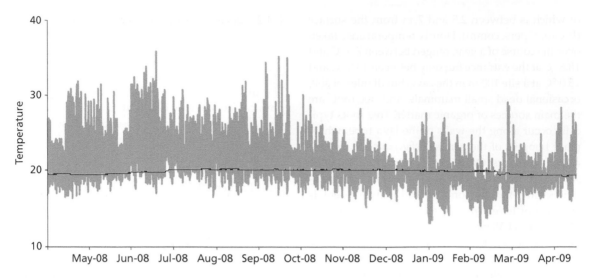

Figure 1.8 Temperature 200 m inside Pahoa Cave, Hawaii, USA (black line) and temperature 10 m outside the entrance (grey line) at hourly intervals from April 2008 to April 2009. Data courtesy of F. Stone.

was nearly as great as the coefficient of variation on the surface (17 per cent compared to 19.5 per cent, Table 1.2, Fig. 1.9). Both the surface and MSS site showed strong seasonal variation, but only the surface site showed daily variation (Fig. 1.9). This MSS site was species-rich, with ten troglomorphic species.

For other SSHs, especially epikarst, the connection between surface conditions and conditions in the SSH are not so linear and immediate as they appear to be in the MSS. For example, Kogovšek (2010) found that temperature in three epikarst drips in Postojnska jama in Slovenia varied less than 0.5 °C over the course of a year and surface temperature was not

Table 1.1 Comparison of hourly temperatures for 8570 hours inside and immediately outside of Pahoa Cave, Hawaii from April 2008 to April 2009. Data courtesy of F. Stone.

	10 m outside	200 m inside	300 m inside near roots
Mean	20.67	19.92	20.01
Standard error	0.032	0.003	0.003
Median	20.21	20.12	20.16
Standard deviation	3.03	0.28	0.30
Coefficient of variation	14.66	1.38	1.50
Sample variance	9.18	0.08	0.09
Kurtosis	0.99	−1.15	−1.28
Skewness	0.88	−0.56	−0.47
Range	23.09	0.97	0.99
Minimum	12.74	19.31	19.34
Maximum	35.83	20.28	20.33

Table 1.2 Comparison of hourly temperatures for 10,555 hours on the surface and 70 cm deep in the MSS in the Teno area of northwest Tenerife, Canary Islands, from May 2008 to June 2009. Data from Pipan et al. (2011).

	70 cm deep	Surface
Mean	12.49	12.91
Standard error	0.021	0.025
Median	12.51	12.59
Standard deviation	2.12	2.51
Coefficient of variation	16.97	19.44
Sample variance	4.51	6.31
Kurtosis	−1.21	−1.1
Skewness	0.28	0.24
Range	7.19	10.5
Minimum	9.24	7.93
Maximum	16.43	18.43

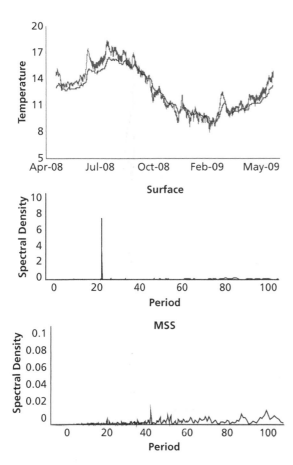

Figure 1.9 Top panel, temperature profiles at hourly intervals for an MSS site (black line) and nearby surface site (grey line) in a laurel forest in Teno in northwest Tenerife, Canary Islands, from April 2008 to May 2009. Centre panel, spectral densities (y-axis) for different cycle periods (x-axis) for cycles up to 100 days for the surface site. Note the strong period at 24 hours. Bottom panel, spectral densities (y-axis) for different cycle periods (x-axis) for cycles up to 100 days for theMSS site. Note the absence of a 24-hour period even at very low spectral densities. From Pipan et al. (2011), used with permission of Taylor and Francis.

a good predictor of drip temperature, because water remains in the epikarst for months or even years. In contrast, drip rates of water were highly variable, ranging from 0 mL min⁻¹ to 3500 mL min⁻¹. The seasonal pattern was extraordinarily complex, dependent on rainfall, amount of water stored, and particularities of the 'plumbing' of each drip (Kogovšek 2010). An example of the complex connection between infiltration from rainfall and discharge rate is shown in Fig. 1.10. Qualitatively, the epikarst seems

to act like a sponge, only discharging water through drips when it is saturated.

The second impact of close surface connections is with the flux of nutrients and organic carbon. An unusual feature of lava tubes compared to other caves is that their walls and ceilings are frequently penetrated by roots, such as those of *Metrosideros polymorpha* in Hawaii, which provide an important source of organic carbon and nutrients. In Hawaii and the Canary Islands, where the system has been extensively studied (Oromí and Martin 1992, Stone et al. 2012), a food web based on the root exudates which are fed on by Homoptera in the family Cixiidae has developed. This represents an additional food source compared to deeper subterranean habitats. Bilandžija et al. (2012) have found roots and cixiids in limestone caves in Croatia but they are in very shallow sections, or near entrances.

Other SSHs also have additional sources of organic carbon and nutrients either not present in deeper subterranean habitats or are present in lower quantities in deeper subterranean habitats. Several SSHs show strong vertical gradients in food availability. This is well known in soils where soil fertility declines with depth (Bardgett 2005). Gers (1998) points out that for MSS habitats there are two primary sources of organic carbon—water percolating through the soil and animals moving from the soil into MSS habitats. Generally, the quantities of both of these factors are greater when the subterranean habitat is closer to the surface, but there are few quantitative data.

In the case of epikarst, organic carbon and nutrients enter the system as dissolved or particulate organic matter in percolating water. We know very little about processing in the epikarst itself but the water that comes out of epikarst in ceiling drips in caves tends to be relatively low in total organic carbon (TOC), often around 1 mg L⁻¹ (Simon et al. 2007, Ban et al. 2008). However, the character of the organic carbon (e.g. low aromaticity) is such that it is more easily assimilated by organisms that other forms of organic carbon (Simon et al. 2010).

The third impact of close surface connections is that they provide pathways for the movement of animals between SSHs and the surface. Almost all species lists of the inhabitants of SSHs contain not only

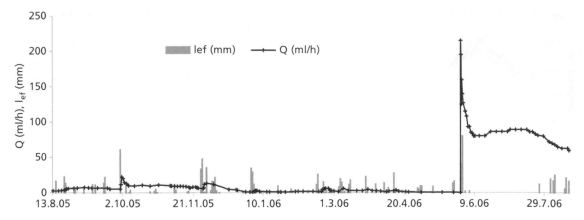

Figure 1.10 Effective infiltration (I_{ef}) in mm, and the hydrograph (Q) in mL h^{-1}, of a drip in Postojnska jama, Slovenia, for the 2005–2006 hydrological year. Note the different responses to similar precipitation events. From Kogovšek (2010), used with permission of ZRC SAZU, Založba ZRC.

stygobionts and troglobionts, but also species that show little morphological modification for subterranean life. Culver and Pipan (2008) provide several examples of species lists for strict sense SSHs. In fact, of the dozens of SSHs we have sampled, only one—epikarst drips in Županova jama in Slovenia—has almost exclusively stygobiotic or troglobiotic species. Fourteen of the 16 copepod species found there are stygobionts (Pipan 2005a). More typical is the list of species from the Teno MSS site in the Canary Islands, where there are 22 edaphobionts (soil specialists), 10 troglobionts, and 43 other species. The 43 other species may include established reproducing populations as well as recent invasions, many of which are destined to become part of the available supply of organic carbon and nutrients. What these numbers show is that there appears to be, in most cases, a pathway for invasion of the subsurface. The exception may be calcrete aquifers, which in some areas may be the only subterranean habitat (see chapter 5). In addition, because the surface environment is extremely arid, there may be little or no surface water for species to live in.

It would be wrong to assume that all of the nonspecialists in SSHs (or deep subterranean habitats for that matter) are destined to die, fail to reproduce, and become part of the food base. Romero (2009, 2011) goes so far as to claim that there is no convergent morphology of cave animals, and by analogy, SSH animals, that is, that there is no convergence or troglomorphy (see chapter 12). While this is an extreme view, it does point to the dangers of assuming that we understand more about adaptation to subterranean life than we do, and that it is easy to fall into a facile circularity—how do we know that a species is subterranean specialist? Because it is eyeless? How do we know this is subterranean habitat? Because it has eyeless species!

There can also be movement from SSHs to the surface, especially in the case of aquatic habitats. Upwelling of the underflow of streams brings not only organic carbon, but also stygobionts to the stream surface. Likewise, during periods of high precipitation or low evapotranspiration, stygobionts in hypotelminorheic habitats move or get pushed out through seepage springs. Nearly all the stygobionts and troglobionts that move from SSHs to surface environments fail to survive for very long, most likely the result of competition and predation (see chapter 9), although the direct harmful effects of light may play a role in some cases (Culver and Pipan 2009).

It should also be noted that connections between shallow and deep subterranean habitats can also promote movement of animals between these two habitats. The dripping of water out of epikarst habitats into drip pools in caves is an example of individuals moving from a shallow to a deep subterranean habitat. Many epikarst species do not fare well in drip pools, and the fauna of drip pools is quite distinct from that of the adjacent epikarst (Pipan 2005a, Pipan et al. 2010). However, it is still a potential colonization route.

1.3.3 Availability of organic carbon and nutrients

Many biologists consider extreme food shortage as the hallmark of the cave environment, and by extension, subterranean environments. Among the best-studied subterranean animals are cavefish in the family Amblyopsidae (Poulson 1963, Poulson and White 1969, Niemiller and Poulson 2010), top predators in relatively deep caves. Poulson provides convincing evidence that chronic food shortage is a driving factor in the evolution of demography, ecology, and morphology. He argues that the evolution of low metabolic rate, low reproductive rate, and elaborated extra-optic sensory structures in the cave-dwelling species of the family only makes sense if food shortage exerted a strong selective pressure.

A number of SSHs have, at least relatively speaking, high nutrient and organic carbon levels and fluxes. Hypotelminorheic sites are in broad contact with the leaf litter layer, and share many features with other detritus-based aquatic habitats. Levels of dissolved organic carbon—DOC (reviewed in chapter 10) are in the range of 5 mg L^{-1}, somewhat less than but similar to that of nearby surface streams. Ban et al. (2008) reports that DOC in epikarst drip waters varies on an annual cycle, but can reach values of 3 mg L^{-1}. In what is a highly anomalous case, Laiz et al. (1999) report organic carbon values from epikarst drips of up to 2200 mg L^{-1} in Altamira Cave, Spain. Organic carbon amounts in hyporheic and groundwater zones below rivers and streams are often approximately the same as that of the river and stream itself. Marmonier et al. (2000) reported DOC levels of approximately 1.5 mg L^{-1} in the surface waters of Vanoise brooks, tributaries of the Rhône River in France, while DOC levels in sediments 40 cm beneath the surface were 1.9 mg L^{-1}. Organic carbon enters the hyporheic zone either through episodic burial of particulate organic carbon following a disturbance or from transport by stream water or groundwater intrusions (Buss et al. 2009).

It is more difficult to quantify organic carbon availability in terrestrial habitats because it is more patchily distributed, and in the case of soil presents considerable technical difficulties (Jones

and Willett 2006). In MSS habitats, organic carbon can enter the system from the movement of water downwards, the active movement of animals in all directions (Gers 1998), and perhaps the action of gravity. In some circumstances, such as the volcanic terrains of the Canary Islands, the surface community is supported by aeolian fallout of organic carbon (Ashmole and Ashmole 2000), with little if any photosynthetic fixation of carbon occurring. MSS habitats in volcanic terrains may also have roots extending into the spaces, as is the case with lava tubes (Stone et al. 2012). The movement of organic carbon by gravity may also be important in lava tubes. We observed small deposits of soil in La Cueva del Viento in the Canary Islands, which entered the lava tube via ceiling cracks and fissures. Organic carbon in the soil varies with soil type, age, local conditions, depth, and especially hydrology and extraction method (Kalbitz et al. 2000, Jones and Willett 2006).

The general question of energy limitation in subterranean habitats, including caves, needs reexamination. A growing number of examples of at least relatively high resource bases and fluxes in subterranean habitats have been reported (Culver and Pipan 2009).

1.3.4 Geographical distribution patterns of SSHs

With the exception of epikarst, the broad-scale distribution of strict sense SSHs is very poorly understood. Hypotelminorheic habitats have been described only from a few sites in Croatia, France, USA, and Slovenia (Meštrov 1962, Culver et al. 2006b). Hypotelminorheic habitats and the associated seepage springs should occur in areas with some topographic relief (in order for a sufficient hydraulic gradient to occur), with sufficient precipitation for runoff to occur, and a layer, usually clay, that impedes the vertical movement of water. It certainly is present in additional countries and regions, and may well be described in the literature, but with different names. For example, seepage springs bear some resemble to very small helocrene springs in the classification of Springer and Stevens (2009). On a local scale, large numbers of seepage springs can occur but they are highly

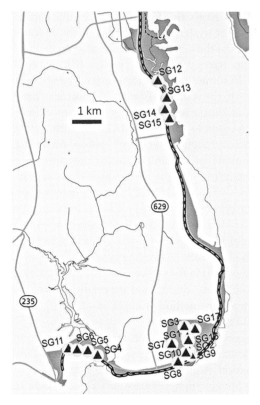

Figure 1.11 Map of southern section of George Washington Memorial Parkway near Washington, DC, USA, showing the location of seepage springs. The watercourse on the right is the Potomac River.

clumped. A striking example of this is shown in Fig. 1.11, which indicates the location of seepage springs along a 20 km stretch of the George Washington Memorial Parkway, a protected area along the Potomac River near Washington, DC. A total of 17 seepage springs are found in three clusters. The absence of seepage springs in most of the area is likely because the clay layer only occurs intermittently.

Calcrete aquifers, at least in the Yilgarn region of Western Australia are widespread but patchy (Fig. 1.12). Except in rare instances, access is only via wells and boreholes, and little direct observation of the aquifer is possible (Fig. 1.4). Genetic differences among amphipods in the family Chiltonidae in a 3.5 km² area of 5–11 m deep boreholes in Sturt Meadows, Western Australia suggest considerable small-scale habitat patchiness (Bradford et al. 2010, see chapter 5).

Since it is a karst feature, epikarst is limited to karst regions, which cover about 15 per cent of the earth's terrestrial surface. Epikarst probably occurs nearly everywhere in karst, except for arid regions such as the calcrete aquifers of Western Australia. Its thickness ranges from a few centimetres to 10 m (Williams 2008). There is considerable heterogeneity in epikarst, even at small scales. Most of what we know about epikarst comes from sampling drips in caves, and the fauna and chemistry of drips even a few metres apart can be quite different (Pipan 2005a, Pipan and Culver 2007a). Pipan et al. (2006) found that the extent of a particular epikarst copepod community was usually less than 1 km. Epikarst is a highly replicated habitat since karst areas are usually many thousands of square kilometres in extent. Caves are also highly replicated, but the scale of replication is larger than for epikarst.

Terrestrial SSHs—the *milieu souterrain superficiel* (MSS)—have been described from a serious of mountainous sites in Europe and Asia. Whether MSS-like habitats occur in regions of low relief is not known, and it is not clear how they would form in such regions since there are no scree or talus slopes. The nature of MSS habitats varied considerably from region to region (see chapter 4) and a general operational definition was the presence of troglobiotic species. Using this definition, MSS sites appear to be more or less continuous in some regions, especially volcanic terrains such as those in northern Tenerife and northcentral Hawai'i (Fig. 1.13). In other sites, even in the Canary Islands, they appeared to be quite local in occurrence, such as on El Hierro. In non-volcanic sites, MSS habitats are more frequent in areas of considerable elevational relief.

Among the broad sense SSHs, lava tubes are the most patchy in their distribution. Volcanic rocks occur along the margin of the Pacific Ocean (the 'ring of fire') as well as scattered sites in the Mediterranean and Macaronesia (Palmer 2007). A rich troglobiotic fauna is known from the Canary Islands and Hawaii lava tubes. On Tenerife, 335 of 8500 0.5 × 0.5 km² quadrats had lava tubes with troglobionts (Izquierdo et al. 2001) The density of lava tubes can be quite high. For example, the 6.92 km² Kiholo Bay lava flow on the island of Hawaii has 98 lava tubes, more than 14 per km² (Medville 2009, see chapter 8).

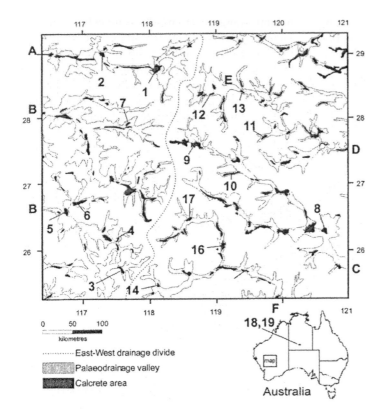

Figure 1.12 A map of the sampled calcrete aquifers and their relative positions in paleodrainages in Western Australia. Letters and numbers refer to the paleodrainages (dotted lines) and calcrete aquifers (in black), respectively. From Leys et al. (2003), used with permission of Blackwell Publishing.

Figure 1.13 Field of a'a lava near Emesine Cave in Hawaii, USA. Photo by T. Pipan. (see Plate 8)

Hyporheic habitats are widespread, occurring in gravel- and stony-bottomed streams throughout the world. The hyporheic zone can extend hundreds of metres laterally from the stream (Stanford et al. 1994). Relatively few areas have been sampled for fauna, with extensive studies done only in the Rhône River basin in France, the Danube River in Austria, the Flathead River in Montana, USA, the Never Never River in Australia, and several rivers draining into the Ljubljana Marsh, Slovenia

(Brancelj and Mori 2005). The presence of stygobionts is highly variable (see chapter 6) due in part to the nature of the connection between the hyporheic and the underlying groundwater (Malard et al. 2000). There is also considerable heterogeneity of habitat within a stream, the classic example being the analyses of Rouch et al. (1989) and Rouch (1991) of a 30 m² area of Lachein Brook near Moulis, France (see Fig. 1.5). Species composition, richness, and physical chemical characteristics varied from metre to metre, and with depth. Overall, there were 20 stygobiotic and 15 other crustacean species.

The final broad sense SSH is the soil, surely the most ubiquitous of any of the SSHs. Except in newly emergent volcanic terrains, high mountains, and glaciers, it is everywhere. Many factors influence species composition and richness, especially the interconnected factors of soil type, geography, and climate.

1.3.5 Habitat size

Habitat, or pore size, varies among and within different SSHs, but is intermediate for strict sense SSHs (Fig. 1.14). In hypotelminorheic habitats, organisms live in the spaces between fallen leaves, and in the dirt and gravel that comprises the habitat—ranging from approximately 1 mm to 1 cm in diameter. Epikarst spaces are created by the solution of limestone and to a certain extent by fracturing. The minimum size of solution tube for turbulent flow and rapid enlargement is approximately 0.2 mm (Dreybrodt et al. 2005) and in epikarst the practical limit to diameter is 5 cm. Calcrete aquifers develop extensive channels below the water table, with dimensions ranging from 0.2 mm to perhaps 1 m (W. Humphreys, pers. comm.). MSS dimensions can be very small (0.1 mm), especially in sediment among the large spaces to a maximum of around 10 cm. Broad sense SSHs bracket the strict sense SSHs in habitat size. Soil has the smallest dimensions, ranging from less than 0.1 mm to perhaps 1 mm. Similarly, aquatic interstitial habitats are the smallest aquatic SSH, with habitat dimensions of between 0.1 mm and 1 mm. Lava tubes bracket the other end of the size spectrum, with dimensions between 10 cm and 10 m.

1.4 Features of SSHs of general ecological and evolutionary interest

1.4.1 Are SSHs ecotones?

The ecotone concept has been applied to the boundary or zone between surface and subterranean

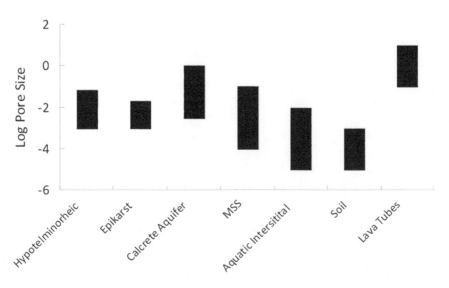

Figure 1.14 Histogram of log pore size of different shallow subterranean habitats.

environments (Gibert et al. 1990, Plénet and Gibert 1995). Ecotones were originally defined by E.P. Odum (1953) as

a transition between two or more communities; it is a junction zone or tension belt which may have considerable linear extent but is narrower than the adjoining community areas themselves.

While the emphasis in Odum's original definition was the community, ecotone studies have often taken a broader view, incorporating energy and nutrient fluxes, filtering effects, and a more detailed spatial context (e.g. Vervier et al. 1992).

Because of their proximity to the surface, SSHs are likely candidates as ecotones. This concept has been applied to the hyporheic, which Buss et al. (2009) define as a transition zone or ecotone:

the saturated transition zone between surface water and groundwater bodies that derives its specific physical (e.g. water temperature) and biogeochemical (e.g. steep chemical gradients) characteristics from active mixing of surface- and groundwater to provide a habitat and potential refugia for obligate and facultative species. (p. 93)

Most perspectives on the hyporheic emphasize its importance in relation to surface streams, such as a habitat for early instar aquatic stream insects (Stanford and Gaufin 1974). A more subterranean-oriented view with respect to hyporheic habitats is that of Gibert and colleagues (Gibert et al. 1990, Plénet and Gibert 1995, Plénet et al. 1995). For example, they document the distribution of stygobiotic species in the hyporheic zone in seven sites in the Rhône and Ain Rivers in France and demonstrate a sharp boundary between surface and subsurface waters. Vervier et al. (1992) point out that as such the hyporheic ecotone acts as a photic filter, a mechanic filter (restricting large particles), and a biochemical filter that affects chemical and biological processes. While nearly all stony-bottomed streams have a hyporheic, not all are directly connected with groundwater because in some situations there is an impermeable layer (e.g. shale) between the hyporheic and the underlying groundwater (Malard et al. 2000).

Epikarst habitats can also be considered from the ecotonal perspective, although the direction of influence is almost exclusively downwards, from the soil to the cave. For example, percolating water is processed in litter and soil layers, enters the epikarst where it is probably the only important source of carbon and nutrients, passes through the vadose zone, and finally exits into deep subterranean habitats, especially drip pools in caves. Carbon and nitrogen from dripping water forms biofilms and supports the aquatic cave community (Simon et al. 2003, Simon et al. 2007). In this case epikarst acts as a filter, by preventing the transport of larger organic particles, and by a complex biochemical processing that ultimately results in fewer aromatic organic compounds and humic and fulvic acids (Simon et al. 2010).

A third situation where SSHs may be ecotones is actually a possible connection between two SSHs—the MSS and lava tubes. Pahoehoe lava, the lava type in which most lava tubes are found, is also rich in MSS habitats, especially as the result of cracks and small tubes that were formed at the time of the initial lava flow. These MSS habitats correspond to Howarth's (1983) 'mesocaverns', which he defines as openings between 0.1 and 20 cm in diameter. These two habitats are often in close connection, and nutrients and animals can move between them, especially in the direction of lava tubes. In nearly all areas, the MSS is in close connection with the soil (another SSH habitat), and the soil is really the matrix in which the MSS is embedded (see chapters 4 and 7). Nutrient movement into the MSS must take place through the soil, but it is stretching the definition of ecotones to call the soil an ecotone for the MSS. In any case its physical dimensions are larger than the MSS.

For other SSHs, they are rarely if ever ecotones, because there is simply no direct or proximate connection with deeper subterranean habitats. In the case of lava tubes, there are no deeper habitats because lava tubes are always close to the surface. As new flows occur, previous lava tubes are obliterated. Calcrete aquifers are typically the only subterranean habitat in the arid regions where they occur. Hypotelminorheic habitats are by definition (see Meštrov 1962, Culver et al. 2006b) isolated from regional water tables. Finally, the soil is not necessarily connected to deeper subterranean habitats, although deep soil dwellers are occasionally found in caves.

Perhaps the most obvious subterranean–surface ecotone is the area around the entrance to a cave, the so-called twilight zone, where there are intermediate environmental conditions, including light intensity. There are often species unique to this typically terrestrial ecotone, such as orb-weaving spiders in the genus *Meta* (Chapman 1993). Prous et al. (2004) proposed a method for the delineation of the extent of the ecotone, based on the pattern of community overlap. In their study, they found that the width of the ecotone for several Brazilian caves was 10–12 m. Moseley (2008, 2009) suggested that an entire cave is a transition zone between epigean and subterranean environments, because of the presence of many species unique to smaller cavities, and because he considered habitat size more definitive than light intensity in delimiting the subterranean environment.

1.4.2 Are SSHs staging areas for colonization of deep subterranean habitats?

The idea that species with less modified morphology, e.g. reduced but not absent eyes, are less adapted has a strong resonance in the study of subterranean life, especially due to the forceful arguments of Poulson (1963) and Christiansen (1961) in their neo-Darwinian treatment of the evolution of cave life. If troglophiles are 'troglobionts in training', then perhaps the training or staging habitat for troglophiles is SSHs. That is, do many deep subterranean species have shallow subterranean species as ancestors? SSHs are in most ways, except for the presence of light, intermediate in environmental parameters between surface and deep subterranean habitats (see chapter 10). A realistic version of this hypothesis is not that all SSH species give rise to deep subterranean species but rather that some do. For example, it is hard to imagine how SSH species limited to the Coastal Plain of eastern North America (Culver et al. 2012a) could give rise to species in deep subterranean habitats such as caves that are 100 km away.

1.4.3 Evolution of eye and pigment loss

The nature of evolution in subterranean habitats is one in which losses of structures, such as eyes and

pigment, seem to dominate. This process of loss of eyes and pigment is often given the name *regressive evolution*, a term not without controversy because of its roots in discredited neo-Lamarckian explanations of evolution (Romero 2009). The phrase *evolution in reverse*, suggested by Porter and Crandall (2003) is better but most authors still refer to the process as regressive evolution. Many species found in shallow subterranean habitats, often exclusively so, are eyeless and without pigment. Two competing, overarching explanations for eye and pigment loss have been put forward (see chapter 9). One is that natural selection and adaptation is not involved, and that the evolution of eyelessness and pigment loss is the consequence of the accumulation of structurally reduced, selectively neutral mutations (Wilkens 2010). The other is that natural selection is involved, usually in the form of either selection for metabolic and developmental economy in a resource-poor environment (Poulson and White 1969), or pleiotropic effects where the same gene acts to decrease the eyes and pigmentation and increase extra-optic sensory structures (Jeffery 2005a).

Because many SSHs are rich in resources relative to deeper subterranean habitats, selection for metabolic efficiency should be reduced, and eye and pigment loss less pronounced. That this seems not to be the case suggests that if selection is important, then pleiotropy, in the sense of multiple effects of a single gene, is likely to be involved. Even more puzzling is that SSH species do not appear to be intermediate in morphological increases, such as relative appendage length and segment number (Culver et al. 2010). Neither neutral mutation theory nor adaptation to harsh, food-poor environments provides a credible explanation for this. We explore the hypothesis in chapter 9 that the absence of light is the predominant selective factor and this explains the similarities of morphology in different subterranean habitats.

1.4.4 What is the geographical pattern of species richness in SSH faunas, and is it similar to that of cave faunas?

The pattern of species richness of stygobionts and troglobionts on a global scale is not one of an increase in diversity towards the tropics. Culver et al.

(2006a) proposed that a ridge of species richness of troglobionts was present along a latitude of approximately 45° N in Europe and 35° N in eastern North America, and that this ridge was at the position of maximum actual productivity. Deharveng et al. (2012) suggest stygobionts follow a similar pattern at least in Europe. Do SSH specialists follow a similar pattern? Although the data are less extensive, the pattern is at least consistent with this hypothesized ridge. For example, the most species-rich epikarst copepod faunas are found in Slovenia, on the biodiversity richness ridge.

1.4.5 Are there special conservation concerns associated with the SSH fauna?

Any subterranean habitat or fauna is hidden from view, and this poses problems in generating a constituency to support its protection. An even more fundamental problem is the identification of SSH sites. The only surface manifestation of hypotelminorheic habitats is a wet spot in the woods, and unless it has been identified as an interesting habitat, it may seem like nothing more than a difficult spot to maintain a trail even if it is in nature park. MSS sites may appear to be insignificant talus slopes, and the banks of streams and rivers may be considered important only because of the sediment load that enters the stream, rather than a habitat per se. Nevertheless, SSHs have been recognized as important habitats in some parks and natural areas, such as: the Danube Flood Plain National Park, a UNESCO International Biosphere Reserve in Austria; in the National Capital Region of the US National Park Service around Washington, DC; and national parks centred around lava tubes and volcanic landscapes, such as Teide National Park in Tenerife, Canary Islands.

1.5 Overview of chapters

Chapter 2 considers the hypotelminorheic and associated seepage springs. The conditions required for their formation are elaborated and what is known of their global occurrence is discussed. The best-studied examples are those from near-sea-level locations in the lower Potomac basin in the vicinity of Washington, DC, and from Nanos Mountain in

Slovenia at an elevation of approximately 700 m. We document the fauna found, including both cave-like and other species. Techniques for sampling this unusual habitat are discussed.

Chapter 3 covers epikarst. The best-studied examples are from dripping water and associated pools in caves in Slovenia, and we enumerate the fauna, both terrestrial and aquatic, including both troglomorphic and other species. We also analyse extensive data for sites in the USA and Romania. The emerging patterns of the relative frequency of troglomorphy are emphasized. The special techniques developed for sampling dripping water and pools are described.

Chapter 4 summarizes information about intermediate-sized terrestrial SSHs, the term MSS (milieu souterrain superficiel) being used to describe some or all of these habitats. The best-studied examples are from the French Pyrenees, the Slovenian Dinaric Plateau, Czech scree lobes, and lava flows on the Canary Islands, and we summarize the species found in these areas. This habitat is also difficult to sample, especially non-destructively, and details of a sampling device that can be left in place for long periods of time are described.

Calcrete aquifers are reported from extensive areas of southwestern Australia, and they likely occur in arid regions elsewhere. In chapter 5, we review the environmental conditions and the fauna of calcrete aquifers in the Yilgarn in Western Australia, describing the remarkable species richness of this fauna.

The sixth chapter introduces a broad sense SSH— aquatic interstitial habitats. This hyporheic and groundwater zone has been sampled in many areas, but especially in: the Rhône River and its tributaries in the vicinity of Lyon, France; the Danube wetlands in the vicinity of Vienna; the Flathead Lake region of Montana; the South Platte River in Colorado; and the Never Never River in Australia. Study of these habitats is made possible by a specialized sampling pump developed by the French biologists Bou and Rouch in 1967, which allows for a detailed analysis of microdistribution of species. An overarching pattern of the hyporheic is its heterogeneity, both spatial and temporal.

In chapter 7 we review soils as an SSH. Soils are a major field of scientific inquiry, and many areas of soil science are not of direct relevance to the study of SSHs, such as soil fertility and formation. Emphasis is on the transfer of carbon and nutrients between the soil and SSHs, especially milieu souterrain superficiel habitats, and morphological features both shared by and different from cave fauna. Nearly all of this work comes from studies in the French Pyrenees. In common with many cave and SSH species, many soil species, especially deep soil species that never encounter light, are eyeless and depigmented, but differ in having greatly reduced appendages. Methods for sampling the soil fauna, such as Berlese extractions, are reviewed.

In chapter 8, we consider lava tubes, their formation and their fauna. Of special interest is the relatively short lifespan of a lava tube, typically measured in hundreds or thousands of years, two orders of magnitude shorter than karst caves. Two regions are relatively well studied and known to harbour a rich specialized fauna—Hawaii and the Canary Islands. We review the fauna, both specialized and non-specialized, and summarize evidence concerning mode of colonization and mode of speciation.

Chapter 9 begins the exploration of the commonalities of SSHs. A defining factor for SSHs is the absence of light. We examine the relationship of light and SSHs in terms of gradients and proximity. While stygobionts and troglobionts of SSHs are rarely found in photic zones, many eyed species are found in SSHs. We offer hypotheses about why stygobionts and troglobionts sometimes survive in photic zones as well as why many eyed species survive in aphotic zones.

Chapter 10 focuses on patterns of environmental variation. SSHs are more variable than caves and less variable than the surface. We summarize hourly records of temperature and other environmental parameters for five types of SSHs—seepage springs, epikarst drips, MSS, lava tubes, and interstitial habitats. The amount of overall variation varies from habitat to habitat, but all share the characteristic that the extremes of temperature are considerably reduced in SSHs compared to nearby surface habitats. We present data on environmental variability and the richness of stygobiotic and troglobiotic species, and conclude that there is little if any connection between environmental variability and species richness.

Chapter 11 examines commonalities in organic carbon and nutrient processes in SSHs. Because of proximity to the surface, fluxes and amounts of carbon are considerably higher in SSHs than in deeper subterranean habitats, but SSHs and deeper subterranean habitats share a dependence on allochthonous organic matter. Different allochthonous sources of organic carbon are summarized. Evidence that the SSH fauna may be carbon limited rather than nutrient limited is reviewed. We place the carbon- and nutrient-rich subterranean guano communities in the context of resource availability in subterranean communities in general.

Chapter 12 takes up the question of morphology. Subterranean species, irrespective of habitat, tend to lose both eyes and pigment. Cave-dwelling species evolve longer appendages while interstitial and soil species evolve shorter appendages, certainly the result of differences in size of the habitat space. We examine studies, especially of the amphipod genus *Stygobromus* in North America and the spider genus *Dysdera* in the Canary Islands, that suggest the absence of light and habitat dimension, but not resource availability or environmental variability, are the primary selective factors driving morphological evolution in subterranean environments.

In chapter 13, we investigate the colonization of and dispersal among SSHs. The close proximity of SSHs to surface habitats and the relatively benign environmental conditions of SSHs suggests that there is likely a steady stream of colonists, accidental or otherwise. Just like troglomorphic species, the frequency of these 'accidentals' is rarely zero. In contrast, most results suggest highly restricted dispersal among SSHs (i.e. subterranean dispersal), less than a few km in extent. The implications of this for speciation are discussed, especially in relationship to parapatric and allopatric speciation, and the two special models of cave colonization—the adaptive shift hypothesis and the climate relict hypothesis.

Phylogenetic relationships between shallow subterranean species and deep subterranean species are the subject of chapter 14. We test the proposition that species in SSHs are more basal and represent an intermediate step in the evolution of deep cave animals. Examples include spiders in the genus

Dysdera in the Canary Islands and amphipods in the genus *Niphargus* in Europe. There is little support for the proposition, and it appears that some SSH species are most closely related to other SSH species, others are most closely related to deep subterranean species, and still others are most closely related to surface-dwelling species.

In chapter 15 we review the critical questions of conservation and management. Both little-studied and insignificant in size and appearance, SSHs have largely been neglected by conservation planners. Protection, to the extent that it occurs, is a by-product of protection of other habitats (caves in the case of epikarst and streams in the case of interstitial habitats). Their proximity to the surface and their inconspicuous nature makes them especially vulnerable. Lava tubes and associated MSS habitats are often protected in national parks. The overall importance of a landscape approach to protection is emphasized.

In the final chapter we review the claim that SSHs, disparate and diverse as they are, should be considered together. This case includes some commonalities of the physical habitats as well as commonalities of the types and morphology of the fauna. The key question of their connection or lack of connection to the morphological evolution of the fauna of deeper subterranean habitats is highlighted. The answer is not yet within our grasp but there are tantalizing clues.

1.6 Summary

We introduce the idea that there are many subterranean habitats close to the surface that are little known and do not fit comfortably into any habitat classification scheme. Four of these shallow subterranean habitats (SSHs) are strict sense shallow subterranean habitats—the hypotelminorheic and seepage springs, milieu souterrain superficiel (including talus and scree), epikarst, and calcrete aquifers—and have intermediate-sized habitat spaces, no light, and close connections to the surface. Broad sense shallow subterranean habitats include habitats with large (lava tubes) or small (aquatic interstitial and soil) spaces. SSHs are generally broadly but patchily distributed across the landscape, although some have restricted physical requirements, such as the presence of a shallow clay layer for hypotelminorheic habitats. Close surface connections have impacts on environmental conditions, nutrient fluxes, and movement of animals through SSHs. While SSHs can be ecotones, they are habitats in their own right, and not necessarily connected with deeper subterranean habitats. They are of general biological interest because of the presence of eyeless, depigmented species, their possible role as stepping stones to adaptation to deeper subterranean environments, their geographical pattern, and conservation issues raised by them. Brief examples of each type of SSH are discussed.

Seepage springs and the hypotelminorheic habitat

2.1 Introduction

The shallowest of groundwater habitats, defining groundwater in the sense of water under the surface not exposed to light, can occur less than a metre beneath the ground. They are miniature subterranean drainage basins, draining less than 1000 m² and rarely more than 10,000 m² (Culver et al. 2006b). Meštrov (1962) applied the term 'hypotelminorheic' to shallow groundwater habitats that are vertically isolated from the water table and are 'constituted of humid soils in the mountains, rich in organic matter, and traversed by moving water' (authors' translation). This groundwater habitat has usually been ignored in overall groundwater classification schemes (e.g. Hahn 2009). Juberthie (2000) included it in his discussion of subterranean habitats, but more as a special case than an integral part of the subterranean realm. The very non-euphonious nature of the word (Greek roots expressed in French by a Croatian biologist) has even led to ridicule (Chapman 1993), but we believe the term is very useful.

Water emerges from these habitats in springs, which take a wide variety of forms (Kresic 2010). Because springs provide access to these habitats, although indirectly, they are often the only places where this groundwater fauna can be sampled. The emergence points of water from these shallow subterranean habitats have been given a series of names, none of them entirely satisfactory, and this has resulted in continuing confusion. Perhaps the earliest name used was 'seep' (e.g. Holsinger 1967), but this term, in American usage at least, often refers to petroleum oozing out of the ground. Less

confusing is the term 'seepage spring'. According to Kresic (2010), a seepage spring is a diffuse discharge of water, when the flow cannot be immediately observed but the land surface is wet compared to the surrounding area. This captures the essence of many spring habitats in the mid-Atlantic region of the eastern USA—wet spots in the woods (Figs. 1.1 and 2.1). Kresic (2010) also provides a useful context for the classification of seepage springs within the general framework of springs. Flows of seepage springs are typically less than 10 cm³ per second, making them eighth-order springs in Kresic's extension of Meinzer's (1923) categorization of springs by discharge. Seepage springs are gravity-fed and situated in sediment.

Kresic (2010) pointed out that variability of discharge is an important hydrological and ecological parameter, and indicated that if the ratio of the maximum to minimum discharge exceeds 10, then the spring can be considered highly variable. Because many seepage springs have little or no flow during hot, dry periods, they would be classified as highly variable.

Seepage springs fit less comfortably in other spring classification schemes. Springer and Stevens (2009) defined 12 'spheres of discharge' or spring types, and seepage springs fall under the category of helocrene springs, i.e. springs that emerge with diffuse flow from low gradient wetlands. However, more typical helocrene springs include soap holes or quicksand (Springer and Stevens 2009)! Seepage springs also have some characteristics of limnocrene springs (see also Danks and Williams 1991), springs that emerge into pools, but the fit to this classification is poor at best.

Shallow Subterranean Habitats. David C. Culver and Tanja Pipan.
© David C. Culver and Tanja Pipan 2014. Published 2014 by Oxford University Press.

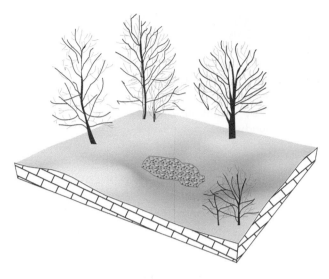

Figure 2.1 Sketch of the hypotelminorheic. From Culver et al. (2006b).

Based on Meštrov's (1962) definition and his sketch of the habitat (redrawn as Fig. 2.2), Culver et al. (2006b) proposed that the term 'hypotelminorheic' be used to describe habitats with the following major features (see also Culver and Pipan 2008):

1. a perched aquifer fed by subsurface water that creates a persistent wet spot;
2. underlain by a clay or other impermeable layer typically 5–50 cm below the ground surface; and
3. rich in organic matter compared with other aquatic subterranean habitats.

Culver et al. (2006b) also indicated that the drainage area of a seepage spring is typically less than 1 ha,

that the seepage spring is in a shallow depression, and that the leaves are characteristically blackened and usually not skeletonized. Without a clay layer, water should tend to move vertically, and there would be no persistent water. The water exits at a seepage spring, although there may not be flow at all times during the year.

For most of the species discussed here, it is the subterranean water of the hypotelminorheic and not the seepage spring, i.e. the groundwater/surface water ecotone (see Gibert et al. 1990 and section 1.4.1), that is their primary habitat. The seepage spring is the point of collecting the fauna (see Box 2.1), but is not the shallow groundwater habitat

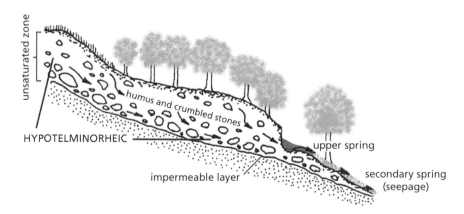

Figure 2.2 Slightly modified sketch of hypotelminorheic habitat by Meštrov (1962). See Fig. 2.3 for a photograph of the same habitat. From Culver et al. (2006b).

itself. A few species are primarily inhabitants of the ecotone itself. The hypotelminorheic and the ecotone are clearly an example of a groundwater-dependent ecosystem (Eamus and Froend 2006). Seepage springs are also isolated wetlands, although highly miniaturized ones.

Clay is a critical component of hypotelminorheic habitats, not only because it acts as a barrier to the downward movement of water, but also because during periods of drought, the water retained by the colloidal clay may serve as a refuge for invertebrates in the hypotelminorheic. According to Ginet and Decu (1977), clay may also have some nutritional value.

2.2 Chemical and physical characteristics of the hypotelminorheic

2.2.1 Hydrology of the hypotelminorheic

Because of its miniature size, most of the standard tools employed by a groundwater hydrologist to measure discharge and to delineate subterranean basins cannot be used. Flow rates are too low to gauge, and water volumes are too small to use colorimetric dyes. Some qualitative features are clear. One is that there is considerable temporal variability. Many of the seepage springs we have studied in the USA and Slovenia have no visible flow during the summer months. A seep with high biodiversity on Nanos Mountain, Slovenia, had no visible flow or even moisture for a period of nearly a year, but began to flow again, with stygobionts still present. A new seepage spring also appeared on Nanos, apparently the result of habitat alteration by wild animals, probably boar. Likewise, D. Fong (pers. comm.) observed the exit of a seep in Pimmit Run in George Washington Memorial Parkway in Virginia (USA) migrate downslope about 2 m after a dry period. Given the very superficial nature of both the hypotelminorheic and its seepage spring, this kind of observation is not surprising in principle but the disappearance of a particular seepage spring is always unexpected. There are also times of intense precipitation that may hydrologically connect seepage springs with above ground flow, i.e. sheet flow, especially since seepage springs are clustered geographically (Fig. 1.11).

Seepage springs are also a type of wetland. Euliss et al. (2004) place all wetlands along two axes—one is precipitation and the other is the relative amount of discharge and recharge from the wetland to the aquifer. Seepage springs are at the end point of the groundwater discharge–recharge axis since they are, by definition, discharge points for shallow groundwater. Known seepage springs occur in temperate zone areas of moderate precipitation, placing seepage springs near the mid-point of the vertical axis. Euliss et al. (2004) suggest that wetlands at this place on their continuum are open, with submergent vegetation and short wetland perennials. Seepage springs studied in relation to hypotelminorheic habitats are usually too small to have any characteristic vegetation, but in the lower Potomac basin, the eastern skunk cabbage, *Symplocarpus foetidus*, is often associated with seepage springs. *S. foetidus* blooms in late winter (February or early March) and seepage springs typically remain unfrozen throughout the winter. Larger wetlands at this position on the continuum are present in the lower Potomac basin, and they often have submergent vegetation. However, these larger wetlands have not been sampled for subterranean invertebrates, and it is not known if any occur.

Vernal pools are often present in the same area as seepage springs. As we demonstrate in section 2.2.3, the physico-chemical signature of the two habitats is quite different, but their physical appearance can be quite similar. Keeley and Zedler (1998) defined vernal pools as

Precipitation-filled seasonal wetlands inundated during a period when temperature is sufficient for plant growth, followed by a brief waterlogged terrestrial stage and culminating in extreme desiccating soil conditions of extended duration.

Except for extreme desiccating soil conditions, the other features can be present in seepage springs. The major difference with vernal pools is of course that in seepage springs and hypotelminorheic habitats, water is continuously present, at least in colloidal clay. There certainly exist habitats with the features of each, e.g. vernal pools whose duration is extended by groundwater discharge. This presents the very interesting possibility that both species specialized for subterranean life and species specialized for life in intermittent water co-occur. The only case we know of this is a special kind of wetland in karst, where lakes and ponds appear intermittently but are groundwater-fed, at the Pivka Intermittent Lake in Slovenia (Mulec et al. 2005). This series of lakes appears sporadically, with some appearing most years and others appearing very occasionally. Pipan (2005b) reported the presence of stygobionts including the copepod *Diacyclops charon* and the isopod *Asellus aquaticus cavernicolus*, vernal pond specialists including the fairy shrimp *Branchipus schaefferi*, and even a species specialized for karst intermittent lakes, the fairy shrimp *Chirocephalus croaticus*. Additional studies of such 'mixed' habitats would be most informative.

2.2.2 Geography of the hypotelminorheic

The original definition of the hypotelminorheic by Meštrov (1962) placed the habitat in regions of moderate to high slopes, no doubt because the first sites reported were on Medvednica Mountain, and subsequent studies in the French Pyrenees foothills were also in mountainous areas (Meštrov 1964). Medvednica Mountain is an isolated mountain in northwestern Croatia with a variable geology, composed of flysch, limestone, and metamorphic rocks. The actual site (Fig. 2.3) is at 927 m asl, underlain by green schist. However, the widespread occurrence of the habitat in the lower Potomac basin, even in the very flat Coastal Plain (Culver et al. 2012a), indicates that the habitat is not limited to mountainous areas. There does need to be enough relief for a least some hydraulic head, but this is a minor elevational restriction. More critical is the requirement that there is some sort of aquiclude, especially clay, that prevents the movement of water downward. Clay is of course widely distributed geographically, but its occurrence in any one area seems to be very sporadic. This would explain the patchy distribution of seepage springs in George Washington Memorial Parkway (Virginia, USA, Fig. 1.11), and the highly sporadic occurrence of seepage springs on Nanos Mountain in Slovenia as well. Given the unimposing physical appearance of hypotelminorheic habitats, it is not surprising that only a few sites are known. Known sites are Medvednica Mountain (Croatia), Ariège (France), Nanos Mountain, the

Figure 2.3 Photograph of Sanja Gottstein at seepage spring on Medvednica Mountain. See also Fig. 2.2. From Culver et al. (2006b).

lower Potomac basin, and a site in Rio Grande do Sul, Brazil (Rodrigues et al. 2012).

2.2.3 Physico-chemistry of the hypotelminorheic

Based on a ten-month monitoring period (March 2007–January 2008) of a hypotelminorheic habitat in Prince William Forest Park (Virginia, USA), the habitat is variable temporally although not to the extent of aquatic surface habitats (Fig. 2.4). From May to September, seepage spring temperatures were depressed compared to a small surface stream less than 5 m away, and approximated surface water temperatures for the rest of the year. In spite of the variability, the amplitude of variation in seepage spring temperatures was less than that of surface waters. The maximum recorded temperature in the hypotelminorheic habitat was 22 °C compared to 28 °C in the nearby stream (Culver and Pipan 2011). The coefficient of variation of stream temperature was 49.8 per cent and the coefficient of variation of seepage spring temperature for the

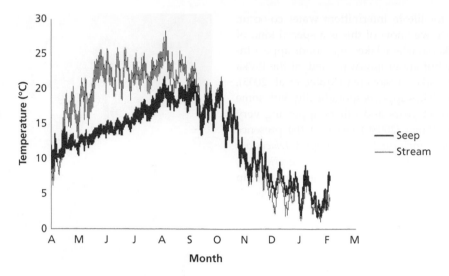

Figure 2.4 Hourly temperature from 7 April 2007 to 4 February 2008 in a seepage spring and adjoining stream in Prince William Forest Park, Virginia, USA. Because of the scale, line thickness indicates the extent of daily fluctuations. From Culver and Pipan (2011), used with permission of John Wiley & Sons.

same period was 38.2 per cent. This is a remarkable difference given the superficial nature of the hypotelminorheic habitat and the close proximity of the two.

A nearly 18-month hourly record of conductivity and temperature from a seep on Nanos Mountain, Slovenia, provides additional insight into the complex pattern of the hypotelminorheic. The datalogger, placed under a 20 cm thick rock at a site of a seep with stygobionts (*Niphargus stygius*), shows a pattern of increasing temperature and conductivity from March 2011 to July 2012 (Fig. 2.5). During this time, the two parameters are strongly correlated (r = 0.82, p<0.001), but with temperature showing more variability (Table 2.1). Both show a 24-hour cycle, based on spectral analysis. The rise in water temperature parallels the rise in air temperature (from approximately 5.5 °C to 20.2 °C) during the summer months, and the rise in conductivity (from approximately 220 µS cm^{-1} to 370 µS cm^{-1}) is likely the result of increased residence time of water in the subsurface and perhaps some evaporation. On 14 August at 1600 (4 pm), conductivity fell from 376 µS cm^{-1} to 27 µS cm^{-1}, the result of the drying of the habitat. Until 25 April 2012 the site was dry, and then conductivity slowly began to rise, indicating re-wetting.

Amphipods were present by May 20, and the site experienced periodic wetting and drying until the datalogger was removed on 10 July 2012. As the site dries, the animals must move to water or at least to burrow into clay to avoid desiccation. While certainly less variable than surface waters, this is a highly variable subterranean habitat, one with stygobionts.

Some basic physical–chemical water measurements are available for three hypotelminorheic sites—George Washington Memorial Parkway (Virginia, USA), Medvednica Mountain (Croatia), and Nanos Mountain (Slovenia, Table 2.2). Average temperatures vary considerably from site to site. The temperature of groundwater approximates the mean annual temperature (assuming it is not thermal water), and the differences reflect this as well as seasonal differences. Both pH and conductivity are consistent from site to site—pH is near neutral and conductivity is moderately high, around 350 µS cm^{-1}. The moderately high conductivity indicates that water has been underground for some period of time (or evaporating on the surface, an unlikely scenario). Dissolved oxygen values also vary, perhaps because of different amounts of decaying organic matter in the habitat.

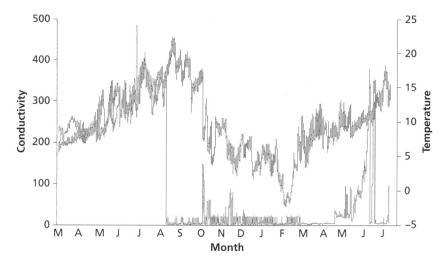

Figure 2.5 Hourly temperature (upper line) and conductivity (lower line) from 10 March 2011 to 10 July 2012 for a seep on Nanos Mountain, Slovenia.

Table 2.1 Summary statistics for conductivity and temperature for a Nanos (Slovenia) seep during the first period when it was wet (see Fig. 2.5).

	Conductivity (μS cm^{-1})	Temperature (°C)
Mean	280.6	11.4
n[1]	3773	3773
Standard Deviation	54.9	3.5
Coefficient of variation	19.6	30.3
10% Quantile	222.8	6.9
90% Quantile	365.7	16.1
Minimum	202.7	5.5
Maximum	486.3	20.2
Range	283.6	14.7

[1] 8,760 hours in a year.

Table 2.2 Comparison of chemical parameters (means only) at three hypotelminorheic sites with stygobiotic amphipods. Data from Culver et al. (2006b).

Site	Temperature (°C)	pH	Conductivity (μS cm^{-1})	Dissolved oxygen (mg L^{-1})
George Washington Memorial Parkway, USA	16.1	6.56	336	6.22
Medvednica Mountain, Croatia	7.3	7.14	384	9.37
Nanos Mountain, Slovenia	7.1	7.32	365	4.97

Since not all or even most small bodies of water on the surface are seepage springs, it is useful to find physical–chemical markers to separate them from other water bodies such as springs, vernal pools, and temporary rainwater pools. Using temperature, conductivity, dissolved oxygen, and pH, Culver et al. (2006b) identified three kinds of hypotelminorheic water emerging on the surface based on their biota:

1. Seepage springs with populations of the stygobiotic amphipod *Niphargus stygius* and *N. tamaninii*.
2. Seepage springs with few or no *Niphargus* but no surface-dwelling amphipods.
3. Seepage springs and small springs dominated by surface-dwelling amphipods such as *Synurella ambulans* and *Gammarus* sp.

Chemical and physical parameters are quite distinct for the three categories (Table 2.3). Temperatures were highest in the seepage springs with stygobionts.

Table 2.3 Comparison of chemical parameters for seeps without clay, hypotelminorheic sites dominated by stygobiotic *Niphargus* species, and those dominated by surface-dwelling species, all on Nanos Mountain, Slovenia. Data from Culver et al. (2006b).

Parameter		Seepage springs and small springs with surface-dwelling species	Seepage springs not dominated by *Niphargus* sp.	Seepage springs dominated by *Niphargus* sp.
Temperature (°C)	Mean	3.2	4.7	7.5
	Range	0.3–6.9	3.0–5.4	
	N	4	4	1
Conductivity (μS cm^{-1})	Mean	348	314	468
	Range	203–433	292–373	
	N	4	4	1
Dissolved oxygen (mg L^{-1})	Mean	1.9	3.9	8.2
	Range		3–5.6	
	N	1	3	1
pH	Mean	7.5	7.4	7
	Range	6.8–8.1	7.4–7.5	
	N	4	4	1

Sampling was done in March 2006 when surface temperatures were close to 0 °C, while temperatures in the first category seepage springs were the warmest, and closer to the mean annual temperature of 9 °C. Conductivity was much higher and pH was somewhat lower than other seepage springs and small springs, consistent with the water emerging from seepage springs with *Niphargus* populations being underground longer than water from the other seepage springs and small springs. Finally, dissolved oxygen was much higher, reflecting in part the greater solubility of oxygen at lower temperature.

In an unpublished study of basic physical and chemical parameters of more than 70 putative seepage springs in the George Washington Memorial Parkway (Virginia, USA, see Fig. 1.11), Chestnut and Culver found similar results. Sites with the stygobiotic genus *Stygobromus* had significantly lower temperatures (the study was done in spring and summer), higher dissolved oxygen, lower pH, and lower NO$_3^-$. The occurrence of *Stygobromus* in water with lower nitrate levels may indicate negative impacts of anthropogenic sources of nitrate. Conduc-

tivity was higher at *Stygobromus* seepage springs, but not significantly so. Using standardized coefficients[1], they found the discriminant function:

0.743 *Temperature* − 0.107 *Conductivity* − 0.405 *Dissolved Oxygen* + 0.972 *pH* + 0.586 *NO$_3^-$*.

Overall, seepage springs without *Stygobromus* had a score of −1.40 while those with *Stygobromus* had a score of +0.27. This function correctly classified 72 seepage springs 73 per cent of the time.

2.2.4 Analogues with other habitats

There are other aquatic shallow subterranean habitats, which do not comfortably fit the definition of the hypotelminorheic, and which occur very close to the surface. For example, there are some seepage springs and small springs emanating from solid rock crevices with very thin soils (Culver et al. 2012a). An artificial analogue of the hypotelminorheic is a tiled field and

---

[1] $z_i = \frac{x_i - \bar{X}}{S}$, where x_i is the original measurement, S the standard deviation, and $\bar{X}$ is the mean.

Figure 2.6 Photograph of a tile drain with the tiled field in the background in Isle of Wight County, Virginia, USA. Photograph courtesy of J.R. Holsinger, © Biological Society of Washington Allen Publishing Services, used with permission.

associated tile drain (Fig. 2.6). Tiling is the laying of pipes at shallow depths (approximately 1.5–2 m) in fields to increase drainage. They are common in the Middle West of the USA in glaciated, poorly drained soils as well as elsewhere. The drains of tiled fields in particular have been productive collecting sites for stygobiotic species (Hubricht and Mackin 1940, Koenemann and Holsinger 2001). The tile drains often dry up during summer but water persists in the pipes, which may act like clay in natural systems. Of course, the stygobionts must have been present in or near the fields before the construction of tile drains. Otherwise the rapid colonization of these artificial habitats is difficult to explain (Culver et al. 2012a).

Near the headwaters of streams, small eighth-order springs (springs with discharge of less than 1 L per second) can issue from naturally occurring tubes in sediments, up to several centimetres in diameter, a feature Kresic (2010) calls gushets, albeit miniature ones. Stream headwaters are often marshy, partly as the result of the lateral movement of water underground. It is an interesting, very rarely studied habitat.

Fišer et al. (2007) studied a very different kind of very shallow subterranean habitat that was inhabited by stygobiotic *Niphargus*. The stream Kolaški potok in southwest Slovenia rises, sinks, and rises again over a distance of approximately 500 m. The depth of the 150 m long underground portion is not known but it is probably less than 1 or 2 m. The stream itself is perched on top of a layer of granulated substrate of clay, marl, flattened stones, and organic debris. There is no gravel or sand in the stream, and thus no hyporheic extension (see chapter 6). The primary geological formations in the area are silicate–clay rich sediments in sandy marl layers in flysch, alternating layers of sandstone and shale.

In addition, there are subterranean habitats that are not as shallow as the hypotelminorheic and its analogues, but nonetheless only a few metres deep. For example, there is no clear distinction between a small spring and a seepage spring, except for the volume of flow, and perhaps the depth of the basin drained. There are seventh- and sixth-order springs present (springs with discharges up to 1 L per second) that are exits for groundwater from fractured rock aquifers. There are also shallow wells, often less than 5 m deep, that intersect groundwater and are also important collecting sites of stygobionts (Fig. 2.7). Animals probably concentrate in such wells because of the organic carbon input from the surface. Throughout most of the developed world, at least, such wells have disappeared. This same habitat, below the hypotelminorheic but above the regional water table, has also been accessed by sampling the water seeping out of the banks of deeply eroded streams, the result of uncontrolled storm water runoff (Hobson 1997).

This range of shallow subterranean habitats, combined with the apparent ease with which

Figure 2.7 Photograph of Biggers Spring/Well near Edsall Road in Fairfax County, Virginia, USA. The length of the line held by Roman Kenk (on the left) is the water depth. Person on the right is William Biggers. Photograph taken in March 1973 by J.R. Holsinger, used with permission.

species colonize drainage tiles and tile drain systems, suggests that there are even more such habitats, some of them perhaps little more than small cavities in water-logged soil. The phrase 'aquatic edaphic' habitats may be a useful one to refer to the totality of these habitats.

2.3 Biological characteristics of the hypotelminorheic

2.3.1 Organic carbon and nutrients in the hypotelminorheic

Much of the available organic carbon is likely in the form of particulate organic carbon, decaying leaves in particular (see Fig. 1.1). Information on organic carbon levels in seepage springs is only available for

some sites on Nanos Mountain, Slovenia, and only for dissolved organic carbon (Table 2.4). In addition to the three types of seeps and small springs listed in Table 2.3, a hyporheic site (see chapter 6) is also included for comparison. Dissolved organic carbon levels in all four types of habitat were quite variable, the result of the very superficial nature of the habitats. We do not know what caused the spikes in organic carbon in any of the sites, but cattle and wild boars have been occasionally observed. The highest average dissolved organic carbon concentration (4.52 mg L^{-1}) was at the site with large populations of *Niphargus stygius* and *N. tamaninii*. The lowest dissolved organic carbon concentrations were observed at those sites with few amphipods, either stygobionts or surface dwellers. Intermediate values of dissolved organic carbon were present in small springs dominated by surface-dwelling species and in the hyporheic habitat of a very small stream. These data are tantalizing in the apparent connection between organic carbon and the presence of stygobionts, and this would be a very interesting line of further inquiry.

Box 2.1 Collecting in the hypotelminorheic

Collecting is primarily accomplished by hand collecting at seepage springs and tile drains. Because the hypotelminorheic habitat is so superficial, it is sometimes possible to sample thoroughly an area of 10 m^2 or more by systematically picking up and examining leaves. However, this is a potentially destructive form of sampling since it disrupts the structure of the habitat even if leaves are returned (see Culver and Šereg 2004 for an example). In some cases, baiting has been performed using raw pieces of shrimp placed in a 500 ml plastic water bottle, which is cut in half and the top inverted into the bottom. This method sometimes yields numerous amphipods, isopods, and planarians, but, in the USA, more frequently the traps are found by raccoons (*Procyon lotor*) and destroyed. The Bou–Rouch pump, widely used to sample the underflow of streams (see Bou and Rouch 1967, chapter 6), is ineffective because there is neither sufficient water nor coarse sediment in seepage springs. Shallow wells have usually been sampled by lowering a jar baited with raw shrimp, with holes punched in the lid, and leaving it for 24 hours or more.

continued

Box 2.1 *Continued*

Leijs et al. (2009) described a device for sampling the invertebrate fauna of groundwater-fed sites, and it holds considerable promise for sampling larger seepage springs as well as small groundwater-fed wetlands. They located groundwater upwelling sites in late summer when they were identifiable by the presence of green vegetation. A 90 mm hand auger with several extensions was used to drill a hole until the underlying rock became too hard. A 90 mm diameter PVC pipe with 4 mm slots in the bottom 40 cm was inserted in the hole. A manual diaphragm pump with 25 mm tubing was used to clean out the bore until the water became clear. Water can be periodically pumped from the bore and filtered through a plankton net or other device. Efficiency of this device increased after several months, apparently because the open space near the end of the pipe was colonized by a number of stygobiotic invertebrates.

2.3.2 History of biological studies of the hypotelminorheic

Meštrov (1962, 1964) was probably the first to sample and certainly the first to describe the hypotelminorheic habitat, one that he thought was important in understanding the colonization of and evolution in subterranean habitats in general. As is often true of first descriptions of new habitats, it was not entirely clear what was unique to the original site and what were critical general features of the habitat. There has been this duality to the definition of 'hypotelminorheic' ever since its discovery—it is not only a site with a set of defined physical characteristics, but it is also, and perhaps more importantly, a shallow subterranean habitat isolated from streams (where the hyporheic occurs) and regional groundwater, both of which harbour stygobionts.

The study of hypotelminorheic and related habitats was not taken up by many other researchers, with the exception of Meštrov's student, R. Lattinger (1988). In the USA, the study of hypotelminorheic habitats was taken up in the mid-1960s, due to the impetus of J. Holsinger. He explored and sampled these habitats in a systematic manner. Rock Creek Park, in the heart of Washington, DC and administered by the National Park Service, was especially important in this regard. Stygobiotic flatworms, snails, isopods, and especially amphipods that were limited to these shallow subterranean habitats were collected and described. An astonishing total of five species of the amphipod genus *Stygobromus* were found in Rock Creek Park (Culver and Šereg 2004, Pavek 2002), signalling that a rich, interesting fauna was present in these little-studied, small habitats. The study of the hypotelminorheic in Europe lay dormant until the recent work of C. Fišer, S. Gottstein, and T. Pipan (Culver et al. 2006b, Culver and Pipan 2011, Fišer et al. 2007, 2010).

2.3.3 Overview of the hypotelminorheic fauna

Amphipods have been the most studied group in hypotelminorheic habitats, both because many are stygobiotic and because they are usually the numerically dominant macro-invertebrates. Based on

Table 2.4 Dissolved organic carbon (mg L^{-1}) for seeps, small springs, and hyporheic sites on Nanos Mountain, Slovenia. Categories, except for hyporheic, are also analysed in Table 2.2.

	Hyporheic	*Niphargus* seep	Other seeps	Small springs
Mean	2.77	4.52	1.63	2.72
Standard Deviation	3.79	3.72	1.40	2.48
Minimum	0.41	0.83	0.4	0.13
Maximum	10.44	9.89	5.53	7.07
N	6	7	11	12

Table 2.5 Amphipod species found in three study areas. Modified from Culver et al. (2006b).

Family	Species	Ecological classification
George Washington Memorial Parkway, USA		
Crangonyctidae	*Crangonyx shoemakeri*	stygophile
	Stygobromus pizzinii	stygobiont
	Stygobromus sextarius	stygobiont
	Stygobromus tenuis potomacus	stygobiont
Gammaridae	*Gammarus fasciatus*	surface/stygophile
Nanos Mountain, Slovenia		
Niphargidae	*Niphargus stygius*	stygobiont
	Niphargus tamaninii	stygobiont
Crangonyctidae	*Synurella ambulans*	stygophile
Gammaridae	*Gammarus* sp.	surface/stygophile
Medvednica Mountain, Croatia		
Niphargidae	*Niphargus foreli*	stygobiont
	Niphargus stygius licanus	stygobiont
	Niphargus tauri medvednicae	stygobiont
Crangonyctidae	*Synurella ambulans*	stygophile

the three areas that have been studied (Croatia, the USA, and Slovenia), four or five amphipod species can occur within a small area of a few km², and half or more are stygobionts (Table 2.5). As in other aquatic subterranean habitats, *Niphargus* dominates in Europe and *Stygobromus* dominates in the USA.

In a broader geographical study of the lower Potomac basin (Culver and Pipan 2008) which included the species from George Washington Memorial Parkway listed in Table 2.6, a total of 15 crustaceans and molluscs were common in seepage springs. Of these, five species were habitat specialists, found only in the hypotelminorheic and seepage springs. Two were stygobionts found in other subterranean habitats, six were stygophiles, and two were accidental inhabitants that may occasionally establish breeding populations in the hypotelminorheic, perhaps entering the habitat during times of flood. For nearly all of the species, it is the hypotelminorheic and not the seepage spring itself

(a groundwater/surface water ecotone) that is their primary habitat. However, the stygobiotic isopod *Caecidotea kenki* is likely concentrated around the ecotone itself rather than the groundwater (Fong and Kavanaugh 2010).

When the fauna of slightly deeper shallow subterranean habitats, such as shallow wells, is included, the situation becomes more complex. Culver et al. (2012a) listed localities and habitats for all 24 described species of stygobionts (4 planarians, 1 snail, 14 amphipods, and 5 isopods) from the Coastal Plain and Piedmont of the District of Columbia, Maryland, and Virginia. Except for some very localized patches of carbonate rocks, there are no major karst areas and only a handful of caves are known. The majority of species (14) are known only from seepage springs and small springs (Table 2.7), and one other species (the planarian *Sphalloplana hypogea*) is known only from the artificial analogue of a seepage spring—a tile drain (see Fig. 2.6). Four other species (the

Table 2.6 Species of amphipods, isopods, and gastropods found in seeps in the lower Potomac River drainage and environs of Washington, DC. Modified from Culver and Pipan (2008).

	Species	Ecological category	Hypotel minorheic specialist	Troglomorphic
Amphipoda:	*Stygobromus sextarius*	stygobiont	yes	yes
	Stygobromus kenki	stygobiont	yes	yes
	Stygobromus hayi	stygobiont	yes	yes
	Stygobromus tenuis potomacus	stygobiont	no	yes
	Stygobromus pizzinii	stygobiont	no	yes
	Crangonyx floridanus	stygophile	no	no
	Crangonyx shoemakeri	stygophile	no	no
	Gammarus minus	stygophile	no	no
	Crangonyx palustris	accidental	no	no
	Crangonyx serratus	accidental	no	no
	Crangonyx stagnicolous	accidental	no	no
	Gammarus fasciatus	accidental	no	no
Isopoda:	*Caecidotea kenki*	stygobiont	yes	weakly
	Caecidotea nodulus	stygophile	no	no
Gastropoda:	*Fontigens bottimeri*	stygobiont	yes	weakly

Table 2.7 Frequencies of species found in caves and shallow subterranean habitats of the Coastal Plain and Piedmont of the District of Columbia, Maryland, and Virginia (USA). Data from Culver et al. (2012a).

Category	Number of species
Seepage and small springs only	14
Tile drain only	1
Shallow well only	4
Cave only	0
Seepage and small springs plus shallow wells	2
Seepage and small springs plus shallow wells plus tile drains	2
Seepage and small springs plus cave	1
Total	24

planarians *Sphalloplana holsingeri* and *S. subtilis* and the amphipods *Stygobromus obrutus* and *S. phreaticus*) are known only from shallow wells which intersect water several metres deeper than that of hypotelminorheic habitats. All of the wells where these species were found have disappeared, although *S. phreaticus* has also been found along the dirt and clay banks of a deeply eroded stream. There are likely other yet undiscovered species living a few metres beneath the surface, but with the disappearance of shallow wells in the region, there are few opportunities to sample the habitat. Only five stygobiotic species were found in both the hypotelminorheic and habitats deeper than the hypotelminorheic. The isopod *Caecidotea pricei* is primarily a cave-dwelling species outside of the Piedmont and Coastal Plain, and has been found in the only well-sampled cave in the Piedmont and Coastal Plain. It has only been found once in a small spring. The remaining four species—the isopod *Caecidotea phreatica* and the amphipods *Stygobromus indentatus*, *S. pizzinii*, and *S. tenuis tenuis*—are widespread and common, and seem to be habitat generalists.

As is true for subterranean species in general, ranges of hypotelminorheic species are small, and even within their ranges, habitat occupancy is low. For example, the amphipod *Stygobromus caecilius* is known from a single seepage spring, and

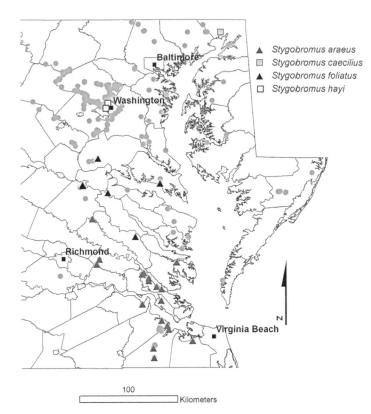

Figure 2.8 Distribution of *Stygobromus araeus*, *S. caecilius*, *S. foliatus*, and *S. hayi* in the Coastal Plain and Piedmont of the District of Columbia, Maryland, and Virginia, USA. Grey dots represent all sampling sites with stygobionts. From Culver et al. (2012a).

S. foliatus is known from five widely dispersed seepage springs up to 100 km apart (Fig. 2.8). There are several more widespread species which are also rather common within their range. The subspecies of *Stygobromus tenuis* range over several hundred km, and are found in hundreds of sites (Fig. 2.9, Culver et al. 2012a).

Several observations point to the idea that the distribution of hypotelminorheic species may actually be much more widespread and frequent within their range. One is the remarkable collection of *Stygobromus tenuis tenuis* in flooded terrestrial pitfall traps near an intradunal pond at the tip of the Delmarva Peninsula, which forms the western boundary of the Chesapeake Bay (S. Roble, pers. comm.). No seepage springs or springs were nearby. Another is the frequent collection of stygobiotic species at tile drains, especially in the thick soils of the glaciated American Midwest (Hubricht and Mackin 1940). They must have colonized these sites from somewhere. Finally, most

of the seepage spring sites in the Fort Hunt area of the George Washington Memorial Parkway in Virginia were the result of leaking buried water pipes, and the sites disappeared when the pipes were relocated (E. Oberg, pers. comm.). Collecting sites may be places where there are concentrations of animals, either because of nutrient inputs or because of relatively large water-filled spaces in the shallow subsurface. This also gives currency to the idea of the hypotelminorheic as the aquatic edaphic zone.

2.3.4 Species richness in hypotelminorheic habitats

Reliable estimates of species richness of hypotelminorheic habitats are especially difficult to obtain both because of the difficulties of collecting, which can often only be done in spring, and because of the very local differences in species composition among seepage springs. In a 5 km long section of Rock

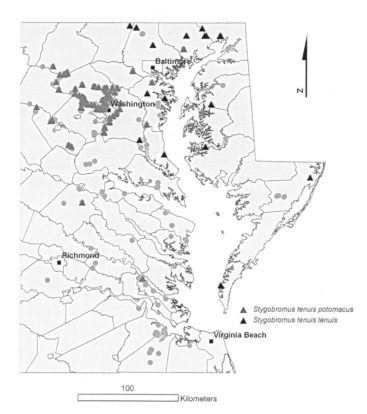

Figure 2.9 Distribution of the subspecies of *Stygobromus tenuis* in the Coastal Plain and Piedmont of the District of Columbia, Maryland, and Virginia, USA. Grey dots represent all sampling sites with stygobionts. From Culver et al. (2012a).

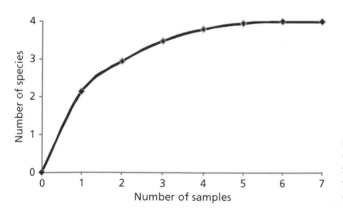

Figure 2.10 Accumulation curve, based on 50 randomly drawn samples, of the amphipod fauna of seven seepage springs in Rock Creek Park, Washington, DC, USA. From Culver and Pipan (2011), used with permission of John Wiley & Sons.

Creek Park in Washington, DC (USA), seven seepage springs were intensively and repeatedly sampled. The species accumulation curve (Fig. 2.10) reached an asymptote of four species of *Stygobromus* after an average of four seeps. Estimates of missing species, such as the Chao 2 estimate (see Box 2.2), also yielded the same result—four species

were present. There are no data available that indicate how many times a given seepage spring needs to be sampled (but see chapter 3 for a case for individual epikarst drips), but at least in the lower Potomac basin, sampling outside of springtime, when discharge rates are high, is unproductive (Culver and Šereg 2004).

Box 2.2 Estimating species richness

There has been a long tradition among subterranean biologists of emphasizing how many undiscovered species there are. In one way, this is appropriate because of the high levels of endemism in both the SSH and cave fauna coupled with the relatively low frequency that available habitat has been sampled. Estimates of species richness or the geographical pattern of species richness have been treated as premature by some speleobiologists. Combined with a general lack of use of multivariate quantitative methods in the discipline (see Herrando-Perez et al. 2008), this has resulted in relatively few attempts to estimate either sampling completeness or total species richness of both known and unknown species. There have however been several attempts (Schneider and Culver 2004, Pipan and Culver 2007b, Zagmajster et al. 2010, Christman and Zagmajster 2012, and the papers in the special issue of *Freshwater Biology* devoted to analysis of European patterns of groundwater biodiversity [Gibert and Culver 2009]), and the techniques employed (see Colwell 2009 for the essential software) deserve wider recognition and use by subterranean biologists.

The easiest way to visualize sampling completeness is to make a graph of species numbers comparing n randomly drawn samples for n = 1,2,3 . . . k, where k is the total number of samples. The software program EstimateS (Colwell 2009) accomplishes this efficiently as well as providing the analytical formulas (called Mao Tau estimates, Colwell et al. 2004) for the expected number of species. In this and other measures, either abundance or incidence (presence/absence) data may be used. Examples of its use for an SSH dataset in this context are in Pipan and Culver (2007b) and Figs. 2.10 and 3.14. Pipan and Culver's study is a rare

example of where, at least at some scales, sampling was complete.

A second set of techniques that has been used with some success is that of estimating the number of missing species. A variety of techniques are available—the most popular among subterranean biologists, and conceptually the simplest, are the Chao1 and Chao2 estimates (Chao 2005). In its simplest version, the formula for incidence-based data (the most frequent type of data available) is:

$$\hat{S}_{Chao2} = S_{obs} + \frac{Q_1^2}{2Q_2}$$

where $\hat{S}_{Chao2}$ is the Chao2 estimate, S_{obs} is the observed number of species, Q_1 is the frequency of unique species (ones found in only one sample), and Q_2 is the frequency of species found in exactly two samples. Deharveng et al. (2009b) prefer the jack-knife 1 estimate:

$$\hat{S}_{jack-knife1} = S_{obs} + \frac{Q_1(m-1)}{m}$$

where m is the number of samples. Both estimators utilize the number or frequency of unique species (single site endemics) to approximate the number of missing species. In most situations, as the number of unique species tends to zero, the observed and Chao2 estimates converge. However, in many subterranean communities, the number of single site endemics does not approach zero as the number of samples increases, and the effect of this on the estimators has not been investigated.

2.3.5 Ecology of hypotelminorheic organisms

Fišer et al. (2010) have done the only study to date about niche separation of different species living in the hypotelminorheic. They investigated the co-occurrence and water chemistry of two species of *Niphargus* that are often found in hypotelminorheic habitats. Both species—*N. sphagnicolus* and *N. slovenicus*—occur not only in hypotelminorheic habitats, but have also been found in surface waters associated with groundwater. The type locality of *N. slovenicus* is a forest ditch near the city of Kranj,

Slovenia, and the type locality of *N. sphagnicolus* is a small *Sphagnum* wetland in the middle of Ljubljana, the capital of Slovenia. Both of these habitats are likely groundwater-fed. They investigated the two *Niphargus* species by examining a seepage spring and a spring in 65 3×5 km grids in the alluvial plain of the Sava River between Kranj and Ljubljana. In each grid, a seepage spring and spring were sampled, although not all grids had both habitats. Of the 110 sites examined, 19 had *N. slovenicus* and 17 had *N. sphagnicolus*. Neither species showed a preference for seepage springs or for springs, and were found in nearly the same number of springs and seepage springs. However, they never co-occurred, even though they were often found in nearby sites. If their distributions were independent, three co-occurrences were expected which significantly differed from observed (Fisher's Exact Test, p = 0.042). Fišer et al. (2010) found differences in pH, oxygen concentration, and conductivity, but not temperature in the sites with the two species. Based on a generalized linear model (GLM) with variables added according to the Akaike information criterion (AIC), they found that a simple model involving only pH correctly distinguished *N. slovenicus* sites from *N. sphagnicolus* sites 84 per cent of the time. *N. slovenicus* is confined to more basic and more oxygenated waters, while *N. sphagnicolus* is found in more acidic, less oxygenated waters. An important point to emerge from their study is that not all seeps are identical, nor can they be expected to harbour the same species.

Fong and Kavanaugh (2010 and unpublished) studied the relationship between temperature and abundance of two species found in a seepage spring along Pimmit Run in the George Washington Memorial Parkway in Virginia, USA. The amphipod *Stygobromus tenuis potomacus* is common in many seepage springs in the area (see Fig. 2.9), and is eyeless and depigmented (Fig. 2.11). The isopod *Caecidotea kenki* is also common in many seepage springs in the area, but typically retains pigment and small eyes (Fig. 2.11) but individuals are variable with respect to both pigment and eye development. Their abundance with respect to temperature is quite different (Fig. 2.12). *S. tenuis potomacus* abundance shows a strong, significantly negative relationship with temperature, indicating that it prefers cooler temperatures. The lowest temperature recorded in the seepage spring was 9.4 °C, and it is possible that if seepage spring temperature gets colder, *S. tenuis potomacus* would retreat back into the hypotelminorheic. *C. kenki* shows a non-linear quadratic relationship, with a maximum abundance at 14 °C. *S. tenuis potomacus* is a true hypotelminorheic inhabitant. When temperatures in the seepage spring are cooler, *S. tenuis potomacus* moves out of the hypotelminorheic into the relatively nutrient-rich seepage spring. This is consistent with the

Figure 2.11 Photographs of the amphipod *Stygobromus tenuis potomacus* (top) and the isopod *Caecidotea kenki* (bottom), both inhabitants of the seepage spring at Pimmit Run, George Washington Memorial Parkway, Virginia, USA. Photos by W.K. Jones, used with permission. (see Plate 9)

hypothesis that *C. kenki* inhabits the seepage spring itself rather than the associated hypotelminorheic, which will be colder than the seepage spring during the warmer months. The mean annual temperature of hypotelminorheic water should be close to 12.1 °C, the mean annual air temperature of Washington, DC. In spite of this microhabitat difference, both are part of the obligate fauna of aquatic SSHs. As was the case for the study of Fišer et al. (2010), inhabitants of seepage springs and the hypotelminorheic showed microhabitat separation.

Fišer et al. (2007) examined the relationship between a surface-stream dwelling amphipod *Gammarus fossarum* and a stygobiotic amphipod, *Niphargus timavi*, inhabiting a hypotelminorheic-like habitat in Slovenia with a sinking stream perched above the water table and unconnected with any

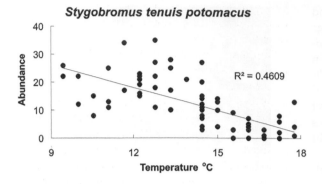

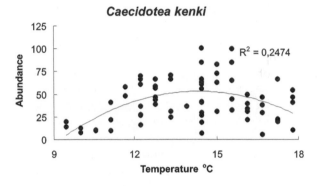

Figure 2.12 Graphs showing relationship between water temperature and abundance of *Stygobromus tenuis potomacus* (upper panel) and *Caecidotea kenki* (lower panel), in Pimmit Run, George Washington Memorial Parkway, Virginia, USA. For *S. tenuis potomacus*, the quadratic term was not significant, but the linear term was ($p<0.001$) and accounted for 46 per cent of the variance in abundance. For *C. kenki*, the quadratic term was significant ($p<0.001$), and accounted for 24 per cent of the variance in abundance. Data from Fong and Kavanaugh (unpublished), used with permission.

deep groundwater (see section 2.2.4). This study looked at the questions both of how surface and subsurface species interact, and what prevents subterranean species from extending much distance in surface habitats. In the study site, the small stream sinks below the primary spring, flows for about 150 m below the surface and resurges again. *G. fossarum* never reached the stretch above the sink, but *N. timavi* occurred along the 250 m surface watercourse. After the re-emergence of the stream, *G. fossarum* dominated but both species were present for a distance of nearly 1 km. Fišer et al. (2007) suggest that the dominance of *G. fossarum* in the part of the stream that is both permanent and continuously connected via surface waters to the Reka River is due to its higher reproductive potential compared to *N. timavi*. More interesting is the persistence of a stygobiotic species in a surface stream. Obviously the presence of light does not prevent it from surviving on the surface, but the presence of a competitor (*G. fossarum* in this case), severely limits its surface distribution. The presence of a subterranean habitat (the underground portion of the stream),

may also be critical for the continued presence of surface populations of *N. timavi*. In other words, the subsurface population may be a source population and the surface-stream population a sink. Gottstein et al. (2010) report on what may be a similar situation with a population of *Niphargus dalmatinus* occurring in surface waters near a spring emptying into the Cetina River in Croatia.

2.3.6 Adaptations to the hypotelminorheic

In its general morphological aspect, hypotelminorheic species are indistinguishable from deeper cave species, according to the preliminary morphological studies of *Stygobromus* amphipods by Culver et al. (2010), also reviewed in chapter 3. Part of the paradigm of adaptation to subterranean life (Poulson and White 1969, Culver and Pipan 2009) is a reduction in number of eggs, increased longevity, reduced metabolic rate, and other modifications to life in food scarce aphotic environments. We know of no life history studies of hypotelminorheic species, but there is one study of the metabolic rate of a hypotelminorheic species. Culver and Poulson (1971) studied *Stygobromus tenuis potomacus*, found in hypotelminorheic habitats in the Middle Atlantic region of the USA (Culver et al. 2012a). Its standard metabolic rate (SMR) was intermediate (2.1 ± 0.6 µg O_2/g/hr) between that of two other subterranean *Stygobromus*—*S. emarginatus* and *S. spinatus*. Both of these species are primarily epikarst species (see chapter 3), although *S. emarginatus* can also be common in small cave streams. Thus there seems to be no difference among species in different subterranean habitats.

2.4 Summary

The shallowest of all subterranean habitats are hypotelminorheic habitats and their associated outlets (seepage springs), which are isolated from lower waters by a clay or other impermeable layer. Inconspicuous wet spots, and an extreme example of an isolated wetland, they appear in clusters in areas with sufficient hydraulic head, and a clay layer. Flows can be zero during warm, dry periods, but water is retained in colloidal clays where the animals probably burrow. Little studied, their geographical distribution is unknown, but there are hypotelminorheic habitats in Croatia, Slovenia, and the USA. Hypotelminorheic water tends to approximate the mean annual temperature of the region, have moderately high conductivity (350 µS cm^{-1}), and near neutral pH. Organic carbon is present in particulate form (decaying leaves) and dissolved organic carbon is moderately high (4 mg L^{-1}). The chemical signature of hypotelminorheic water is distinct from that of other surface or near-surface water.

The agricultural practice of tiling fields creates an artificial hypotelminorheic-like habitat, and hypotelminorheic animals can be collected at the outlet of the tiling—tile drains. Other shallow subterranean habitats can be present in areas where hypotelminorheic habitats are present, but at slightly greater depths. The most interesting of these are shallow wells, extending less than 10 m vertically. Unfortunately, nearly all of these wells have been destroyed, leaving nothing but the historical records.

Sampling animals in the hypotelminorheic is difficult because there are no pumps or continuous sampling devices that have proved effective. Hand sampling is most productive during periods of higher water flow in the spring. Multiple seepage springs (four in the case of one study) need to be sampled in order to collect all the species. The first biologist to both describe the habitat and report the presence of stygobionts was the Croatian biologist Meštrov (1962). Work in the USA was initiated by J. Holsinger, working in habitats in the lower Potomac River basin near Washington, DC.

In Croatia, Slovenia, and the USA, the macroscopic fauna of seepage springs is dominated by amphipods, but stygobiotic isopods and molluscs are also known. Regionally, the fauna can be quite rich. Culver et al. (2012a) report 24 stygobiotic species from the hypotelminorheic and other similar shallow subterranean habitats of the Coastal Plain and Piedmont of the mid-Atlantic region of the USA. The majority of these species are known only from hypotelminorheic habitats. Ranges of hypotelminorheic species tend to be small, but some are quite common within their range, such as the amphipod *Stygobromus tenuis*. The occurrence of *S. tenuis* in artificially created habitats such as tile drains and in intradunal sands suggests it is widespread in the shallow subsurface, a kind of aquatic edaphic species, that becomes abundant when suitable food and habitat are available.

Potentially competing hypotelminorheic species tend to occupy water with different pH and oxygen concentration, based on a study of two amphipod species by Fišer et al. (2010). Fong and Kavanaugh (2010) found that the isopod *Caecidotea kenki* occupied the seepage spring itself while the amphipod *Stygobromus tenuis potomacus* occupied the hypotelminorheic itself. Nevertheless, both species are specialists for their respective habitats. Finally, hypotelminorheic stygobionts can occur in surface habitats when no competitors are present.

CHAPTER 3

Epikarst: the soil–rock interface in karst

3.1 Introduction

Wherever there is soil covering a base rock, there is a zone of contact between the rock and the soil, typically consisting of an unconsolidated layer of rock mixed with soil—the regolith. This zone often has spaces larger than the soil above. When the base rock is water soluble at the pH of water in the area, this area is greatly enhanced by the dissolution of rock into small channels and cavities. These are karst areas. It is the shallow part of karst areas where stress release, climate, tree roots, and karst processes fracture and enlarge rock joints and cracks, creating a more porous zone over the carbonate rock in which only a few vertical joints and cracks occur (Bakalowicz 2012). This zone, the epikarst, is important as a site of cave formation, water storage, and a habitat for many species.

The word epikarst came into widespread use in the 1990s following the definition by Mangin (1973) of an epikarst aquifer as a perched saturated zone within the superficial part of the karst that stores a part of the infiltrated water (Bakalowicz 2012). Hydrogeologists, such as Mangin (1973), Bakalowicz (1995, 2012), and Williams (1983, 2008), have studied epikarst intensively, and not surprisingly have emphasized its role in storage of percolating water as a result of lateral movement of water. Typically 3–10 m thick, epikarst overlies the water infiltration zone, which is itself intersected by occasionally enlarged vertical fractures and conduits. Because of this, the base of the epikarst acts as an aquitard, a layer of low permeability, resulting in a local perched water table and a perched aquifer. Williams (1983) used the phrase subcutaneous karst

to describe the epikarst zone, but he too now uses the term epikarst (Williams 2008). Epikarst is often easy to visualize (Fig. 3.1) but is rather difficult to precisely define. Participants in a workshop on epikarst (Jones et al. 2004) decided upon the following definition of epikarst:

Epikarst is located within the vadose zone and is defined as the heterogeneous interface between unconsolidated material, including soil, regolith, sediment, and vegetative debris, and solutionally altered carbonate rock that is partially saturated with water and capable of delaying or storing and locally rerouting vertical infiltration to the deeper, regional, phreatic zone of the underlying karst aquifer.

The boundaries of epikarst are not always as clear as those in Fig. 3.1, and some biologists (Fong 2004, Bichuette and Trajano 2004) even include the upper parts of some cave passages in the epikarst.

The storage capacity of the epikarst zone is clear from the often reproduced ground penetrating radar diagram (Fig. 3.2) of Al-fares et al. (2002). According to Williams (2008) the typical porosity (per cent open space) of unweathered limestone is 2 per cent while that of epikarst typically exceeds 20 per cent. More generally, water storage in epikarst is the reason why cave streams typically have water for long periods of drought.

3.2 Chemical and physical characteristics of epikarst

3.2.1 Hydrology of epikarst

The connection of outflow from epikarst, which is measured by the output from drips in caves coming from epikarst, has a complex connection with

Figure 3.1 Photo of a quarry in Salem, Indiana, USA showing a distinct epikarst layer. Photo by A.N. Palmer, by permission.

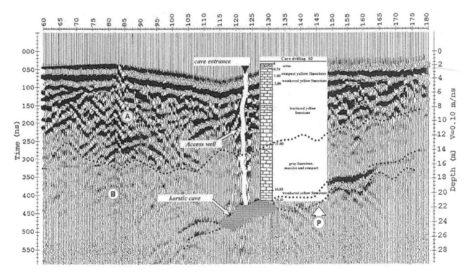

Figure 3.2 Ground penetrating radar profile through the Hortus field test site (Herault, France). A cave, dipping rock, a local fault, and the epikarst are shown. From Al-Fares et al. (2002).

precipitation and the infiltration zone (Kogovšek 2010, see Fig. 1.10). Typically, output spikes after several precipitation events, which cumulatively fill the cavities in epikarst and the infiltration zone. Based on continuous monitoring of three epikarst drips for three years in Postojnska jama (Slovenia), Kogovšek (2010) was able to estimate total surface catchment area of an individual drip using precipitation and drip rate data (Table 3.1). Even the

largest catchment area of a drip (I in Table 3.1) was quite small, approximately 200 m². These catchment areas are to a certain extent virtual since the actual connections between drips and the surface are complex and likely to be overlapping with at least some mixing. Additionally, the apparent catchment areas changed from year to year, and the relative contribution of each drip changed as well. For drip I in Postojnska jama, discharge was much more

Table 3.1 Yearly volume of outflow, in m³, through three drips (I, J, and L) in Postojnska jama, Slovenia with catchment area, in m², defined by dividing outflow by annual precipitation for that year. Data from Kogovšek (2010).

Drip	2003		2004		2005	
	m³	m²	m³	m²	m³	m²
I	299	244.8	175	158.5	216	204.9
J	12.2	10.0	7.3	7.3	10.4	9.9
L	0.16	0.1	0.06	0.1	0.186	0.2
Total	311.4	255	182.4	165.9	226.6	215

variable than epikarst temperature, which remains relatively constant throughout the year (Table 3.2), with temperature of the drips varying between 9 °C and 10 °C, and discharge varying from near zero to over 4000 mL min⁻¹. Outside temperature was intermediate in variability between drip discharge and drip temperature (Table 3.2), varying from –8.6 °C to 23 °C. Interestingly, both temperature profiles were platykurtic (thin-tailed) while discharge was thick-tailed, as well as having a long-tailed distribution to the right (positive skew). Positive skew is the

Table 3.2 Drip and outside temperatures, discharge for Drip I in Postojnska jama (Slovenia) from October 2003 to July 2004. Data courtesy of J. Kogovšek.

	Drip temperature (°C)	Discharge (mL min⁻¹)	Outside air temperature (°C)
Mean	9.61	833.70	8.41
Standard error	0.002	11.80	0.41
Median	9.6	373.68	8.9
Standard deviation	0.16	1071.45	7.52
Coefficient of variation	1.69	128.52	89.42
Kurtosis	−0.44	1.47	−0.96
Skewness	−0.66	1.61	−0.01
Range	0.7	4119.25	31.6
Minimum	9.1	0	−8.6
Maximum	9.8	4119.25	23
n	8408	8242	343

result of flood events, albeit on a small scale in this study. All in all, temperature and discharge patterns of drip I are a striking example of the complexity and heterogeneity of epikarst.

The storage capacity of the epikarst is determined by the thickness and connectivity of the epikarst, the average porosity, and the relative rate of inflow and outflow of water (Williams 2008). Inflow depends on precipitation and outflow depends on the characteristics of the infiltration zone. Water storage and movement in epikarst can be placed in a more general context of the hydrology of a karst aquifer. Rainwater recharges the karst aquifer and emerges at its springs by the following pathways:

1. A part of the water infiltrates directly and quickly through wide fractures and vertical conduits, from dispersed infiltration at the karst surface or from point infiltration from sinkholes (sinking streams), bypassing the epikarst and to some extent the infiltration zone, and
2. The other part is stored in the epikarst where it contributes to different processes. Some of the stored water is consumed by plants (Huang et al. 2011) and some percolates through the fine cracks (slow infiltration). During heavy rains, water stored in epikarst is flushed away into vertical conduits of the infiltration zone of the dolina-shaft system.

Bottrell and Atkinson (1992) found by direct observation of water soluble dye in White Scar Cave, England, that there were three epikarst flow components:

1. a rapid through-flow with a residence time of 3 days;
2. a short-term storage of 30–70 days; and
3. a long residence time of 160 days or more, water flushed out only during periods of high flow.

Clearly, the extent of the long residence time interval depends on rainfall variation. Kogovšek (2010) calculated overall residence times in two drips in Postojnska jama, based on oxygen isotope composition (Kogovšek and Urbanc 2007) of 2.5 months to over a year. Some precipitation may also contribute to soil moisture rather than epikarst storage (Tooth and Fairchild 2003).

3.2.2 Epikarst evolution

Epikarst is also an important site of dissolution of $CaCO_3$. Ford and Williams (2007) point out that about 70 per cent of the dissolution takes place in the top 10 m of limestone, the typical extent of epikarst. Organic carbon has a dual role, as a source of food for heterotrophic organisms in epikarst, and as the source of CO_2 and ultimately H_2CO_3 (carbonic acid). Epikarst represents the vertical extension of the soil and acts as a reservoir for the accumulation of organic matter (Bakalowicz 2012). Organic carbon is transformed by biological activity into CO_2, which, when dissolved in water, is the main solvent of carbonate rocks. Epikarst thus acts as a CO_2 reservoir, recharging the infiltration zone as water moves vertically downwards. This process drives the CO_2-rich air in soil down into the system of cracks and fissures of epikarst and disperses it throughout the whole infiltration zone. The epikarst gradually gives way to the main body of the infiltration zone, which comprises largely unweathererd bedrock with a porosity of less than 2 per cent.

This zone functions mainly as a transmission zone with minimal storage (Bakalowicz 1995).

Gabrovšek (2004) presents an elegant mathematical model of the evolution of the widening of fractures in epikarst as a result of the dissolution kinetics of the H_2O–CO_2–$CaCO_3$ system. He suggests that a constant recharge, rather than a constant hydraulic head (basically the elevation difference in the aquifer) assumption yields more realistic results. Clemens et al. (1999), also using a modelling approach, explored the connection between epikarst evolution and evolution of the karst aquifer itself. They concluded that epikarst development led to an acceleration of enlargement of cave passages in the karst aquifer.

The connections between the soil, epikarst, and the unsaturated zone below are shown in Fig. 3.3. Most of the water, and hence most of the available habitat, is in the epikarst zone, rather than in the infiltration zone below. Sampling of the fauna (see Box 3.1) is done beneath both the epikarst and the infiltration zone, and a few biologists still hold that there is no distinct epikarst fauna, and consider it

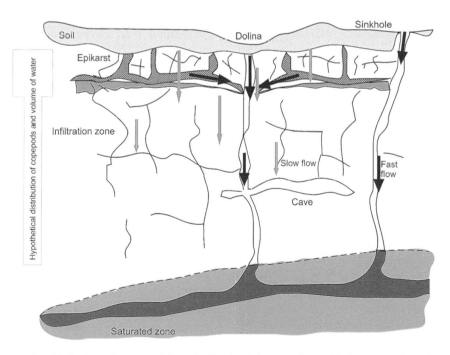

Figure 3.3 Conceptual model of epikarst. Grey arrows indicate the direction of slow water flow and black arrows are faster flow paths. From Pipan (2005a), used with permission of ZRC SAZU, Založba ZRC.

part of the upper vadose fauna (Sket et al. 2004). Our view, based on both hydrogeological evidence and evidence that drips from thin ceilings (indicating a thin unsaturated zone) yield more animals than drips from thick ceilings (see section 3.3.6), is that the epikarst zone is the primary habitat, but animals certainly occur in the infiltration zone as well.

A contrarian view of the geological role of epikarst is that of Šušteršič (1999). Based on his geological studies of the Dinaric karst where there is both rapid uplift and rapid denudation of karst, he sees epikarst as a zone of destruction of karst, which he calls speleothanatic space. There is little doubt that in circumstances of both rapid uplift and erosion caves become unroofed and a destruction zone can be identified. Even in this case, epikarst can still have major hydrological function for water storage, even if epikarst is transitory in a geological time frame. Šušteršič's work does point out the ephemeral nature of epikarst, more so than the underlying caves.

Kresic (2013) offers two objections to the widespread use of the term epikarst. Firstly, he argues that is absent in many places. This is an empirical question, and its presence in flank margin caves and hypogenic caves, for example, is little studied (Jones 2013). Curiously, Kresic (2013) uses the presence of vertical shafts and sinkholes as evidence for epikarst, but this is not an integral part of epikarst, and Bakalowicz (2004) even argues that it not even part of epikarst, since it represents rapid transmission routes through upper karst layers. Secondly, he quite correctly points out that epikarst is not a usable aquifer by humans, but that speaks to its size, not its presence.

3.2.3 Geographical variation of epikarst

The extent of epikarst development varies with climate, history, geological context, and elevation. Carbonate rocks do not always have a functioning epikarst. The weathered epikarst 'skin' can be scoured off by glaciation (Williams 2008). In many alpine areas where the carbonate rock has been tectonically stressed and deformed during uplift, and then later on exposed by rapid erosion and valley incision, relatively wide fissures can be opened and extended by dissolution of $CaCO_3$. This process

simultaneously extends the epikarst zone and reduces its storage capacity. Sometimes epikarst never develops, especially in coralline limestone and chalk, where primary porosities are so high (20–45 per cent) that little dissolution occurs because of the rapid downward movement of water. Examples of this include the 'stone forest' regions of China, Sarawak in Borneo, and elsewhere (Ginés et al. 2009). The opposite conditions of very low primary and secondary porosity in the uppermost zone of carbonate rock can occur under tropical climate because of 'case-hardening', a consequence of secondary deposition of carbonate in primary pores immediately beneath the surface (Ireland 1979). A similar effect occurs in arid zones, including calcrete aquifers (see chapter 5). Together, these variations suggest that epikarst aquifers and, by implication, the epikarst habitats are best developed in temperate regions with moderate rainfall, a hypothesis borne out by what is known of the distribution of species limited to epikarst (see section 3.3.5).

3.2.4 Physico-chemistry of epikarst

While drip rate is highly variable (Table 3.2), water temperature is not, at least as measured in drip pools. For example, water temperature measured at hourly intervals from 27 March 2007 to 29 July 2009 in the Pivka River in Pivka jama (Slovenia), varied between 0.5° C and 17.1° C while temperature in an adjacent drip pool only varied between 8.9° C and 9.4° C. The lower variability of the drip pool is because there is a longer underground residence time of water in drips compared to water in cave streams (Table 3.3).

Table 3.3 Summary of hourly temperatures in a drip in Postojnska jama and the river in Pivka jama from 27 March 2007 to 29 July 2009. Both sites are in the Postojna Planina Cave System (Slovenia).

	Drip pool	Cave stream
Mean	9.21	9.42
Coefficient of variation	1.17	32.58
Standard deviation	0.11	3.07
Range	0.47	16.60
Minimum	8.90	0.51
Maximum	9.37	17.11
n	19,344	19,344

There have been a number of reports on the inorganic chemistry of water dripping from epikarst, and geochemists have been especially interested in concentrations and kinetics of the Ca^{2+}–CO_3^{2-}–$CaCO_3$ system because secondary carbonate deposits result from precipitation of $CaCO_3$ from epikarst drip water. The mechanism of precipitation is the outgassing of CO_2 when drip water is exposed to air (Palmer 2007). Pipan (2003, 2005a) provided extensive data on inorganic chemistry of drips in her study of the epikarst fauna of six Slovenian caves. These data are summarized in Table 3.4. As expected, conductivity was high, largely because of the high concentration of Ca^{2+} ions. Škocjanske jame had somewhat lower values of Ca^{2+} and Županova jama had somewhat higher values of Ca^{2+} (Table 3.4). The ceiling of Škocjanske jame has few if any stalactites and the ceiling of Županova jama was especially rich in stalactites, probably a consequence of the differences in Ca^{2+} concentration, with $CaCO_3$ deposition occurring in Županova jama. As is also typical for carbonate waters, pH was slightly basic, typically around 7.8. Other cations (NH_4^+, K^+, Na^+, and Mg^{2+}) were typically low, and varied little from cave to cave, with two exceptions. Cation concentrations, except for NH_4^+, were highest in Pivka jama, probably because there is a campground and associated structures above the cave. Dimnice had elevated levels of Na^+ (and Cl^-), perhaps because of the presence of a salt block to attract deer above one of the drips in an area where the ceiling was thin—less than 10 m (Pipan 2003). Among the anions, NO_3^- and NO_2^- were elevated only in Pivka jama, most probably related to the impact of the campground. None of the six caves are in agricultural areas, and this is somewhat atypical. Postojnska jama and Črna jama were especially unremarkable in their inorganic chemistry, with no indications of abnormal values for either means or coefficients of variation in any parameters. Nonetheless, Postojnska jama had the fewest copepod species in drips and Črna jama was average in its species richness (Pipan 2005a, Pipan and Culver 2007b). There were no cases with concentrations of PO_4^{3-} above the level of detectability of 0.05 m L^{-1} in any of the drips.

Meleg et al. (2011a) did a similar analysis for three Romanian caves for pH, conductivity, and NO_2^- as well as the concentrations of several heavy metals—Al^{3+}, Cr^{3+}, and Fe^{3+} (Table 3.5). Compared to Slovenian caves (Table 3.4), pH was consistently higher in the Romanian caves, above 8.0; conductivity, except in Peştera Vadu Crişului, was lower and nitrites were much higher in Romanian caves than in Slovenian caves. Differences in parent rock and anthropogenic impacts are likely the reasons for the discrepancy between the two countries. Concentrations of aluminum and chromium ions were similar in the three Romanian caves for which data are available, but Peştera Ciur Izbuc had elevated iron concentrations relative to the other two caves, possibly due to differences in composition of the parent rock.

Musgrove and Banner (2004a, 2004b) studied spatial and temporal variability of epikarst drips in Natural Bridge Caverns, Texas (USA), using not only concentrations of ions but also stable isotope ratios of strontium. An example of both spatial and temporal variability of Sr stable isotope ratios is shown in Fig. 3.4, for six drips between 25 and 150 m apart. Note that some drips are quite stable in their isotope ratios and others are highly variable. The authors suggest that the differences in ratios were the result of different groundwater residence times and water–rock interactions with overlying soils and rock. Kogovšek (2010) also provides detailed information about temporal patterns of ionic concentrations for three drips in Postojnska jama, Slovenia.

Two consistent themes emerge from all of the studies of geochemistry of dripping epikarst water. One is that drip water has high concentrations of the ions associated with $CaCO_3$ dissolution, e.g. Ca^{2+}, the result of water being in contact with carbonate rock for significant periods of time, i.e. weeks to months. The second theme is that there is considerable temporal and spatial variability in geochemistry, even at scales on the order of 10 m. As we show in section 3.3.6, the inhabitants of epikarst are affected by this heterogeneity.

3.2.5 Non-karst analogues of epikarst

With respect to its vertical position in the landscape, epikarst is the regolith of karst, albeit solutionally modified (White 2004). The air- and water-filled

Table 3.4 Average values and coefficients of variation (standard deviation × 100/mean) for pH, conductivity, and nine ions for six Slovenian caves. Ionic concentrations are in mg L^{-1}. Data from Pipan (2003, 2005a).

Cave		pH	Conductivity (μS cm^{-1})	NH_4^+	K^+	Ca^{2+}	Na^+	Mg^{2+}	NO_3^-	NO_2^-	SO_4^{2-}	Cl^-
Črna jama	Mean	7.84	376.67	0.08	0.39	36.64	0.94	0.71	0.64	0.004	4.91	1.14
	Coeff. var.	2.55	22.89	34.43	37.91	68.13	222.43	69.36	150.80	236.52	59.78	62.51
	n	70	70	25	25	25	25	25	25	25	25	25
Dimnice	Mean	7.70	330.04	0.11	0.42	36.86	2.16	0.95	3.16	0.001	5.17	5.54
	Coeff. var.	2.55	25.94	74.83	52.18	40.42	91.25	63.14	190.16	345.22	80.84	135.08
	n	45	45	45	45	45	45	45	45	45	45	45
Pivka jama	Mean	7.84	416.36	0.11	0.83	42.93	2.17	1.14	11.81	0.012	5.44	2.37
	Coeff. var.	3.39	27.73	69.95	50.41	61.69	94.82	60.56	122.62	399.73	46.81	69.84
	n	69	69	24	24	24	24	24	24	24	24	24
Postojnska jama	Mean	7.81	342.57	0.07	0.42	39.00	0.99	0.86	3.09	0.001	5.73	1.26
	Coeff. var.	2.38	27.28	52.63	32.31	36.42	42.80	31.25	74.57	433.17	47.79	36.00
	n	219	219	69	69	69	69	69	69	69	69	69
Škocjanske jame	Mean	7.87	308.40	0.12	0.45	23.64	1.40	0.88	1.05	0.004	6.40	0.37
	Coeff. var.	2.14	28.43	104.46	53.11	68.55	77.70	52.56	95.29	655.74	43.66	54.90
	n	43	43	43	43	43	43	43	43	43	43	43
Županova jama	Mean	7.65	371.29	0.10	0.46	47.83	0.88	1.09	0.31	0.001	8.74	1.40
	Coeff. var.	3.09	20.89	44.75	45.85	30.05	68.72	17.10	47.64	343.60	21.26	46.69
	n	45	45	45	45	45	45	45	45	45	45	45

Table 3.5 Average values and coefficients of variation (standard deviation × 100/mean) for pH, conductivity, NO_2^-, Al^{3+}, Cr^{3+}, and Fe^{3+} in three caves in Romania. Ionic concentrations are in mg L^{-1}. Data from Meleg et al. (2011a).

Cave		pH	Conductivity (µS cm^{-1})	Al^{3+}	Cr^{3+}	Fe^{3+}	NO_2^-
Peştera Ungurului	Mean	8.70	256.13	0.42	0.01	0.63	0.25
	Coeff. var.	4.02	28.88	169.05	200.00	115.87	96.00
	n	12	12	12	12	12	12
Peştera Vadu Crişului	Mean	8.36	380.84	0.11	0.01	0.31	0.46
	Coeff. var.	2.75	15.88	63.64	100.00	203.23	76.09
	n	12	12	12	12	12	12
Peştera Ciur Izbuc	Mean	8.49	265.31	0.54	0.01	2.94	0.44
	Coeff. var.	2.59	32.24	135.19	200.00	211.22	261.36
	n	12	12	12	12	12	12

spaces of non-karst regolith may also provide aquatic and terrestrial habitats (see chapter 4 for a discussion of some of these terrestrial habitats and chapter 6 for a discussion of some of these aquatic habitats).

From a hydrogeological point of view, the signature feature of epikarst is that it stores water, not enough for human consumption (Kresic 2013), but enough to be both aquatic habitat and to be an important geochemical agent. Although they are often much smaller in areal extent and volume, the typically non-karst hypotelminorheic habitats discussed in chapter 2 are also perched aquifers.

Although it is the aquatic epikarst habitat that has been studied most, epikarst is above the permanent saturated zone, and therefore a terrestrial habitat as well. The *milieu souterrain superficiel* (MSS) described by Juberthie et al. (1980b), and discussed in detail in chapter 4, is the non-karst analogue of terrestrial epikarst. Some MSS habitats are part of the regolith, such as erosional MSS, but others, such as scree slopes, are not.

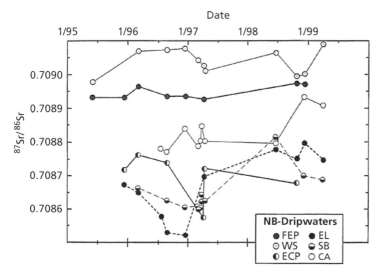

Figure 3.4 Temporal variations in $^{87}Sr/^{86}Sr$ values for epikarst drip waters sampled from Natural Bridge Caverns, Texas, USA, between 1995 and 1999. See Musgrove and Banner (2004b) for details about the locations of the six drips.

3.3 Biological characteristics of epikarst

3.3.1 Organic carbon and nutrients in epikarst

Organic carbon in drip water is especially interesting because the fauna of epikarst and caves is likely carbon- rather than nutrient- (nitrogen or phosphorus) limited (Simon and Benfield 2002, Simon et al. 2007), and because fluctuations in organic carbon in drip water may limit the utility of laminae in stalactites as a climate record (Ban et al. 2008, Dreybrodt 2011, Tooth and Fairchild 2003). The source of organic carbon in epikarst water is the soil. Rainwater does not contain organic carbon, but because of biological activity in the soil, water leaching from the soil into epikarst cavities has organic carbon.

Ban et al. (2008) did an extensive series of temporal and spatial measurements of organic carbon in three drips between 70 m and 100 m apart in Shihua Cave, China. All three drips exited from stalactites or other secondary $CaCO_3$ deposits. The temporal pattern of dissolved organic carbon (DOC) is shown in Fig. 3.5. The general shape of the pattern is the same for all three drips, but the amount of DOC varies from drip to drip. The highest DOC

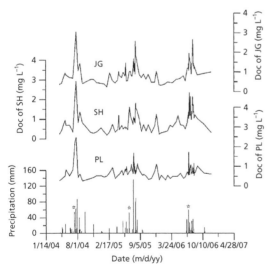

Figure 3.5 Rainfall amount and intra- and inter-annual variation of DOC in three drip sites at Shihua Cave, China, during the period from April 2003 to December 2006. The important rainfall events causing the increase in DOC concentration in drip water are marked with hollow stars. From Ban et al. (2008), used with permission of John Wiley & Sons.

value was recorded in drip JG, with a value of 2.76 mg L^{-1}. Mean values ranged from 1.06 mg L^{-1} at JG to 0.73 mg L^{-1} at PL. For two of the drips (JG and SH in Fig. 3.5) there was a positive relationship between DOC concentration and drip rate (discharge). Ban et al. (2008) suggest that the differences in the relationship between drip rate and DOC concentration are either due to differences in the length (and time) of the flow paths, with longer flow paths losing more DOC during transport, or to differences in the size of the reservoir feeding the drip, with larger reservoirs being diluted during major rainfall events.

Simon et al. (2007) found similar DOC concentrations in drip water in both Organ Cave, West Virginia, USA and Postojna Planina Cave System (PPCS) in Slovenia to that found in Shihua Cave. They also put epikarst carbon in a more general context of a karst aquifer with sinking streams, cave streams, and resurgences (Fig. 3.6 and Table 3.6). In Organ Cave, mean DOC concentrations in epikarst were 1.10 mg L^{-1} while in PPCS they were 0.70 mg L^{-1}. The differences are likely the result of different land uses. The land above Organ Cave is mostly pasture and the land above PPCS is forest. In addition to water entering cave passages through percolating water, both Organ Cave and PPCS had sinking streams. DOC concentrations in sinking streams averaged at least five times higher than in percolating water (Table 3.6). However, many conduits in both caves had no stream, and the only source of carbon was percolating water. While most organic carbon entering a cave from dripping water was dissolved organic carbon, the steady rain of organisms coming in through drips contributed to the standing crop of organic carbon. However, this contribution was minuscule compared to DOC, even though up to 1 copepod per drip per day entered Organ Cave (Pipan et al. 2006a). The amount of DOC in cave streams depended in large part on the relative contribution of sinking streams and epikarst drips. In PPCS, the sinking stream was much larger than in Organ Cave, and DOC concentration in the PPCS stream was accordingly higher.

The studies of Ban et al. (2008) and Simon et al. (2007) only considered the amount of organic carbon. An equally interesting question is the type, or quality, of organic carbon in epikarst. The source

Organic Carbon in Organ Cave

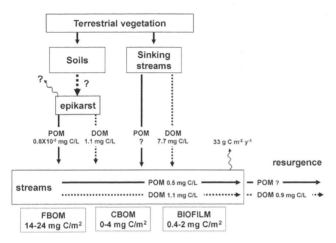

Figure 3.6 A conceptual model of energy flow and distribution (as organic carbon) in a karst basin with estimates of fluxes and standing crops for Organ Cave, West Virginia, USA. Standing stocks are particulate (POM) and dissolved (DOM) organic matter in the water column and fine (FBOM) and coarse (CBOM) benthic organic carbon and microbial films on rocks (epilithon). Solid and dashed arrows represent fluxes. Data are standing stocks of carbon except for respiration flux. Values for FBOM, CBOM and microbial film are taken from Simon et al. (2003); the whole-stream respiration rate (wiggly line) is from Simon and Benfield (2002); and the remaining values are from Simon et al. (2007). Modified from Simon et al. (2007), used with permission of the National Speleological Society (www.caves.org).

of organic carbon in the epikarst ultimately comes from the decomposition of material at or near the soil surface, e.g. leaf litter. There are two hypotheses about the nature of organic carbon that reaches the epikarst. One hypothesis is that, as a result of the action of microbial and other biological activity in leaf litter and soil, the remaining organic carbon leaching into epikarst cavities is difficult for organisms to break down, e.g. cellulose. The other hypothesis, and seemingly less likely, is that the intensive microbial processing of organic matter in leaf litter and soil results in more easily assimilated organic carbon, e.g. glucose.

Simon et al. (2010) examined these hypotheses for Organ Cave and Postojna Planina Cave System (PPCS). They measured specific UV absorbance

(SUVA) at 254 nm, a standard measure of the frequency of aromatic compounds. Higher SUVA values tend to mean the compounds are less reactive and less easy to metabolize but there are numerous caveats (Weishaar et al. 2003). They used three other measures of carbon quality using information from fluorescence spectroscopy:

1. Humification index (HIX) which measures the proportion of humic compounds, the higher the value, less metabolic availability.
2. Fluorescence index (FI) which distinguishes terrestrial (low values) from aquatic sources (high values).
3. Biological index (BIX) which indicates the relative importance of recent microbial contributions relative to terrestrial sources of DOC.

The pattern, which is very similar for both caves and for all measures of carbon quality, is shown in Fig. 3.7. Soil, one of the sources of DOC in epikarst, had relatively high SUVA and moderate HIX values. In contrast, epikarst drips had relatively low SUVA and low HIX values. Both SUVA and HIX values suggest that the organic carbon is more metabolically accessible than that of the soil, or at least with lower percentages of aromatic and humic compounds. FI values were lowest for soils and higher for epikarst drips (and all other samples). This indicates the microbial processing is important in epikarst water. BIX values were high in epikarst

Table 3.6 Estimates of dissolved organic carbon in mg L^{-1} from Organ Cave, West Virginia (USA) and Postojna Planina Cave System (Slovenia). From Simon et al. (2007), used with permission of the National Speleological Society (www.caves.org).

	Organ Cave	Postojna Planina Cave System
Input: sinking streams	7.67 ± 1.03	4.36 ± 0.46
Input: percolation water	1.10 ± 0.15	0.70 ± 0.04
In cave: streams	1.08 ± 0.32	4.75 ± 1.57
Output: resurgence	0.90 ± 0.17	2.67 ± 0.80

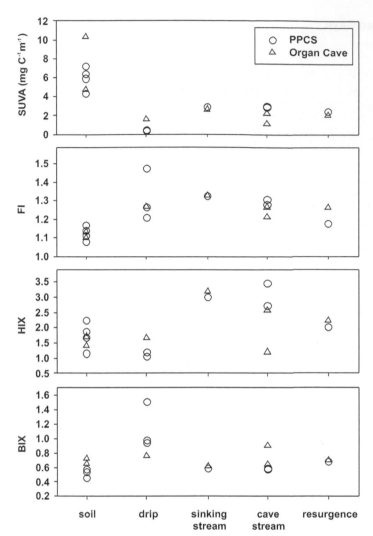

Figure 3.7 Specific UV absorbance (SUVA), fluorescence index (FI), humification index (HIX), and biological index (BIX) values for DOC samples from soil extracts, epikarst drips, sinking streams, cave streams, and resurgences at PPCS (Slovenia) and Organ Cave (West Virginia, USA) karst aquifers in September 2007. From Simon et al. (2010), used with permission of E. Schweizerbart'sche Verlagsbuchlandlung OHG (www.schweizerbart.de).

drips relative to soil, also indicating the importance of microbial processing. Thus their results make the second hypothesis more likely—carbon in epikarst is more available metabolically than in the soil, but more research is needed. The situation is more complicated with respect to carbon in sinking streams, but this is not of direct relevance to epikarst.

3.3.2 History of biological studies of epikarst

Beginning with Racoviţă's 1907 classic 'Essai sur les problèmes biospéologiques' (Moldovan 2006), biologists have recognized that much of the fauna observed in cave passages accessible by humans often occurs more frequently in cracks and crevices. In Racoviţă's time, the epikarst zone was completely unknown to hydrogeologists, and he can scarcely be faulted for not identifying it as a separate habitat. By the mid 20th century, several biologists, such as Petkovski (1959), became aware that there were stygobiotic copepods in caves with only percolating water. He recognized that the accumulation of water from above depended on fractured rock, and that there was water in tiny fissures and cracks which slowly flowed down from the ceiling. He believed that this habitat was the realm of some copepods

Box 3.1 Methods for collecting epikarst fauna

The epikarst habitat has rarely if ever been sampled directly. Instead, biologists have had to rely on indirect samples. Direct methods for sampling fauna in the epikarst are still being developed. In theory, drilling vertically into the epikarst zone and collecting water from voids is possible. A metal or plastic tube, closed at the bottom and with a series of holes some centimetres above the sealed end that is left in position for some time should act as a pitfall trap for fauna there—but that's in theory. In practice, there are two possibilities for sampling the fauna. Pipan and Brancelj (2001, 2004) achieved a breakthrough in sampling epikarst. Rarely if ever is it possible to sample this habitat directly, so they concentrated on a technique for indirectly but quantitatively sampling water exiting the epikarst through dripping water. In their sampling device, water from a ceiling drip is directed via a funnel into a 500 mL rectangular filtering bottle fitted on two sides with plankton netting of 60 μm mesh size, and the filtering bottle is placed within a sampling container (Fig. 3.8). Each sampling container has a drain 3 cm from its base such that collected animals and a small amount of water remain in the filtering bottle while most of the water passes through the filtration unit. Samples, which typically have a volume of about 50 mL, are preserved in either formalin or alcohol, formalin having the advantage of requiring less volume of liquid. Samples are then sorted and identified in the laboratory. Samples need to be collected at weekly or monthly intervals to reduce mortality effects in the sample, especially since larger cyclopoid copepods may prey on the generally smaller harpacticoid copepods. The device has also proven to be effective for the collection of terrestrial species living in the epikarst (Pipan et al. 2008). Pipan and Culver (2007b) give guidelines for how many such filters need to be used and for how long in order to sample completely the epikarst fauna. This sampling device can be combined with others that continuously measure discharge and water chemistry to provide a more comprehensive analysis.

The second method is to collect and filter water from small pools on calcareous slopes or on the bottom of galleries, which are filled with water from the trickles. The volume of the pools can vary from a few millilitres to a litre or more. Water from the pools can be collected by means of a pump (Fig. 3.9), which is very efficient in small rimstone pools or in the deep and narrow cracks on stalagmites that are filled with water. During the sampling, vigorous agitation of the water is recommended to collect particles from the bottom of the pools where most animals are attached. The aspirated water is filtered through either a plankton net or filtering bottle with a mesh size of 60 μm. The sample is preserved in either formalin or alcohol, and sorted in the laboratory.

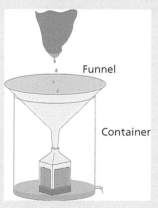

Figure 3.8 Diagram of filtering device to continuously collect invertebrates from drips. From Pipan (2005a), used with permission of Založba ZRC, Ljubljana, Slovenia.

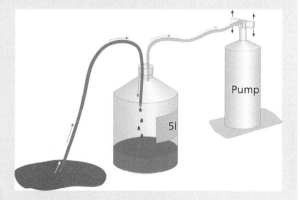

Figure 3.9 Device for collecting fauna from small amounts of percolating water in pools and cracks. From Pipan (2005a), used with permission of Založba ZRC, Ljubljana, Slovenia.

like *Speocyclops* and many harpacticoids. Thus he didn't recognize the infiltration zone as a habitat per se, but as a source of water that filled small depressions in walls, the 'realm of Parastenocarida'. Holsinger (1971) came to similar view with respect to a population of the amphipod *Crangonyx antennatus* living in Molly Waggle Cave in Virginia. Part of the population was in an old trough used for saltpetre mining during the American Civil War. He concluded that the only way the individuals could have got there was via what we would now call epikarst, but he reviewed it more as a dispersal corridor than a habitat with a sustainably reproducing population.

In his study of copepods in the Baget karst basin in France, Rouch (1968, pers. comm.) recognized that the small number of individuals occurring in pools in Grotte de Sainte-Catherine was much too small to constitute a viable population and concluded that there were populations in perched, i.e. epikarst, aquifers. Rouch had the advantage of collaborating with Mangin, one of the discoverers of epikarst (Mangin 1973). Other French biologists, notably Delay (1968) and Gibert (1986) studied the fauna of percolating waters, and included terrestrial species in their study, but did not distinguish different components of the zone of percolation.

Brancelj's discovery of a rich copepod fauna in drip pools in the shallow Slovenian cave Velika Pasica, which has no other water, but with relatively few reproducing individuals, led Brancelj (2002) to conclude that reproduction was occurring in 'small cracks around the cave'. Although he did not use the phrase epikarst, the cave is so shallow that most of the ceiling is epikarst. Unlike Petkovski and Holsinger, he held that reproduction was not occurring in the cave but in crevices in the cave ceiling and walls. Pipan (2003, 2005a) championed the idea that there was an epikarst habitat and fauna distinct from other subterranean habitats, and developed innovative sampling techniques (Pipan and Brancelj 2001). Pipan and Brancelj (2004) also demonstrated that the fauna of epikarst-fed pools was distinct from that occurring in drips themselves, a theme elaborated on by Pipan et al. (2010). Culver and Pipan (2011) argued that epikarst was one of several aquatic shallow subterranean habitats each of which harbours a unique, troglomorphic, stygobiotic fauna.

3.3.3 Faunal differences between drips and drip pools

While sampling drip pools is much easier than sampling drips, the sampling from pools is a biased sample of drip water. A striking example of this is the epikarst and drip pool fauna in Organ Cave, West Virginia, USA (Pipan and Culver 2005). Generic diversity of copepods was much lower in pools, and pools were dominated by *Bryocamptus* species (Fig. 3.10). Pipan et al. (2010) specifically

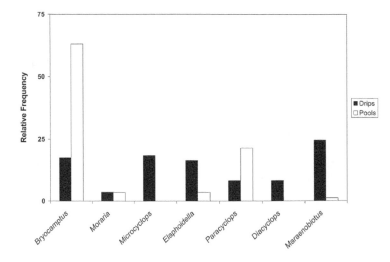

Figure 3.10 Relative abundance of different copepod genera in drips and in pools in Organ Cave, West Virginia, USA. Only the seven most common genera are shown. From Pipan and Culver (2005), used with permission of the National Speleological Society (www.caves.org).

addressed the question of whether the fauna of drip pools reflected the drip community, analysing data for 35 drips and associated pools in six Slovenian caves, with a total of 37 copepod species, including 25 stygobionts. Overall, the frequency of stygobionts was 1.5 times higher in drips than in pools, and the frequency of stygobionts that were epikarst specialists was three times higher in drips compared with pools. Both of these differences were statistically significant. The frequency of immature individuals, suggestive of reproduction at the site, was also lower in pools, with the exception of one artificially enlarged pool in Škocjanske jame. Pipan et al. (2010) suggest that there is increased juvenile mortality in pools and reduced reproduction, indicating that pools are not 'source populations', populations that are self-sustaining in the absence of migrants (Pulliam 1988).

With the exception of the work by Pipan, Meleg, and their colleagues, nearly all papers discuss samples of the epikarst fauna from pools, and hand-collected samples at that, e.g. Holsinger (1969). In spite of the bias of these samples, numerous new species have been found in this way, especially larger invertebrates such as amphipods (Culver and Pipan 2009, Culver et al. 2012b). The habitat of many of these species is listed as 'drip pools', largely because the species were described before epikarst was described or because the authors were unaware of the habitat. (See also Box 3.2.)

3.3.4 Overview of the epikarst fauna

Largely based on reports of species found in drip pool habitats in Botosaneanu's (1986) compendium of subterranean aquatic species, Culver et al. (2012b) list the following genera as having more than ten species known from epikarst habitats:

- Copepoda Cyclopoida—*Speocyclops* and *Diacyclops*.
- Copepoda Harpacticoida—*Elaphoidella* and *Parastenocaris*.

Box 3.2 Why do animals wash out of epikarst habitats?

Just as mineralized particles can be mobilized into the water column, so can copepods and other epikarst inhabitants. It is instructive to think of epikarst animals, not as living organisms, but as organic particles of varying sizes and shapes. A

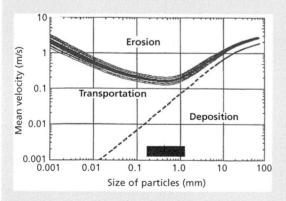

Figure 3.11 Hjulstrom curves for mobilization (erosion), transportation, and deposition of sediments. The body size range of epikarst copepods is shown as a black bar.
Modified from Gordon et al. (1999), used with permission of John Wiley & Sons.

great deal is known about the mobilization of mineralized particles, e.g. Hjulstrom curves (Fig. 3.11). Intermediate-sized particles between 0.3 and 0.6 mm are the easiest to mobilize. The body size of adult copepods ranges from 0.2 mm to 2 mm. Mineral particles 0.3 mm in diameter can be mobilized (erosion in Fig. 3.11) at a velocity slightly greater than 0.1 m s^{-1} and transported at velocities greater than 0.05 m s^{-1}. Organic 'particles', e.g. copepods, are easier to mobilize. Smith (1975) states that the critical velocity for organic matter with a density of 1.05 g cm^{-3} is about 1/6 that of an equivalent-sized mineral particle, suggesting that a passive copepod in the epikarst could be mobilized at velocities of less than 0.05 m s^{-1}. Smart and Friederich (1987) report that epikarst flow rates reach 4.5×10^3 m day^{-1} (0.05 m s^{-1}). It seems reasonable that such rates are often reached in the epikarst. Indeed it would be hard to explain the nearly continuous washout of copepods from the epikarst via drips if this were not the case. It is this property of copepods, i.e. their frequent occurrence in the water column as well as their small size, that makes them potential water tracers (Pipan and Culver 2007a).

- Isopoda—*Caecidotea* and *Proasellus*.
- Amphipoda—*Stygobromus*.
- Syncarida—*Iberobathynella*.

This is a broad brush picture that is highly incomplete but several features of the epikarst fauna emerge. Firstly, it is largely aquatic. No terrestrial genera are known with ten species found in epikarst habitats (Culver et al. 2012b). Secondly, at least numerically, copepods predominate with half of the genera listed above being copepods. Thirdly, the macroscopic species are dominated by isopods and amphipods.

Body sizes of epikarst inhabitants range up to approximately 5 mm, although a few amphipods reach up to 10 mm. Although rarely measured, the diameter of epikarst animals probably is less than 2 mm. The fish *Ituglanis epikarsticus*, described by Bichuette and Trajano (2004), inhabits drip pools rather than the epikarst itself.

For the most part, epikarst habitats are unrecognized, uncollected, or both, so accurate global numbers of species are hard to come by. An exception is the North American amphipod genus *Stygobromus*, intensively studied by Holsinger. Although it is known as a 'cave' genus, only 34 of 125 species known from central and eastern North America are known from cave habitats, primarily cave streams and phreatic lakes. The rest of the species are all known from shallow subterranean habitats (Fig. 3.12). In addition to 38 species found in epikarst habitats, *Stygobromus* species were collected from hypotelminorheic (chapter 2), hyporheic, and other interstitial habitats (chapter 6).

As was the case for hypotelminorheic habitats (Table 2.5), not all species found in epikarst are epikarst endemics, or even stygobionts. For the 35 drips in six caves studied by Pipan (2005a), the number of epikarst endemic species, non-epikarst endemic stygobiont species, and non-stygobiont species was about the same, ranging between 10 and 15 (Fig. 3.13). For all Slovenian records, non-epikarst endemic stygobionts and non-stygobiont species were approximately equal in number, while the number of epikarst endemics was approximately half of the other two categories. The reduced relative frequency of epikarst endemic species is not

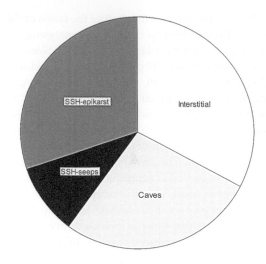

Figure 3.12 Subterranean habitats of 125 species of the exclusively subterranean amphipod genus *Stygobromus* from eastern and central North America. From Culver and Pipan (2009), based on data from J. Holsinger, used with permission of Oxford University Press.

surprising since only seven caves have been thoroughly investigated.

It is instructive to consider the relative abundance of the different categories of species—epikarst endemic, non-epikarst endemic stygobiont, and non-stygobiont—because one could hypothesize that the three categories represent different stages of adaptation, with non-stygobionts being the least adapted. For the six caves where drips were sampled directly (Pipan 2003, 2005a), it is possible to test this directly. Using overall abundance of each species in drip samples as the variable, the mean abundance of non-stygobionts was the lowest of the three groups, as expected (Table 3.7), but the presumably specialized epikarst endemics were less than half as abundant as the other stygobionts. Overall, the ANOVA was not significant (p = 0.0516, Table 3.7), indicating no significant differences in abundance among the ecological groups. The topic of adaptation to SSHs will be considered in more detail in chapter 12.

We use the phrase epikarst endemic for those species that have not been found outside epikarst drips and epikarst-fed drip pools. Of course, because of the method of collection, if for no other reason, it may be that species we have called epikarst endemics also occur in the infiltration

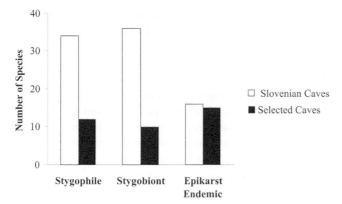

Figure 3.13 Histogram of number of copepod species reported from all Slovenian caves (open bars) and copepod species reported from epikarst habitats in six intensively studied Slovenian caves, according to the categories *stygophile, stygobiont,* and *epikarst endemic.* Epikarst endemics are also stygobionts, but are only listed under the *epikarst endemic* category. Data from Pipan (2005a) and Culver et al. (2009).

(unsaturated) zone of caves, a point raised by Sket et al. (2004). Figure 3.3 makes connection between the habitats clear. As the pioneering work of the hydrogeologists Bakalowicz, Mangin, and Williams demonstrates, most of the water is stored in the epikarst zone. This volumetric relationship suggests most individuals are in the epikarst rather than other areas, especially since the upper vadose zone is a region of vertically moving water. In addition, Pipan et al.'s (2006a) finding that the abundance of all stygobiotic copepod species except *Elaphoidella cvetkae* was negatively correlated with ceiling thickness. If the upper vadose were the major habitat, then most species should have a

pattern like that of *E. cvetkae.* Because all sampling is indirect, it cannot be known with certainty what the distribution of species in the epikarst and upper vadose is. While it is highly likely for the reasons given above that the major habitat is epikarst, many species may well occur in other sites in the upper vadose. Nevertheless, epikarst endemic is a very useful phrase to describe these species just as riparian species may occasionally occur away from stream margins.

Since the study of epikarst is just beginning, it is difficult to say much about the geographical extent of the epikarst, beyond the observation where a fauna exists. As a result of direct sampling of drips, an epikarst fauna, with stygobiotic copepod species, has been reported from Romania (Meleg et al. 2011b, 2011c) and West Virginia, USA. (Pipan and Culver 2005), as well as Slovenia. As a result of sampling of drip pools, stygobiotic copepods have been reported from China (Pipan et al. 2011b), Spain (Camacho et al. 2006), Sicily (Cottarelli et al. 2012), and Thailand (Brancelj et al. 2010). The description of a species of stygobiotic harpacticoid from epikarst pools in Thailand is particularly significant since epikarst development is likely less extensive in the tropics (Williams 2008). While no caves studied entirely lack an epikarst fauna, some drips and drip pools do. Some of these have been extensively sampled, but the reasons for the absence of any individuals at that site are varied, including the presence of ice, moonmilk, and thick ceilings (Papi and Pipan 2011, Pipan et al. 2006b).

Table 3.7 ANOVA and comparison of mean abundance of copepods collected in 35 drips in six caves in central Slovenia. Data from Pipan (2003).

Sources	Sum of squares	Mean squares	F-ratio	p
Ecological category	11,648.73	5824.37	3.239	0.0516
Error	61,141.70	1796.29		
Total	72,790.43			
Ecological category	**Mean**	**n**	**SE**	
Epikarst endemic	20.1	15	10.9	
Other stygobionts	52.9	9	14.1	
Non-stygobionts	6.6	13	11.8	

3.3.5 Copepod species diversity and richness in epikarst

Malard et al. (2009) analysed data on stygobiotic groundwater species on a European-wide scale and found that local α-diversity (in their case a local aquifer) contributed less than ten per cent to overall species richness, and suggested that among sites β-diversity was characteristically high in subterranean systems. Sampling data on epikarst drips allows for the analysis of even finer spatial scales. Instead of a local sample site being on the scale of a few km², as was the case for the data Malard et al. (2009) analysed, the local sample for epikarst data is a single drip. The question is whether α-diversity is also a minor component at this finer geographical scale of individual drips.

Sampling of drips yield data that can be hierarchically arranged according to their spatial scale. Progressing from smaller to larger spatial scales, and thus from finer to coarser detail, there are repeated samples of the same drip through time at the finest scale; there are samples of different drips in the same cave; and finally there are samples from different caves, a still coarser scale. Pipan

and Culver (2007b) provided estimates of species richness of Slovenian epikarst copepods at different scales, allowing the partition of diversity into different spatial scales. For the 35 drips sampled in six caves, an average of 3.20 ± 0.47 species per drip was found. The intensity of sampling was sufficient to collect nearly all the species present drips and in caves, as assessed by species accumulation curves and Chao estimates based on the ratio of doubleton and single species (Table 3.8, Fig. 3.14, and Box 2.1). For the three richest drips analysed (Črna jama #5, Pivka jama #1 and #2), it appears that between zero and two species have yet to be found. At the level of caves, four of six curves appear at or near their asymptote (Table 3.8), while two caves—Postojnska jama and Škocjanske jame—are not completely sampled. For both of these caves, this is likely the result of the large distance between drip samples (>400 m). The average number of observed species per cave was 8.67 ± 1.23, and if Chao estimates are used instead, the expected number of species per cave was 10.83 ± 1.66. In all, 27 species were observed in the 35 drips in six caves, and both the accumulation curve and the Chao estimate of total species richness suggests a number of species

Table 3.8 A comparison of the observed number of copepod species, the number of samples required to find 90 per cent of the total number of observed species, and the Chao estimates of total diversity, sampled and unsampled, in drips, caves, and the central Slovenian region. Sampling was monthly. Data from Pipan and Culver (2007b).

	Observed no. of species	No. of samples for 90% of species	Chao estimate of no. of species
Individual drips			
Črna jama #5	6	3 of 5	6
Pivka jama #1	6	4 of 5	7
Pivka jama #2	10	4 of 5	12
Individual caves			
Črna jama	8	3 of 5	8
Dimnice	7	3 of 5	7
Pivka jama	9	3 of 5	9
Postojnska jama	5	4 of 4	9
Škocjanske jame	9	4 of 5	15
Županova jama	14	4 of 5	16
Total	27	5 of 6	45

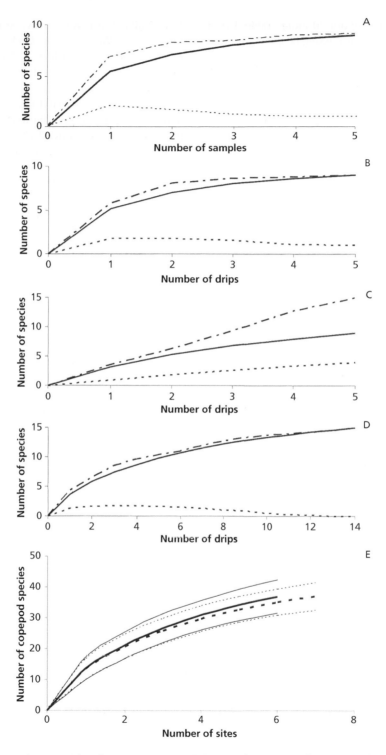

Figure 3.14 Species accumulation curves based on Mao Tau estimates (solid lines), Chao estimates of total diversity (dashed lines), and number of singletons (dotted lines) for A—an individual drip in Pivka jama, B—Pivka jama, C—Škocjanske jame, and D—Postojna Planina Cave System (PPCS, Črna jama + Pivka jama + Postojnska jama), and E—the six study caves, all in Slovenia. Analysis was done using EstimateS (Colwell 2009). See Pipan and Culver (2007b) for more details.

remain to be found—about 20 more (Table 3.8). It is interesting to note that Pipan (2003, 2005a) found an additional 10 species in 35 adjacent epikarst drip pools, lending support to the estimate of 20 missing species in drips in the six caves. Nonetheless, even at the regional scale, sampling of drips provides much more complete sampling than for other estimates of the subterranean fauna. For example, using similar estimation procedures, the estimated total number of cave (as opposed to epikarst) stygobionts in the south-central Slovenian region was 170; yet only 78 species had been reported at the time of Pipan and Culver's (2007b) study. It is a recurring theme of studies of cave biodiversity that sampling is very incomplete because repeated sampling of the same sites are needed (e.g. Zagmajster et al. 2010).

The same pattern of low α-diversity and high β-diversity demonstrated by Malard et al. (2009) for the stygofauna as a whole on a European-wide scale was found for epikarst copepods for smaller geographical scales—individual drips to the south-central region of Slovenia (Pipan and Culver 2007b). The three components of total epikarst copepod diversity for 35 drips in six Slovenian caves were partitioned as follows:

- 12 per cent (3.20 species) of the total species richness was within-drip diversity (α-diversity);
- 20 per cent (5.47 species) of the total species richness was among-drip diversity within a cave (a component of β-diversity); and
- 68 per cent (16.33 species) of total species richness was among-cave diversity (a component of β-diversity).

We can suppose that if more caves had been sampled, β-diversity among caves would likely have been even higher.

The other prism with which to view these data are that diversity at a single drip can be astonishingly high. Ten species were found from repeated samples in a single drip in Pivka jama (Table 3.8), and these populations sampled in this drip and other drips are unlikely to extend for more than a few hundred metres in any direction, a point we will discuss further in section 3.3.6 below.

No other epikarst fauna has been studied as thoroughly as the one in Slovenia but a few estimates are available elsewhere based on sampling of drips.

In a similar but less extensive study conducted in Organ Cave, West Virginia, USA, Pipan and Culver (2005) found that sampling of 13 drips in three sections of the cave for a period of only 35 days in early summer was sufficient to collect nearly all the species present in individual drips, sections of the cave, and in the entire cave. The total number of copepod species collected in all 13 drips in Organ Cave was ten, slightly higher than the average number of epikarst copepods per cave in Slovenia (Table 3.8), although the frequency of stygobionts (0.30) was lower in Organ Cave.

Meleg et al. (2011c) report between 3 and 6 epikarst copepod species per cave in drips and drip pools in Romania (Table 3.9) and a total of 11 species among their five study caves, much lower numbers than those recorded for the Slovenian caves. The best estimate of total species richness was 16 species for the five Romanian caves (Meleg et al. 2011a). In an expanded study, Meleg et al. (2011b), using data from 17 caves in the Pădurea Craiului Mountains of northwest Romania, recorded 17 epikarst species, and estimated that 64 species should have been present in the epikarst, based on Chao 1 estimates. While this number is higher than the total of 45 species estimated for the six Slovenian caves (Table 3.8), it is also based on more caves. Based on this limited evidence, it seems likely that Slovenia and perhaps the Dinaric karst in general is a hotspot for epikarst copepod species, as it is for other subterranean aquatic groups (Sket 1999).

Not only are species numbers higher in general in the Dinaric karst, but nearly all the species are

Table 3.9 Species richness of epikarst copepods for five Romanian caves. The estimated total includes observed and unobserved species, using Chao 1 estimate in EstimateS. Data from Meleg et al. (2011a).

Cave	Number of species
Peștera Ungurului	6
Peștera Vadu Crișului	5
Peștera cu Apă din Valea Leșului	5
Peștera Ciur Izbuc	5
Peștera Doboș	3
Total	11
Estimated total (Chao estimate)	15.5

stygobionts (Table 3.10). An interesting and unusual situation occurs in two caves in Slovenia which had very low diversity (as opposite to the general high diversity recorded for the Dinaric karst): in drips in both Huda luknja, occurring in an isolated karst area, and Snežna jama na planini Arto, an ice cave in the Kamnik-Savinja Alps, only two copepod species were found in each cave, but all were stygobionts. In both Romania and West Virginia, frequencies of stygobionts relative to the total number of copepod species were lower than in Slovenia (Table 3.10).

3.3.6 Ecology of epikarst fauna

Given the high levels of variation in chemical and physical parameters in the epikarst, it is interesting to investigate the extent to which epikarst animals occur throughout such range of physico-chemical conditions. One approach is to consider the differences, if any, between the chemistry of epikarst water where animals were collected, and the one of water without animals, taking into account temporal differences and differences among caves. The analysis is further complicated by the presence of strong correlations among variables as well as

non-linearities. In order to address this problem in the most general way, Christman, Culver, and Pipan (unpublished) used a recursive partitioning approach, known as random forests (Strobl et al. 2009) to determine individual variables of predictive value by minimizing effects of correlated variables. The important variables determined by the random forest were then used to construct classification trees, which repeatedly partition the predictor space.

In her study of epikarst copepods in six Slovenian caves, Pipan (2003) measured NO_3^-, NO_2^-, NH_4^+, SO_4^{2-}, K^+, Ca^{2+}, Na^+, Mg^{2+}, Cl^-, pH, temperature, and conductivity. Of these, temperature, conductivity, Cl^-, NO_3^-, Ca^{2+}, and Mg^{2+} appeared to affect the presence or absence of copepods (Fig. 3.15). The best model from the classification tree is shown in Fig. 3.16. Overall, the tree classified 192 of 252 samples (76 per cent) correctly with regard to the presence or absence of copepods. The presence of copepods was correctly predicted 69 per cent (94 of 136) of the time and the absence of copepods was correctly predicted 84 per cent (98 of 116) of the time. The first branch was for temperature and 57 of 136 occurrences of copepods could

Table 3.10 Frequency of stygobiotic copepods taken from epikarst water in drips in eight Slovenian caves, five Romanian caves, and one West Virginia cave.

Site	No. of species	Per cent troglobionts and stygoboints	Source
Črna jama, Slovenia	8	100	Pipan (2005a)
Dimnice, Slovenia	8	100	Pipan (2005a)
Huda luknja, Slovenia	2	100	Pipan et al. (2008)
Snežna jama na planini Arto, Slovenia	2	100	Papi and Pipan (2011)
Županova jama, Slovenia	14	93	Pipan (2005a)
Škocjanske jame, Slovenia	9	89	Pipan (2005a)
Postojnska jama, Slovenia	5	80	Pipan (2005a)
Pivka jama, Slovenia	11	73	Pipan (2005a)
Peştera Doboş, Romania	3	67	Meleg et al. (2011b)
Peştera Ciur Izbuc, Romania	5	60	Meleg et al. (2011b)
Peştera cu Apă din Valea Leşului, Romania	4	50	Meleg et al. (2011b)
Organ Cave, W.Va., USA	10	40	Pipan et al. (2006b)
Peştera Ungurului, Romani	6	33	Meleg et al. (2011b)
Peştera Vadu Crişului, Romania	5	20	Meleg et al. (2011b)

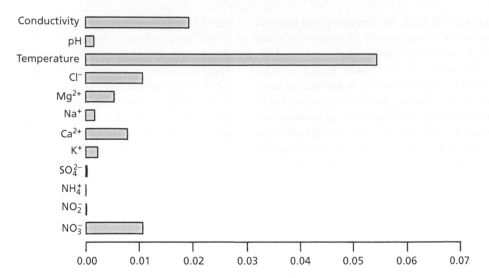

Figure 3.15 Bar graph of a 'random forest' for a series of chemical parameters for the presence/absence of copepods in drips in Slovenian caves, showing the relative importance of different variables in generating 2000 classification trees, when five variables are repeatedly drawn at random. Data from Pipan (2003).

be predicted solely on the basis of temperature being less than 8.2 °C. There was no further branching on the left but for temperatures greater than 8.2 °C, Ca^{2+} concentration was the next partition with high concentrations also being associated with the presence of copepods. The remainder of the partitions involved Cl^- (twice), NO_3^-, Mg^{2+}, and Ca^{2+} (again). Of the eight terminal leafs, three were associated with absence, and the most important of these (81 of 116 samples without copepods) were samples with high temperature, intermediate Ca^{2+}, low Cl^-, low Mg^{2+}, and high NO_3^-.

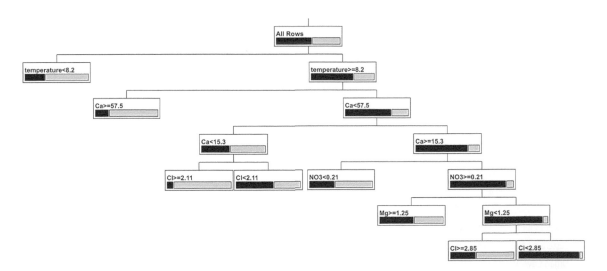

Figure 3.16 Classification tree for presence or absence of epikarst copepods in Slovenian caves. The size of the black rectangle is the relative number of empty samples and the size of the grey rectangle is the relative number of samples with copepods. Variables were chosen from the random forest analysis (Fig. 3.15).

Conductivity itself was unimportant in the classification tree, presumably because some its major components, especially Ca^{2+} and Mg^{2+}, were exposed by the random forest analysis, which teases apart correlated variables. Copepods tend to be in samples with lower temperatures, and so the connection between temperature and copepod presence is likely a seasonal one. It is unlikely, but possible, that the epikarst community itself changes seasonally, but it is more likely that copepods get washed out of epikarst in greater numbers in winter. Since the drip samples of copepods are samples of individuals washed out of their habitat, it is more likely that seasonal effects are the result of differences in washout rather than community changes.

Of the other variables shown to be important in the classification tree, Ca^{2+} has a strong connection with the biology of copepods. It is critical in the moulting process, and some crustacean species, like the amphipod *Gammarus minus*, are limited to carbonate springs (Glazier et al. 1992). Also of interest is that copepods tend not to be found in water with temperature greater than 8.2 °C and Ca^{2+} concentrations greater than 57.5 mg L^{-1}. Water in epikarst can be supersaturated with respect to Ca^{2+} (part

of the mechanism of deposition of $CaCO_3$ in caves [e.g. stalactites]) and this may cause physiological problems for animals in this water. While carbonate geochemists have long focused on the $Ca^{2+} - HCO_3^-$ system, we suggest it deserves more attention from biologists working in the same systems. Mg^{2+} is also a critical nutrient, and it is possible that it is limiting in some contexts.

Pipan et al. (2006b) used Canonical Correspondence Analysis to search for patterns and connections between individual species and environmental parameters. In this study, it was only the samples with copepods that were used, and the emphasis was on niche differences among species. Fig. 3.17 is a two-dimensional plot of the 12 chemical and physical parameters of epikarst drips from five Slovenian caves (Postojnska jama was excluded because species were rare and more than half of the drips had no fauna). The most important parameters which separated the different drips were NO_3^- concentration and ceiling thickness. Each cave formed a relatively compact cluster, with Županova jama both having the largest cluster and being the most distinct. The clusters for Pivka jama and Dimnice were also quite distinct. Ceiling thickness was the

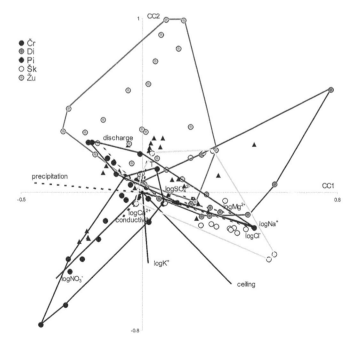

Figure 3.17 Ordination diagram based on species composition and abundance data in drips in five Slovenian caves. Lines indicate the environmental variables and their orientation on the canonical axis. Triangles indicate different species and dots represent individual drips. Convex hulls enclose the drips for individual caves. From Pipan 2005a, used with permission of ZRC SAZU, Založba ZRC. (See Plate 10)

most important axis in separating Županova jama, while NO₃⁻ was the most important axis separating Pivka jama. Škocjanske jame and Črna jama had very compact clusters very similar to each other. It is interesting to observe that each cave has a relatively compact cluster and that drips in three of the five caves could easily be separated from the others. When species were superimposed on the two-dimensional plot (Fig. 3.18), three clusters of species could be distinguished. One is represented by the single species *Parastenocaris* cf. *andreji*, and largely separated by low concentrations of NO₃⁻ and high concentrations of Na⁺ and Cl⁻, and was only found in Dimnice drips. A second cluster—*Moraria varica, Maraenobiotus* cf. *brucei, Bryocamptus dacicus*, and *Bryocamptus* sp.—was separated by high concentrations of NO₃⁻. These species were only found in Pivka jama. A third cluster comprised all of the other species (Fig. 3.18). Because the first two clusters were related not only with particular environmental conditions but also with a particular cave, it was impossible to distinguish which factor (physical-chemical or geographical separation) was important. Nevertheless, each species occupied a distinct set of sites, and even species within a cluster slightly differed in their preferential conditions.

Given the highly fragmented character of epikarst and the high levels of heterogeneity of physical and chemical conditions, there are many possibilities for niche separation, both along geochemical and spatial axes. It may be this heterogeneity makes possible the high β-diversity of epikarst habitats (see section 3.3.5).

Of course, not all potentially important environmental parameters were measured, and organic carbon in particular may be very relevant in determining the distribution of copepods, not because copepods are utilizing dissolved organic carbon directly, but because they feed on micro-organisms that are directly utilizing dissolved organic carbon.

Meleg et al. (2011a, 2011b, 2011c) and Pipan et al. (2006a) report parallel, but less extensive analyses for epikarst communities in Romania and USA, respectively. Using pH, temperature, conductivity, and precipitation as predictor variables, Meleg et al. (2011a) produced correspondence plots for the entire epikarst community, including amphipods, isopods, ostracods, and copepods, the latter being the dominant taxa as number of species and individuals. They found that, unlike Pipan et al. (2006b), differences in conductivity were correlated with abundance of different species and that the

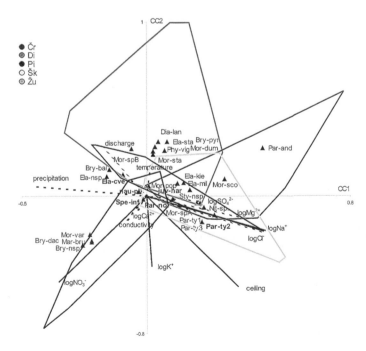

Figure 3.18 Ordination diagram shown in Fig. 3.17 with species indicated by abbreviations as follows: CYCLOPOIDA, *Diacyclops languidoides* Dia-lan, *Speocyclops infernus* Spe-inf; HARPACTICOIDA, *Bryocamptus balcanicus* Bry-bal, *B. dacicus* Bry-dac, *B. pyrenaicus* Bry-pyr, *Bryocamptus* sp. Bry-nsp, *Elaphoidella cvetkae* Ela-cve, *E. kieferi* Ela-kie, *E. stammeri* Ela-sta, *E. millennii* Ela-mil, *Elaphoidella* sp. Ela-nsp, *Maraenobiotus* cf. *brucei* Mar-bru, *Moraria poppei* Mor-pop, *M. stankovitchi* Mor-sta, *M. varica* Mor-var, *Moraria* sp. A Mor-spA, *Moraria* sp. B Mor-spB, *Morariopsis dumonti* Mor-dum, *M. scotenophila* Mor-sco, *Nitocrella* sp. Nit-sp., *Parastenocaris nolli alpina* Par-nol, *P.* cf. *andreji* Par-and, *Parastenocaris* sp. 1 Par-ty1, *Parastenocaris* sp. 2 Par-ty2, *Parastenocaris* sp. 3 Par-ty3, *Phyllognathopus viguieri* Phy-vig, cf. *Stygepactophanes* sp. Sty-nsp From Pipan et al 2006b, used with permission of ZRC SAZU, Založba ZRC. (See Plate 11)

stygobiotic taxa were usually found in water with higher conductivity. Higher conductivity implies longer residence time of the water in epikarst as the water becomes saturated with $CaCO_3$ (Covington et al. 2012).

Moldovan et al. (2011) have made important steps in understanding why different epikarst communities are different, not just that they are different. Working in Peştera Ciur Izbuc in Romania, they used a combination of stable isotopes, drip rates, and species composition and abundance, to understand differences between the epikarst community in two sections of the cave 300 m apart. They concluded that the downstream epikarst section had smaller, well-connected voids that allowed for rapid transport of animals and water from the surface. The upstream section had lower secondary porosity but larger spaces (and larger animals). Finally, they also demonstrated that the surface-dwelling copepod *Bryocamptus caucasicus* was able to penetrate the epikarst and had a negative impact on the stygobiotic copepods, even though *B. caucasicus* probably did not maintain permanent populations in subterranean habitats.

Pipan et al. (2006a) used temperature, pH, oxygen concentration, drip rate, and ceiling thickness as predictor variables for the copepod community of Organ Cave, West Virginia, USA. They found that oxygen concentration, drip rate and ceiling thickness all had statistically significant correlations with the copepod communities, and the other variables did not. This is similar to the results of Pipan et al. (2006b) where ceiling thickness was also important (see Fig. 3.17). Of all the copepod species found in Organ Cave epikarst drips, only *Bryocamptus zschokkei alleganiensis* tended to be found where ceiling thickness was greater. Perhaps this species typically lives in vertical cracks and fissures of the infiltration zone, rather than in the epikarst itself. The study by Pipan et al. (2006b) was the only one that reported values of dissolved oxygen, and their results indicate that different species may occupy areas with different oxygen concentrations. Possibly there are different redox zones in the epikarst, with different faunas. Pipan et al. (2006b) also found that the three different areas of the studied cave formed a compact cluster of points each, and that only two of the areas had any overlap in the 2-dimensional ordination plot. As was the case for the analysis of the Slovenian data, environmental variables and geographical separation were conflated.

A further analysis of the relationship between geographical distance and community structure in the Organ Cave drip community highlights the importance of spatial heterogeneity and patchiness. The highly dissected nature of epikarst (Figs. 3.1 and 3.3), may constrain dispersal. If this is the case, then community similarity should decline rapidly with distance. The Jaccard index (J) measures the degree of similarity between two communities, and (1–J) measures the difference.[1] The relationship between (1–J), a measure of community difference and distance between drips in Organ Cave is shown in Fig. 3.19. The pattern is quite striking. There is an increase in (1–J) with distances up to 100 m, after which point there is an increase in the variability of (1–J). Thus, drips within several hundred metres of each other tend to have similar composition and there is an expected decline in similarity with distance. This relationship between distance and similarity breaks down when drips greater than 1 km apart were considered, and communities are substituted with new ones. The 'new' communities may or may not be similar to the adjoining communities, as evidenced by the high variability of the Jaccard coefficients. Epikarst copepod communities thus form a patchwork of communities, with variation on a scale of 100 m.

3.3.7 Terrestrial epikarst communities

Since epikarst organisms are sampled either in drip pools or in dripping water itself, it may seem like a contradiction in terms to consider a terrestrial epikarst community. Since epikarst is well above the water table, and the amount of water storage in epikarst varies, then it follows that it includes some air-filled habitat. Very little is known about this habitat because the only terrestrial animals collected are those which are flushed out by moving epikarst water. Terrestrial species are usually much less common in epikarst habitats although

---

[1] J = c/(a + b) where c is the number of species shared between the two communities, and a and b are the total number of species in the two communities.

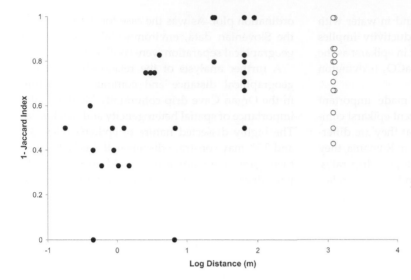

Figure 3.19 Semilog plot of geographical distance (in m) against (1−J) for epikarst copepods in Organ Cave, West Virginia, USA., where J is the Jaccard indices. Closed circles are pairs of drips on the same side of the syncline that is the major structure determinant of cave passage position; open circles are on opposite sides of the syncline. From Pipan et al. (2006b).

all studies of the drip fauna have found terrestrial species (Delay 1968, Gibert 1986, Pipan 2003, Pipan and Culver 2005, Pipan et al. 2008, Moldovan et al. 2007, Papi and Pipan 2011). Only in the cave Huda luknja in Slovenia is the terrestrial fauna richer in species and more abundant than the aquatic epi-karst fauna (Table 3.11). Terrestrial animals were rarely identified to species, probably because aquatic biologists were making the collection, but at least one troglobiont was noted in two studies— a Collembola in the genus *Arrhopalites* in Grotte du Cormoran in France (Gibert 1986) and the beetle *Otiorhynchus anophthalmus* in Huda luknja in Slovenia (Pipan et al. 2008). The frequency of terrestrial

Table 3.11 Number of aquatic and terrestrial animals collected in dripping water, arranged by frequency of terrestrial species. Caves are in Slovenia unless otherwise indicated.

	Terrestrial	Aquatic	Frequency of terrestrials	Source
Huda luknja	78	46	0.63	Pipan et al. (2008)
Organ Cave, USA	176	212	0.45	Pipan and Culver (2005)
Postojnska jama	11	27	0.29	Pipan (2005a)
Grotte du Cormoran, France	755	2391	0.24	Gibert (1986)
Peştera Vadu-Crişului, Romania	2	11	0.15	Moldovan et al. (2007)
Peştera Ungurului, Romania	29	326	0.08	Moldovan et al. (2007)
Snežna jama na planini Arto	11	198	0.05	Papi and Pipan (2011)
Peştera Vântului, Romania	3	55	0.05	Moldovan et al. (2007)
Dimnice	8	158	0.05	Pipan (2005a)
Črna jama	16	348	0.04	Pipan (2005a)
Planinska jama	10	335	0.03	Moldovan et al. (2007)
Škocjanske jame	8	525	0.02	Pipan (2005a)
Županova jama	7	462	0.01	Pipan (2005a)
Pivka jama	9	795	0.01	Pipan (2005a)

individuals, relative to aquatic individuals, varies considerably. In some cases, especially Postojnska jama and Peştera Vadu-Crişului, Romania, few aquatic individuals were found relative to the numbers in other caves in the study. In other caves, such as Huda luknja, Organ Cave, and Grotte du Cormoran, both aquatic and terrestrial components were abundant.

Overall biodiversity of the epikarst in Huda luknja, a cave in an isolated karst area in northeast Slovenia outside the Dinaric karst, is shown in Table 3.12. The reasons for the rich terrestrial

epikarst found in this cave are not clear, but the geographical position of Huda luknja is such that overall, the subterranean terrestrial fauna is likely richer than the aquatic because the main climate factor forcing species into aquatic caves—the Messinian salinity crisis—was not at play in this cave. In contrast, Pleistocene effects, which often force the terrestrial fauna into caves (Culver and Pipan 2010), were likely to be strong. More examples of epikarst communities dominated by terrestrial species are needed before their hypothesis can be tested.

Table 3.12 List of taxa found in 12 epikarst drips during a one year sampling in Huda luknja in northeast Slovenia. From Pipan et al. (2008).

Higher group	Class	Order	Genus and/or species	Terrestrial T or Aquatic A
Nematoda		Rhabditida	Unidentified	A
Annelida	Clitellata	Oligochaeta	Unidentified Enchytraeidae	A
			Unidentified Lumbricidae	T
Arthropoda	Arachnida	Palpigradi	*Eukoenenia* cf. *austriaca*	T
		Araneae	*Troglohyphantes diabolicus*	T
		Acarina	*Ixodes vespertilionis*	T
		Mesostigmata	Unidentified	T
		Oribatida	Unidentified	T
	Crustacea	Amphipoda	*Niphargus scopicauda*	A
		Ostracoda	Unidentified	A
		Copepoda	*Parastenocaris nolli alpina*	A
			Bryocamptus balcanicus	A
	Chilopoda		Unidentified	T
	Diplopoda	Acherosomatida	Unidentified	T
			Polyphematia moniloformis	T
	Insecta	Collembola	Unidentified Poduridae	T
			Unidentified Sminthuridae	T
			Unidentified Entomobryidae	T
		Diplura	*Plusiocampa* sp.	T
		Coleoptera	cf. *Atheta* sp.	T
			Laemostenus schreibersi	T
			Otiorhynchus anophthalmus	T
		Diptera	Unidentified Sciaridae	T
			Unidentified Trichoceridae	T
			Unidentified larva	T

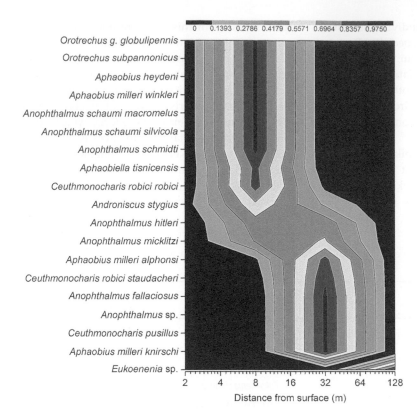

Figure 3.20 Comparative normalized spatial density map of 19 troglomorphic taxa in small caves in north-central Slovenia. From Novak et al. (2012). (See Plate 12)

Novak et al. (2012) indirectly sampled epikarst communities in 54 small caves in north-central Slovenia, in much the same geological setting as Huda luknja. Rather than sampling drips (not all of the caves even had dripping water), they set pitfall traps and took Berlese samples at 3.5 m intervals in the cave. They found a very striking bimodal pattern of distribution of troglobionts with respect to distance from the entrance and vertical distance to the surface (Fig. 3.20). In fact the majority of troglobiotic species analysed (10 of 19) were typically found at sites 8 m below the surface. Seven others were found at depths of 32 m, and two had a broad distribution, ranging between 6 and 20 m below the surface. The authors suggest that the deeper group are the true cave species while the shallow group are epikarst species. They also point out that the epikarst species may also occur in voids in the soil, such as around tree roots, and they call rhyzophaerobionts the organisms living in these microhabitats! If this is the case, then these species overlap in their distribution between epikarst and the MSS (see chapter 4).

3.3.8 Life-history characteristics of epikarst species

Relatively little is known about life history characteristics of epikarst species, especially when compared to other subterranean-dwelling species. What little is known concerns copepods. In general, subterranean copepods show features of K-selected species, with increased longevity, fewer and larger eggs, and delayed reproduction (Rouch 1968). In surface-dwelling species, fertilized eggs are extruded into an egg sac, but in subterranean species, egg sacs are usually not present (Dole-Olivier et al. 2000) and thus, at least in some subterranean species, eggs are released directly onto the substrate (Pipan 2005a). Rouch (1968) provides some quantitative information on life cycle characters for three species of copepods—the surface-dwelling *Bryocamptus zschokkei* (also occasionally found in drip pools [Pipan 2005a]), the groundwater-dwelling *Nitocrella subterranea*, and the epikarst-dwelling *Bryocamptus pyrenaicus*. *B. pyrenaicus* has a life history intermediate between

Table 3.13 Life history characteristics of three copepod species. Data from Rouch (1968).

Species	Habitat	Mean female length (mm)	Egg volume per brood (mm³)	No. of broods	Lifetime egg volume (mm³)
Bryocamptus zschokkei	Surface	0.63	6.9×10^{-4}	14	9.7×10^{-3}
Bryocamptus pyrenaicus	Epikarst	0.47	4×10^{-4}	16.7	6.7×10^{-3}
Nitocrella subterranea	Deep groundwater	0.50	3×10^{-4}	9.2	1.2×10^{-3}

that of *B. zschokkei* and *N. subterranea* (Table 3.13). It is similar to the surface-dwelling *B. zschokkei* in lifetime total egg volume and in number of broods produced, but similar to the deep groundwater species *N. subterranea* in egg volume per brood and in body size. These intermediate characteristics may be the result of greater amounts of available organic carbon in epikarst than in deeper groundwater.

3.3.9 Morphological characteristics of epikarst species

Given the general small diameter of epikarst channels and openings, it is not surprising that epikarst animals tend to be smaller than related species in other subterranean habitats with large spaces available. The amphipod genus *Stygobromus* occurs in a wide variety of subterranean habitats throughout North America (Holsinger 1978), making a good candidate for the analysis of the effect of habitat on body size. Culver et al. (2010) analysed

morphological information on type specimens of 56 species of *Stygobromus* occurring in subterranean habitats in eastern North America. The species were found in four subterranean habitats—epikarst, cave streams, deep groundwater, and the hypotelminorheic (for definition, see chapter 2). Mature females (male specimens were not available for all species) found in epikarst were significantly smaller than females from other habitats (Fig. 3.21). Overall, habitat accounted for about 20 per cent of the overall variance in female body size.

Culver et al. (2009) found a similar pattern with respect to Slovenian subterranean copepods. Copepods that were epikarst specialists were roughly half the size of stygophilic copepods and 3/4 the size of stygobiotic copepods found in other subterranean habitats (Table 3.14). They also found that body size and species range were unrelated, which might be expected if size were a major determinant of dispersal capacity (see section 3.3.10). Both of these morphological studies strengthen a

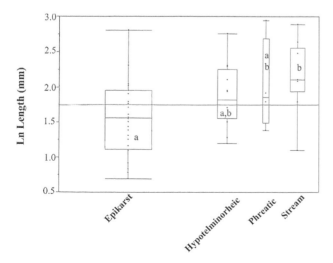

Figure 3.21 Box and whiskers plots of ln female body length of species of the amphipod genus *Stygobromus* for epikarst, hypotelminorheic, cave-stream, and phreatic habitats in North America. Overall mean represented by the line across the entire. Boxes: 50 per cent of the data, line across each box: group median; whiskers: minimum and maximum values. The widths of the rectangles are proportional to sample size. Dots are individual data points. Plots with the same letter (a or b) do not differ according to the Tukey–Cramer HSD test. From Culver et al. (2010), used with permission from Brill.

Table 3.14 Analysis of copepod size in relation to ecological classification, for all copepod sizes known from Slovenian caves. For HSD (honestly significant difference) groups, those not connected by the same letter are significantly different. For ANOVA, $F_{2,52} = 7.58$, p<0.002. From Culver et al. (2009).

Group	n	Mean (mm)	S.E.	Tukey-Kramer HSD group
Epikarst stygobionts	4	0.495	0.061	a
Other stygobionts	25	0.652	0.045	b
Stygophiles	26	0.930	0.071	b

recurring theme in studies of the morphology of subterranean animals—size of the available habitat has a major effect on body size (see chapter 12).

The general features of shape and structure of appendages have been little examined with respect to specialization to epikarst habitats. The most quantitative study is that of Culver et al. (2010), and they found no differences in relative antennal length of *Stygobromus* living in different subterranean habitats (see Fig. 3.21 for a comparison). Pesce and Galassi (1986) emphasized the importance of reduction of spine length on proximal segments of groundwater cyclopoid copepods which move in sandy and muddy sediments.

Brancelj (2007, 2009) gave a number of intriguing suggestions about uniquely convergent features of epikarst copepods. Working with the genera *Morariopsis* and *Paramorariopsis,* he pointed out that animals living in the epikarst must have some morphological adaptations to prevent or minimize their transport downwards. Combined with what he considers a low supply of organic carbon, he proposes that the following convergent features are present in specialized epikarstic copepods to avoid being displaced by water flow.

- Reduction in endopodal segmentation to two or one.
- Reduction in number of spines and setae on the terminal segments of both endopods and exopods.
- Reduction in spines and setae on the caudal rami to one terminal seta.

- Tips of the terminal setae of the caudal rami are far apart.
- Short and robust setae on the endopodal lobe of fifth leg (P5) as well as very strong spinules at the base of the caudal rami.

Brancelj (2009) demonstrated that the genus *Elaphoidella* shows similar convergent features, especially the last two listed. He also proposed that there is convergent reduction in length of the antennules and that robust setae are probably an adaptation for moving through small spaces in fractured rock as well as a protection against washout.

3.3.10 Species ranges and dispersal ability

It might be expected that the habitat is less heterogeneous for smaller stygobiotic copepods than for larger stygobiotic ones. There is no detectable relationship between body size and geographical range (Culver et al. 2009), but there is evidence for dispersal of stygobiotic copepods, albeit indirect. Per cent occupancy of the epikarst habitat was estimated for two different scales—drips within a cave and caves within a small region of Slovenia. If dispersal among epikarst habitats is important (and it need not be if species become isolated in caves with little subsequent dispersal [see chapter 13]), then occupancy of the habitat at the two scales should be related. In fact, there is a striking relationship between the two (Fig. 3.22, Table 3.15), suggesting that distributions of species, at least at small scales, is highly dynamic.

The pattern of distribution of number of caves occupied by stygobiotic copepods is also interesting. Using the six Slovenian caves extensively sampled by Pipan (2005a) and Brancelj (2002), the number of caves occupied by stygobiotic epikarst copepod species is shown in Figure 3.23. Of the 25 species, 16 were only found in one or two caves, and many of the species known from two caves were from the closely adjoining Črna jama and Pivka jama. Only one species—*Speocyclops infernus*—was found in all caves. This pattern of many geographically rare and a few geographically widespread species is likely a recurrent pattern in the subterranean fauna in general.

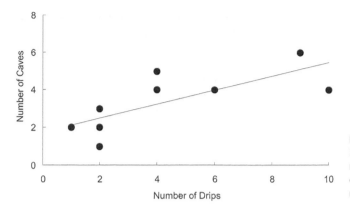

Figure 3.22 Relationship between number of drips in the Postojna Planina Cave System in Slovenia (x-axis) and the number of Slovenian caves occupied by stygobiotic epikarst copepods (y-axis). See Table 3.15 for details of the linear regression. From Culver et al. (2009).

Table 3.15 Linear regression of relationship between number of drips occupied by stygobiotic copepods in Postojna Planina Cave System and number of caves occupied in Slovenia. From Culver et al. (2009).

Term	Estimate	Standard error	t ratio	p
Intercept	1.75	0.434	4.02	0.002
Slope	0.37	0.092	4.05	0.002

$R^2_{adj} = 0.583$, n = 12.

3.4 Summary

Epikarst is a nearly ubiquitous component of karst areas, which cover about 12.5 per cent of the earth's surface. Epikarst is the site of most of the water storage above the water table in karst, and an important shallow subterranean habitat. It is highly dissected, with many miniature (<100 m^2) catchment areas, which in turn often differ in their water chemistry. Organic matter in the soil is broken down in epikarst, both making organic carbon available and releasing CO_2 that produces carbonic acid. Water exiting epikarst typically contains dissolved organic carbon at concentrations of approximately 1.0 mg L^{-1}. It is particularly important biologically, because it is both widespread and relatively labile in terms of metabolic uptake.

Until recently, epikarst fauna was known from drip pools in caves where epikarst water collected. Direct sampling of dripping water allows for a relatively unbiased sample of the epikarst fauna, and the fauna of drip pools is relatively distinct from that of drips. The epikarst fauna is typically species-rich, often with more than ten species known from

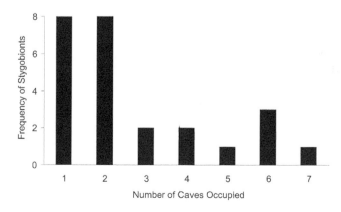

Figure 3.23 Frequency distribution of the number of caves occupied by stygobiotic epikarst copepods, based on seven thoroughly studied caves in Slovenia. Data from Pipan (2005a) and Brancelj (2002).

drips in a single cave. In the Dinaric karst, the epikarst fauna is predominately stygobiotic. Copepods generally dominate, but many other groups of crustaceans, such as amphipods, isopods, and syncarids are present. The epikarst fauna may also be found in other places with cracks and fissures above the water table, but this appears to be largely a secondary habitat. As is usually the case with subterranean fauna, local diversity is a small component of regional diversity, about 12 per cent in the case of six Slovenian caves which were extensively sampled. In those cases when water dripping from the epikarst does not contain animals, it is water low in conductivity and high in pH, suggesting that it has been underground for a very short period of time. As expected because of the highly dissected nature of the habitat, individual species and communities have very limited spatial extent, less than 100 m in linear distance. In some caves, there is evidence of a diverse terrestrial epikarst community. In the cave Huda luknja in north-central Slovenia, terrestrial organisms predominate among the specimens collected in dripping water, perhaps because of strong effects of the Pleistocene on the terrestrial fauna relative to other sites. Based on extensive sampling at different depths in caves, including deep sites and sites close to or in epikarst, Novak et al. (2012) suggests that the epikarst terrestrial fauna may be more diverse than the cave terrestrial fauna.

The limited available information on life histories of epikarst species (Rouch 1968) suggests that their life history may be intermediate between that of surface dwellers and of deep cave dwellers. Body size in epikarst species is smaller than that of species in other subterranean habitats, likely because the habitat dimensions in epikarst are smaller. Some studies of shape have failed to find any consistent morphological difference between epikarst and other subterranean species. However, Brancelj (2007) suggests that there are a series of morphological adaptations related to reducing washout and increasing metabolic efficiency.

CHAPTER 4

Intermediate-sized terrestrial shallow subterranean habitats

4.1 Introduction

Several groups of biologists have investigated the ecology and fauna of aphotic terrestrial habitats, places where the habitable spaces are greater than 1 mm but less than 25 cm, too small for humans to penetrate but larger than the spaces between soil particles. These habitats are above the solid rock layer, and take a variety of shapes and sizes. In chapter 3, we discussed one of these—the epikarst, and that portion of it that is dry is an intermediate-sized SSH.

Perhaps the most obvious and relatively widespread example of an intermediate-sized SSH is an open talus[1] slope (Fig. 4.1). Talus slopes, typically the result of fracturing and subsequent movement of rock as a result of a freeze-thaw cycle, and other thermal and topographical stresses, are in many ways an extreme environment, not only because of lack of light and scarcity of organic matter, but also because the habitat is unstable. Working in the Czech Republic, Růžička (1990, 1999a, 2011) investigated the possibility that open talus slopes harboured some spider species that were characteristic of boreal forests. One of the reasons he thought that such spiders were likely to be present was the generally cool temperature regime in the summer such at that encountered in northern boreal forests. After designing a special long-term pitfall trap that could be placed safely on these unstable slopes (see

Box 4.1), Růžička (1988b, 2011) found a number of species with disjunct distributions, such as *Bathyphantes eumenis*, a boreal species also found in talus slopes in Germany and the Czech Republic. Not only did he find disjunct populations of typically boreal spider species, he and colleagues also found a number of troglomorphic subspecies and species of beetles (J. Růžička 1998) and spiders (V. Růžička 1998).

Studying cave beetles in the Pyrenees, Juberthie et al. (1980a, 1980b) became interested in intermediate-sized terrestrial SSHs from a very different perspective. The distribution of some cave beetles, such as *Speonomus hydrophilus*, was disjunct and occurred in two karst areas separated by a region without caves. Juberthie, based on this distributional anomaly, looked for subterranean habitats (and *S. hydrophilus*) outside of cave regions. Where he looked was not in the open talus slopes Růžička studied, but in stable talus slopes and weathered rock protected and buffered by a layer of mossy covered soil (Fig. 4.2). He chose this habitat because it was likely better environmentally buffered, more isolated, and more physically stable than open talus slopes. Juberthie also found what he was looking for—non-cave subterranean populations of *S. hydrophilus*. He drew attention to this habitat by coining the phrase *milieu souterrain superficiel* (*MSS*), which has been translated as superficial underground compartment (SUC) and transliterated into mesovoid shallow substratum. The phrase was initially limited to covered talus slopes and covered weathered and fissured rock (Juberthie et al. 1980a, 1980b), but later expanded (Juberthie 2000) to include 'cracked lava flow or by scoriaceous [i.e.

---

[1] Scree and talus are nearly interchangeable terms although scree sometimes refers to smaller rock fragments. We use talus and scree interchangeably throughout the chapter to mean the accumulation of broken rock fragments, whatever their size, at the base of cliffs and steep valleys.

Shallow Subterranean Habitats. David C. Culver and Tanja Pipan.
© David C. Culver and Tanja Pipan 2014. Published 2014 by Oxford University Press.

Figure 4.1 Prof. Miloslav Zacharda standing on an open talus slope on Plešivec u Jinců, Czech Republic. Several troglomorphic spiders occur in the talus. Photo by T. Pipan.

Figure 4.2 A covered weathered MSS site in Ariège Province (France) in the Pyrenees. Numerous troglomorphic species occur in this and similar nearby sites. Photo by D. Culver.

clinker] layers' in lava and even aquatic habitats (Juberthie 1983), presumably the hypotelminorheic (see chapter 2).

Talus slopes, fissured rock, and clinker in lava are not the only shallow subterranean terrestrial habitats that harbour either disjunct populations of boreal species, or troglobiotic populations, or both. Ueno (1977) reported that troglobionts in Japan were found in a variety of voids in many different, loose-fissured rock types, and that they can also be collected in mines sampled in these rocks. Novak and Sivec (1977) found troglobionts in shallow Slovenian caves developed in pegmatite, a coarse crystalline igneous rock. Růžička et al. (2010) discovered disjunct populations in a sandstone labyrinth with some aphotic habitats. There are certainly other examples.

These habitats differ in a variety of features, such as lithology, slope, altitude, and latitude, but two features—cavity size and connections to the soil—are especially important. Gers (1992, 1998) stressed the importance of soil proximity both for movement of organic carbon and dispersal of MSS organisms. Gers (1992) suggested that MSS habitats are macropores.[2] Macropores increase the hydraulic conductivity of soil, allowing shallow groundwater to move relatively rapidly via lateral flow. The degree of connection with the soil varies from tree roots and spaces around rocks in the soil to open talus with little soil in the habitat at all (Fig. 4.3). Cavity size varies from 1 mm to perhaps 1 m, with small cavities constraining the size and shape of inhabitants and large size having lower relative humidity than smaller cavities (Howarth 1983). Seven major terrestrial shallow subterranean habitats are shown in Fig. 4.3. In addition to the larger cavity sites discussed above (erosional MSS, lava clinker, covered talus, and open talus), at least three other habitats often harbour a troglomorphic fauna. These are epikarst (discussed in chapter 3, especially Fig. 3.20 and Tables 3.11 and 3.12), spaces around tree roots, and the spaces around rocks in the soil (see chapter 7). Gers (1992) also argued

[2] Cavities in the soil that are larger than 75 μm (Soil Science Glossary Terms Committee 2008).

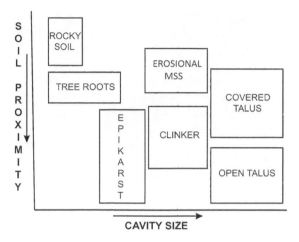

Figure 4.3 Conceptual diagram of different terrestrial shallow subterranean habitats. The position of the rectangles indicates the typical range of pore sizes and connectivity to the soil. Epikarst is covered in more detail in chapter 3 and soil is considered in more detail in chapter 7.

that, as MSS evolves, it becomes more connected with the soil.

Howarth (1983), with the typical North American emphasis on caves to the exclusion of nearly all other subterranean habitats, classified subterranean habitats into microcaverns (<1 mm in diameter), mesocaverns (1 mm–20 cm), and macrocaverns (>20 cm in diameter). He used these categories primarily for terrestrial habitats, and notes that microcaverns correspond to spaces between soil particles. Macrocaverns correspond to caves in his classification. He claimed that cavities of mesocaverns were too large for microcavern species, but without data. In general, his mesocaverns correspond to the habitats shown in Fig. 4.3.

4.2 Chemical and physical characteristics of intermediate-sized terrestrial SSHs

4.2.1 Representative intermediate-sized terrestrial SSHs

The open talus slope on Plešivec u Jinců in the Czech Republic (Fig. 4.1) is one of the most open talus slopes for which fauna are reported. Either its relatively young age, or its steep slope, or both,

have resulted in the near absence of soil. One food source for such communities is aeolian (Ashmole and Ashmole 2000), and it is also likely that falling leaves from nearby forests are a main source of organic carbon and nutrients (V. Růžička, pers. comm.). A troglomorphic subspecies of the spider *Bathyphantes eumenis* (Růžička 1988a, 2011) was collected at this site in pitfall traps specially designed for talus slopes and left in place for over a year (Růžička 1982, 1988b).

The covered talus site (Fig. 4.4) at Kamenec hill, Czech Republic, is a basaltic talus site situated above the Ploucnice River, at an elevation of 330–360 m asl. Ice forms in the lower margin and persists through much of the summer. Spider species with boreal distributions, such as *Bathyphantes simillimus* and *Diplocentria bidentata*, were found exclusively at sites with persistent ice, such as Kamenec Hill (Růžička 1999b).

The partially covered talus slope on Massif de l'Arize in the Ariège valley of the French Pyrenees is at an elevation of approximately 1300 m asl (Fig. 4.5). While this site does not have a complete covering of moss visible on the surface, it is much more stable than the talus site shown in Fig. 4.1. It is possible to rather easily walk on this slope, which is not true for Plešivec u Jinců site in the Czech Republic. The Massif de l'Arize site is dominated by the troglobiotic beetle *Speonomus hydrophilus*, with other Coleoptera and Diptera making

Figure 4.4 An icy covered talus site at Kamenec Hill, Czech Republic. Several boreal spider species have disjunct populations near the ice. Photo by T. Pipan.

Figure 4.5 Partially covered talus slope on Massif de l'Arize in the Ariège valley of the French Pyrenees, at an elevation of approximately 1300 m asl. The site is rich in troglobiotic beetles (Crouau-Roy 1987). Photo by T. Pipan.

Figure 4.6 Erosional MSS site in lava at Teno on the Island of Tenerife, Canary Islands. Ten troglobiotic species are known from the site. Photo by T. Pipan.

up most of the community (Crouau-Roy 1987, Crouau-Roy et al. 1992). Many of the Pyrenean sites, especially at lower elevations in the mountain valleys, had relatively small rocks (< 10 cm in diameter) with considerable soil involvement (Fig. 1.6). These sites are at the final stage of Gers' schemata for the evolution of MSS.

A covered, weathered MSS site near Mašun, Slovenia is shown in Fig. 1.3. While this site is on limestone, there are no caves in the immediate vicinity, and there is little obvious solutional modification of the upper layer of limestone so it is not epikarst by the usual definition (see chapter 3). The forest above is dominated by beech (*Fagus sylvatica*). In appearance it is very much like the original sketches of Juberthie et al. (1980a, 1980b) of MSS, except that it is in calcareous rather than non-calcareous rock. The entire fauna has not been systematically studied but three troglobiotic species are known from the site— one millipede and two beetles (Pipan et al. 2011a).

Although it is in a very different geographical context, the physical structure of an erosional MSS site in lava near Teno on the Island of Tenerife in the Canary Islands (Fig. 4.6) has superficial similarities to the Mašun site. It is in a 'laurisilva' forest at 940 m asl, with at least four species of Lauraceae, and is comprised of fragmentized basaltic rocks several million years old, and covered by 40–60 cm of soil. The tree covering is dense, with low exposure to the sun and rather high humidity. Ten troglobionts are known from the site (Pipan et al. 2011a).

While most epikarst sites are sampled from below (Box 3.1) and contain a mixture of organisms from aquatic and terrestrial sites, with aquatic animals typically predominating (see Table 3.11), it was possible to sample directly a terrestrial epikarst site immediately above the cave Jama v Kovačiji (Fig. 4.7). The site is in a steep dolina in a montane beech (*Fagus sylvatica*)—silver fir (*Abies alba*) forest at an elevation of 1000 m asl. During sampling in the summer, cool air, presumably from the cave immediately below, could be felt at the site. Two troglobionts were known from the site, much smaller than the number known (11) from an MSS site 20 m away but not directly connected with the cave (Pipan et al. 2011a).

The final representative site is La Guancha, a clinker MSS site in a pine (*Pinus canariensis*) forest next to Sima de la Perdiz on Tenerife, Canary Islands, at an elevation of 1580 m asl. It is recent

Figure 4.7 Epikarst site immediately above Jama v Kovačiji in Slovenia. It is directly connected with the cave and has two troglobiotic species. Photo by T. Pipan.

terrain, being only a few tens of thousands of years old. The habitat, comprising basaltic lava flow, was formed at the time of cooling of the lava. Surface conditions are characterized by an open forest, moderate humidity, and high exposure to the sun.

4.2.2 Overall differences and similarities among intermediate-sized terrestrial SSHs

Given the relatively small number of places where terrestrial SSHs have been carefully studied (the French Pyrenees, the Czech Republic, Canary Islands, and Slovenia), it is difficult to make generalities, especially since the different places typically have very different types of MSS sites. Gers (1992, pers. comm.) suggested that several factors are important in determining both species richness and species composition. His focus was on the importance of MSS in terms of connections with other subterranean sites (especially caves) and with community dynamics and energetics. Firstly, above ground vegetation may be important, and both C. Gers in France (pers. comm.) and S. Polak in Slovenia (pers. comm.) noted that MSS sites under *Fagus* forests are often rich in species and numbers of individuals. Gers pointed out that *Fagus* is calcium-rich, and the soil may be rich in cations. A second

factor is lithology. For example, granite makes bad MSS because its cracks and fissures are small and few in number. MSS on a'a[3] lava is often extensive because of the propensity of this kind of lava to form cracks and fissures. Thirdly, high relative humidity is critical for many troglobionts (Howarth 1983), and this is why Gers suggested that covered talus is richer than open talus. Fourthly, age of the habitat is important because the lifetime of any individual terrestrial SSH habitat is usually measured in the thousands of years, although it is likely this depends very much on location. As it ages, it fills with soil.

4.2.3 Evolution of intermediate-sized terrestrial SSHs

Terrestrial shallow subterranean habitats in general, and the MSS in particular, can originate in a variety of geological and lithological settings, including tree roots, cinders from volcanic eruptions, clinker, erosional (and solutional) enlargement of cracks and fissures, and rock falls. Whatever the origin, the lifetime of any particular site is likely to be thousands or tens of thousands of years. A possible exception to this is epikarst, although it is certainly eroded and corroded away, as Šušteršič (1999) pointed out, but likely at a slower rate.

Gers (1992) devoted considerable attention to the evolution of MSS, especially talus slopes. The first stage is the creation of the talus slope, and he gave the name 'vesicular macropores' to the spaces between the rocks. This is the open talus slope studied by Růžička and colleagues (e.g. Růžička 1990, 1999b, Růžička et al. 1995, Růžička and Zacharda 2010). The second stage is a stabilization of the vesicular macropores by soils and moss covering of the soils, the MSS of Juberthie et al. (1980a, 1980b). The third and final stage is clogging of vesicular macropores with soil. At this point, no subterranean habitat remains except for the soil and perhaps a few larger spaces around rocks. This view can be expanded to a broader geographical context, and Gers (1992) suggested how the geographical extent of the MSS changes through time in the 200 km^2 mountain

---

[3] A'a is a rough and blocky lava with many gas bubbles. It is composed of broken lava blocks called clinker.

valley, La Ballongue, near Saint Girons, France. In the Riss-Würm interglacial (approximately 150,000 years bp), MSS would only have existed as fissures in the limited area of limestone, possibly seven such MSS islands. At the end of the Würm glaciation (approximately 30,000 years bp), there was a period of development of open talus slopes, with only the valley floor choked with sediments. Except for one isolated limestone block, open talus and later MSS occur in a wide area, allowing for population interchange. From the Holocene (2000 years bp) to the present, the MSS created after the Würm has become choked with sediment, resulting in 19 isolated MSS sites. Gers (1992) analysed MSS habitats in a general context of soil ecosystems and the development of soil itself. Other perspectives can be taken, e.g. the connection with deeper subterranean habitats, but his general point about the transitory nature of SSHs is important.

4.2.4 Geographical extent

The original conception of MSS was that it was a habitat of mountains, especially talus slopes. Such MSS habitats have been reported from the major mountain chains of Europe, including the Alps, Balkans, Carpathians, Pyrenees, and Rhodopes, and in mountain chains in Japan and Taiwan (Juberthie 2000). Largely due to the efforts of P. Oromí and his colleagues, a number of volcanic MSS sites in the Canary Islands have been investigated (Medina and Oromí 1990, Pipan et al. 2011a). Similar MSS habitats likely occur in other volcanic areas, such as Hawaii and the Azores (Borges 1993). Juberthie (2000) suggested the MSS in tropical areas is absent because the voids ('vesicular macropore' of Gers) are filled with laterite and clay, but this has not been documented. If the entire range of terrestrial shallow subterranean habitats is considered (Fig. 4.3), at least epikarst, rocky soil, and tree root habitats should occur in the tropics as well as nonmountainous temperate areas. The problem is that few biologists have collected in these sites. For example, because of the apparent restriction of MSS sites to mountain areas, no one has looked for terrestrial shallow subterranean habitats in the large low-lying karst areas of the USA, particularly the Interior Low Plateau, an area of high diversity of

cave troglobionts. When such habitats have been examined, biologists have often been surprised. Bilandžija et al. (2012) report finding planthoppers on tree roots penetrating very shallow sections of Croatian caves, a habitat and fauna thought to be restricted to lava tubes.

4.2.5 Physico-chemistry of intermediate-sized terrestrial habitats

An essential part of Juberthie's definition of MSS was that it was covered with soil, which he thought was important in part because of the buffering effects of temperature. Růžička (1990) measured the temperature throughout the day for two sections of a talus slope, one covered and one uncovered, on Lovoš Hill in the České Středohoří mountains, Czech Republic (Fig. 4.8). Temperatures in the talus slopes from depths of 10–120 cm were very similar in the two sections, ranging from 15 °C to 25 °C, with lower temperatures in the lower depths. Air temperatures were nearly identical, ranging from 20 °C to 27 °C. In contrast surface temperatures were quite different in the two sections, reaching 50 °C in the open talus slope and only 25 °C in the covered talus slopes. The rocks themselves provide considerable temperature buffering, and thus the physical habitat of open

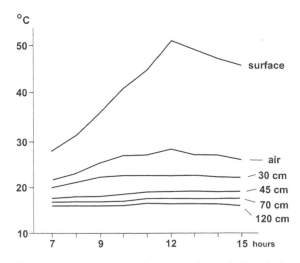

Figure 4.8 Daily temperatures for the air, surface, and various depths in open and covered sections of a talus slope on Lovoš Hill in the České Středohoří mountains, Czech Republic. Data from Růžička (1990).

and covered talus slopes may not be that different, except at the surface.

A special case is that of freezing talus slopes (Růžička 1999b, Růžička and Zacharda 2010, Zacharda et al. 2007). If the topographical orientation results in a cold air trap at the bottom of the talus slope, the lower part of the talus slope is much colder than the surface. Temperature patterns in two freezing talus slopes are summarized in Table 4.1. This temperature regime is a major factor in the survival and reproduction of species from boreal habitats in talus slopes in more southern latitudes.

The most extensive data on environmental patterns and variation in shallow subterranean sites is temperature data reported by Pipan et al. (2011a) for a series of MSS sites in Slovenia and the Canary Islands. Although there are more interesting environmental parameters than temperature, especially relative humidity and perhaps CO_2, temperature is the easiest (or at least most inexpensive) parameter to measure, and multiple frequent measurements can be made. They asked several questions about MSS habitats:

- How does temperature and its cyclicity differ from surface and the MSS?
- How does temperature and its cyclicity differ between MSS and nearby caves?
- How does temperature and its cyclicity differ among MSS types?

- How does temperature and its cyclicity differ with depth in the MSS?

The question of cyclicity, analysed using spectral analysis, is especially interesting both because of the claim that subterranean environments often lack at least daily and often even annual cyclicity, and because of the potential physiological difficulties organisms face adapting to acyclic environments (Poulson and White 1969).

All data were based on Onset (Tidbit)™ dataloggers programmed to record temperature every hour with an accuracy of $+0.2\ °C$ for periods between 412 and 419 days from May 2008 to June 2009. Cyclicity was investigated by examining the density of spectral frequencies up to a period of 100 hours in order to look for the existence of a 24-hour cycle. Spectral analysis decomposes a time series into a spectrum of cycles of different lengths, and the density measures the strength of a cycle at a given period.

How does temperature and its cyclicity differ from surface and the MSS?

We compared temperature patterns (Table 4.2) at 70 cm below the surface and at the surface for the Teno site in Tenerife, Canary Islands (see Fig. 4.6). Given the large sample sizes, all differences were statistically significant. Mean temperature was slightly lower in the MSS site, the result of yearly temperature variation or because 14 months were

Table 4.1 Annual mean, maximum, and minimum temperatures (ºC) for the interior of talus slopes near their base and external temperature on the surface of the slope for two sites in the Czech Republic. From Zacharda et al. (2007).

Locality	Site	Mean	Maximum	Minimum	Total thawing degree days	Total freezing degree days
Kammená Hůra	External	7.5	29.8	−22.3	3030	280
	Interior	−1.6	3.9	−17.0	191	664
Klíč	External	6.8	27.2	−14.8	2820	347
	Interior	0.1	8.0	−11.2	459	490

Table 4.2 Basic statistics for temperature for the Teno MSS and surface site on Tenerife, Canary Islands. Data from Pipan et al. (2011a).

Description	n[1]	Mean	Standard deviation	Coefficient of variation	10% Quantile	90% Quantile	Minimum	Maximum	Range
Surface	10,055	12.90	2.51	19.45	9.80	16.67	7.93	19.43	10.50
Lava MSS—70 cm	10,055	12.50	2.12	17.00	10.02	15.63	9.24	16.43	7.19

[1] 8760 hours in a year.

used, with an overlap of May and June. The reduction in coefficient of variation, which is independent of the mean, was 14.5 per cent, indicating that the MSS is less variable than the surface site. This difference is especially evident in the range which is reduced by 31 per cent, and the reduction comes more at the high end than the low end. The maximum MSS temperature is reduced by 2.0 °C while the minimum MSS is increased by 1.3 °C. The 10 and 90 per cent quantiles show a similar asymmetry (Table 4.2).

Both the surface and the MSS sites exhibited a strong seasonal cycle with a summer high temperature of around 16–18 °C and a winter low temperature of 8–9 °C (Fig. 1.9). The highs were lower and the lows higher at the MSS site, as is also indicated in Table 4.2. The other difference between the two sites is that there was a strong daily temperature cycle on the surface but not in the MSS, as indicated by spectral analysis (Fig. 1.9). Thus, this MSS site had a seasonal but not a daily cycle of temperature.

How does temperature and its cyclicity differ between MSS and nearby caves?

The lava clinker site at La Guancha, Canary Islands (see Fig. 4.9) was compared to nearby cave, Cueva del Mulo, although 400 m lower in elevation (1110 m asl). Not surprisingly, in the comparison between

Figure 4.9 Clinker site at La Guancha, Tenerife, Canary Islands. Four troglobiotic species are known from the site. Photo by D. Culver.

Cueva del Mulo and La Guancha MSS in the pine forest in Tenerife, the MSS site was more variable (Table 4.3). Mean temperature at the MSS site was 1.1 °C lower, the result of the higher elevation of the MSS site. Overall variability in La Guancha MSS was much higher compared to Cueva del Mulo. The coefficient of variation for Cueva del Mulo was only 31 per cent of that of the MSS site, and the range was only 36 per cent of that of the MSS.

There was also a strong seasonal cycle although the pattern was quite different at the two sites (Fig. 4.10). At the MSS site, there was a consistent summer high of 14–15 °C but the winter lows fluctuated from 4 °C to 8 °C at intervals of about one week. The reason for this is unknown but it seems likely that surface temperatures fluctuated as well. Conditions may fluctuate more at higher altitudes in the Canary Islands, where north-oriented intermediate altitudes are usually cloudy but altitudes over 1200 m asl are more exposed to sunlight. There was no evidence of a daily cycle in the MSS site. There was also a seasonal cycle in Cueva del Mulo with a summer high of 13 °C and a winter low of 10 °C (Fig. 4.10), an amplitude much less than the Teno MSS site. There was a very weak daily temperature cycle, largely the result of daily temperature fluctuations in the winter. The presence of a daily temperature cycle is not surprising since light penetrates to the back of the cave, but it is surprising that it is not stronger. Cueva del Mulo is one of the very few cases in the Canary Islands where troglobionts can be found in the presence of light (Oromí and López, unpublished).

How does temperature and its cyclicity differ among MSS types?

Two types of shallow subterranean terrestrial habitats occur near Jama v Kovačiji in Slovenia—MSS and epikarst (see Fig. 4.7), all in a steep dolina with ice present most of the year. Ice at the bottom of the dolina was present both at the time of installation (23 April 2008) and removal (9 June 2009) of the dataloggers. In addition to a data logger 10 m inside the cave in constant darkness, one was placed at the top of the epikarst at a depth of 30 cm, immediately above the cave. A third datalogger was on

Table 4.3 Basic statistics for temperature for MSS at La Guancha and Cueva del Mulo on Tenerife, Canary Islands. Data from Pipan et al. (2011a).

Description	n[1]	Mean	Standard deviation	Coefficient of variation	10% Quantile	90% Quantile	Minimum	Maximum	Range
MSS—clinker—70 cm	10,029	10.38	3.24	31.17	5.71	14.55	3.99	15.02	11.03
Cueva del Mulo	10,029	11.49	1.13	9.8	10.2	13.3	9.73	13.61	3.88

[1] 8760 hours in a year.

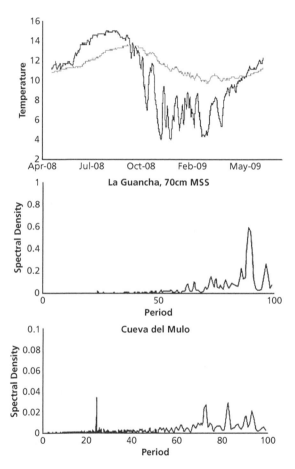

Figure 4.10 Top panel: temperature profiles at hourly intervals for the Sima de la Perdiz MSS site (solid line) and Cueva del Mulo (dotted line), Canary Islands. Centre panel: spectral densities (y-axis) for different cycle periods (x-axis) for cycles up to 100 days for the MSS site. Note the absence of a 24-hour period even at very low spectral densities. Bottom panel: spectral densities (y-axis) for different cycle periods (x-axis) for cycles up to 100 days for the cave sites. Note the weak 24-hour period.

limestone covered with soil, stones, leaf litter, moss, and ferns at a depth of 30 cm, about 25 m from the cave, with no direct connection to the cave.

The cave had the coldest mean temperature because of the persistence (but not permanence) of ice in the cave for most of the year (Table 4.4). The cave and the epikarst site (accessed from the surface immediately above the cave) had nearly identical temperature profiles, with high coefficients of variation and similar 10 and 90 per cent quantiles. Only the maximum temperature differed, being 2.2 °C in the epikarst site. The MSS site, located in the same dolina as the cave was less variable overall—the coefficient of variation was only 29 per cent that of the cave. On the other hand, the temperature range at the MSS site was 22 per cent larger. Quantiles showed a similar pattern, being larger at the MSS site.

All three habitats showed a seasonal pattern (Fig. 4.11). The MSS site had the simplest pattern with summer highs around 9 °C and winter lows around 0 °C. Unlike MSS sites on the Canary Islands, there was a strong diurnal cycle. The cave showed a complex pattern of slowly rising temperatures (up to about 6 °C) until October when the temperature dropped to 0 °C or below until the following June, undoubtedly the result of the formation of ice in the cave. The epikarst site showed a pattern similar to that of the cave although temperatures were generally higher, reflecting the greater distance of epikarst from the ice. The cave had a very weak daily cycle, largely limited to late summer, while the epikarst site had a stronger daily cycle, present through most of the summer.

Table 4.4 Basic statistics for temperature for MSS and epikarst site near Jama v Kovačiji, Slovenia. Data from Pipan et al. (2011a).

Description	n[1]	Mean	Standard deviation	Coefficient of variation	10% Quantile	90% Quantile	Minimum	Maximum	Range
MSS—30 cm	9889	5.38	3.46	64.27	0.16	9.24	−0.16	11.27	11.43
Epikarst—30 cm	9889	1.23	2.71	220.33	−2.63	4.42	−4.65	6.76	11.41
Jama v Kovačiji	9889	0.98	2.17	220.05	−2.06	3.88	−4.76	4.41	9.27

[1] 8760 hours in a year.

How does temperature and its cyclicity differ with depth in the MSS?

At the Mašun MSS site in Slovenia (see Fig. 1.3), probes were placed at depths of 10, 20, and 50 cm. There is a decline both of temperature and the co-efficient of variation with increasing depth (Table 4.5). The higher temperatures closer to the surface suggest that the sample year was warmer than the historic average temperature. The difference in mean temperature at 10 cm in depth compared to 50 cm in depth was 0.5 °C. The coefficient of variation declined 22 per cent and the range declined 27 per cent. The drop in the maximum and 90 per cent quantile was greater than the drop in the minimum and 10 per cent quantile temperature. Only in the 10 cm data logger did the temperature drop below 0 °C.

All three MSS sites, ranging in depth from 10 to 50 cm, showed a very similar seasonal cycle (Fig. 4.12) with summer high temperatures of 14–15 °C for the upper two sites and 13 °C for the lower site, and winter low temperatures of between 2 °C (lower) and 0 °C (upper). Only at the uppermost MSS site did temperatures ever drop below 0 °C (Table 4.5). The uppermost datalogger showed a strong daily cycle and the intermediate datalogger showed a weak one. At the intermediate site, daily cycles were apparent during the summer months.

Very little is known about relative humidity in MSS or other terrestrial shallow subterranean habitats. Gers (1992) reported values of between 5.2 per cent and 48.7 per cent with a mean of 20.0 per cent (based on 27 monthly measurements), slightly lower values than he reported for the B-horizon of the soil, which had a mean of 26.2 per cent. These contrast sharply with those reported by Nitzu et al.

(in review) for covered and uncovered talus slopes in Piatra Craiului National Reserve in the southern Carpathians of Romania. Based on eight samples at each of two sites, they found that relative humidity was always greater than 80 per cent. Gers reported difficulties in measuring relative humidity, and his method (involving soil dessication at 105 °C) measures soil moisture rather than relative humidity. It is interesting to note that Nitzu et al. (in review) found high relative humidity values in both open and closed talus slopes, and this corresponds to the results for temperature in the two kinds of habitat reported by Růžička (1990, see Fig. 4.8).

Gers (1998) has provided the only available data on chemistry of the MSS (Table 4.6). He analysed ionic concentration in water samples taken both from soil layers and the MSS. Among the cations, MSS had the highest concentrations of Ca^{2+} and Mg^{2+}, and among the lowest values for NH_4^+ and K^+. Among the anions, NO_3^- was highest in the MSS and Cl^- and SO_4^{2-} were intermediate.

4.2.6 Analogues with other habitats

There is a confusing state of affairs with regard to names given to terrestrial SSHs, and more terminology would only confuse matters further. Based on existing usage, there are a number of terms that can be used to distinguish among these habitats, including clinker, covered talus, epikarst, macropores, MSS, and open talus. MSS is perhaps the most problematic, because it can be restricted to covered non-calcareous rocky hillsides, the original habitat described by Juberthie et al. (1980a, 1980b), or expanded to cover almost all intermediate-sized terrestrial SSHs. Gers (1992) advocated

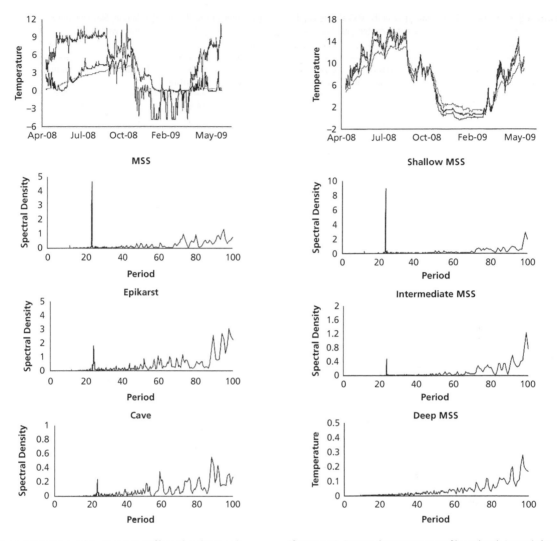

Figure 4.11 Top panel: temperature profiles at hourly intervals for an MSS (thick dotted line), epikarst (solid line), and cave (thin dotted line) site in the dolina where the entrance of Jama v Kovačiji (Slovenia) is located. Upper centre panel: spectral densities (y-axis) for different cycle periods (x-axis) for cycles up to 100 days for the MSS site. Note the strong 24-hour period. Lower centre panel: spectral densities (y-axis) for different cycle periods (x-axis) for cycles up to 100 days for the epikarst site. Note the weak 24-hour period. Lower panel: spectral densities (y-axis) for different cycle periods (x-axis) for cycles up to 100 days for the cave site. Note the weak 24-hour period.

Figure 4.12 Top panel: temperature profiles at hourly intervals for an MSS site at Mašun, Slovenia, at depths of 20 cm (heavy dotted line), 50 cm (solid line), and 80 cm (thin dotted line). Upper centre panel: spectral densities (y-axis) for different cycle periods (x-axis) for cycles up to 100 days for the upper MSS site. Note the strong 24-hour period. Lower centre panel: spectral densities (y-axis) for different cycle periods (x-axis) for cycles up to 100 days for the intermediate MSS site. Note the weak 24-hour period. Lower panel: spectral densities (y-axis) for different cycle periods (x-axis) for cycles up to 100 days for the deep MSS site. Note the absence of any 24-hour period, even at low spectral densities.

the use of the term macropore, in particular vesicular macropore, to designate patches of intermediate-sized SSH, both to denote their size and to denote the connection with soil. There can be no doubt that in many cases, there is an intimate connection with the soil, but not all terrestrial SSHs share this close connection (see Fig. 4.3). Even the distinction between open and covered talus slopes may be blurred because many talus slopes are partially covered. Open talus slopes

Table 4.5 Basic statistics for temperature of the MSS at different depths at two different sites (A and B) in Mašun, Slovenia. Details of temperature patterns are slightly different at the two sites because of differences in placement of the dataloggers with respect to a vertical bank. Data from Pipan et al. (2011a).

Description	n[1]	Mean	Standard deviation	Coefficient of variation	10% Quantile	90% Quantile	Minimum	Maximum	Range
MSS-A—10 cm	9889	7.46	5.31	71.29	0.14	14.85	−0.33	16.59	16.92
MSS-A—30 cm	9889	5.38	3.46	64.27	0.16	9.24	−0.16	11.27	11.43
MSS-B—20 cm	9889	7.38	4.76	64.54	0.84	14.1	0.52	15.37	14.85
MSS-B—50 cm	9889	6.91	3.84	55.61	1.64	12.39	1.01	13.32	12.31

[1] 8760 hours in a year.

Table 4.6 Cations, anions, and pH for La Ballongue MSS site in France. All concentrations are in mg L^{-1}. Data from Gers (1998).

Layer	pH	NH_4^+	Ca^{2+}	Mg^{2+}	K^+	Cl^-	NO_3^-	SO_4^{2-}
Litter	5.61 ± 0.36	0.29 ± 0.96	3.98 ± 2.86	0.32 ± 0.17	1.58 ± 1.28	1.46 ± 0.89	1.33 ± 1.07	3.03 ± 3.46
AO soil horizon	6.35 ± 0.15	0.04 ± 0.07	12.46 ± 0.97	0.43 ± 0.04	0.43 ± 0.45	0.87 ± 0.34	7.52 ± 2.30	2.82 ± 2.60
B soil horizon	6.51 ± 0.10	0.01 ± 0.003	14.34 ± 1.29	0.53 ± 0.04	0.26 ± 0.16	0.87 ± 0.20	10.79 ± 4.88	1.29 ± 0.45
BC soil horizon	6.50 ± 0.13	0.011 ± 0.003	14.67 ± 0.78	0.66 ± 0.03	0.075 ± 0.13	1.25 ± 0.47	8.31 ± 2.14	1.63 ± 0.61
MSS	6.63 ± 0.25	0.02 ± 0.01	15.46 ± 0.90	0.84 ± 0.002	0.02 ± 0.004	1.38 ± 0.34	11.80 ± 3.34	2.71 ± 0.61

often have little soil and while they may eventually develop such connections, there is not at present an intimate connection with the soil. The general term, intermediate-sized SSHs may be useful but it conveys none of the distinctions among such habitats.

There are clear analogies with aquatic habitats, and in the case of epikarst, aquatic and terrestrial animals are often collected at the same drip (see chapter 3). Since epikarst is above the permanent water table, but yet is an important site of water storage (Williams 2008), a mosaic of water-filled and air-filled pockets exists. As the amount of water stored changes, there is a shift between the water-filled and air-filled cavities, and it is when air-filled cavities become water-filled that the terrestrial epikarst fauna ends up washing out of drips. Hypotelminorheic habitats can be thought of as aquatic edaphic habitats, ones whose only permanent water is associated with clay layers. In a similar vein, Juberthie (2000) included the hypotelminorheic as an aquatic MSS.

4.3 Biological characteristics of intermediate-sized terrestrial SSHs

4.3.1 Organic carbon and nutrients in intermediate-sized terrestrial SSHs

Gers (1998) pointed out that there are two potential sources of organic carbon in the MSS. One is from organic carbon dissolved in water that percolates into the MSS. The analysis of organic carbon in water found in the MSS at the La Ballongue site, indicated that the amount of organic carbon was below detectable levels, i.e. <0.2 mg L^{-1}. On the other hand, he argued that the movement of animals between the MSS and the soil resulted in a significant transfer of organic carbon, although he was not able to measure this quantitatively. What he was able to show was that both soil animals and MSS animals moved between the two habitats. For example, he showed that MSS-dwelling beetles in the genus *Speonomus* came within 20 cm of the surface, well above the MSS level of approximately 1 m at the La Ballongue site. Likewise, MSS-dwelling phorid flies came within 0.4 m of the surface. Soil species also moved downwards, and soil-dwelling species are routinely found in the MSS. Gers noted that many soil Collembola were found in the MSS sites. Pipan et al. (2011a) also reported soil-dwelling beetles at MSS sites in the Canary Islands.

4.3.2 History of biological studies of terrestrial SSHs

E.G. Racoviţă, in his 'Essai Sur les Problèmes Biospéologiques', originally published in 1907 (Moldovan 2006), emphasized that cave animals also lived in cracks and crevices adjoining the parts of the cave that humans could enter. This and other ideas of Racoviţă were enormously influential among European speleobiologists, and he remains a figure of near reverence to some European biologists (Sket 2006). On the other hand, he was and is largely unread by and unknown to North American speleobiologists. There is no doubting his insight that cave animals did not require large passages in order to survive in the subterranean realm. This resulted in a conceptual divide between speleobiologists on the two continents, with North Americans being much more cave-focused and Europeans and others being more focused on the subterranean realm in general. It is not surprising that most of the work on terrestrial SSHs came from Europeans.

Among the first speleobiologists to look for troglobionts not only outside of caves but outside of cave regions were Novak and Sivec (1977) and Ueno (1977). Novak and Sivec inventoried the fauna of Jama pri Votli peči in Slovenia, a small 20 m cave in pegmatite (an igneous rock with large crystals). With some exceptions, they found species there that occurred in limestone caves in the region, although the closest cave was more than 2 km away. They clearly indicate that these species are living in fissures and cracks in non-carbonate rocks, in spite of the fact that the underlying chemistry and geology is very different. Novak and Sivec (1977) state that 'If there are fissures . . . in non-carbonates the hypogeic [subterranean] fauna can also enter them permanently . . . ' Ueno (1977) reported that there was a troglobiotic fauna in lava tubes and other non-calcareous formations, but even in a variety of mine adits, although in lower frequency than in

caves. He pointed out that the richest adits were those in more fractured rock, such as mudstones. Ueno (1977) concluded by stating that 'Thus, most of the so-called troglobionts, with the possible exception of ultra-evolved ones, are nothing but such animals as live deep in the earth or in fissures of rocks beneath the soil.' Ueno, as well as Novak and Sivec, used mines and tectonically formed caves as sampling places for the terrestrial SSH fauna, and they did not think that these large cavities were their primary habitat.

As a counterpoint to these early observations of a terrestrial SSH fauna is the even earlier paper by Barr (1967), a prominent North American speleobiologist. Of the over 100 species of *Pseudanophthalmus* beetles known, all were found in caves, and Barr searched for non-cave species. He finally found one in a covered talus slope in West Virginia tens of km from the nearest limestone area. However, he considered this to be an epigean species even though it was found deep in moss-covered talus. He also found several other troglobionts at the site, including the springtail *Pseudosinella gisini*. Barr did think that this 'epigean' covered talus slope was likely similar to the ancestral habitat for cave *Pseudanophthalmus*.

The emergence of the general concept of a *milieu souterrain superficiel* (MSS) was the result of the publication of two papers by Juberthie and colleagues (1980a, 1980b), in which they formally named these SSHs, described their characteristics, enumerated some of the fauna, and pointed out their biogeographical importance. Initially at least, Juberthie and others saw the MSS as a dispersal corridor, a way of explaining disjunct distributions. Soon it became apparent that not only did it harbour troglobionts previously known from caves, but new troglobiotic species as well (e.g. Bareth 1983, Delay et al. 1983, Genest and Juberthie 1983a, 1983b, Gers and Najt 1983). A number of investigators reported MSS habitats outside of France (e.g. Kroker 1983, Medina and Oromí 1990, Rendoš et al. 2012). Two graduate students of Juberthie's wrote dissertations on the MSS (Crouau-Roy 1987, Gers 1992). Crouau-Roy (1987) focused on population genetics and demography of an important MSS species—*Speonomus hydrophilus*—and Gers (1992) focused on ecosystem dynamics of the MSS. With the exception of a monograph on the MSS in Italy (Giachino and Vailati 2010), and continuing work on the MSS in the Canary Islands as part of larger subterranean studies (Arnedo et al. 2007, Frisch and Oromí 2006, Gilgado et al. 2011), work on the MSS by speleobiologists nearly ceased. There are undoubtedly several reasons for this, but one is not that the questions raised by the existence of a troglomorphic and troglobiotic MSS fauna have been answered.

On the other hand, the work of the Czech biologist V. Růžička, which began shortly after Juberthie's studies (e.g. Juberthie 1982), has continued up to the present. Originally interested primarily in boreal spiders with disjunct distributions in talus slopes (e.g. Růžička 1988a), he expanded his interest to include the evolution of troglomorphy and the connection of the talus fauna with a deeper subterranean fauna (e.g. Růžička 1998, 1999a).

Box 4.1 Collecting in intermediate-sized terrestrial SSHs

Two problems make collecting in terrestrial SSHs especially difficult. The first is that almost any trapping design requires the destruction and recreation of the habitat. In order to sample the aphotic zone of terrestrial SSHs, rocks and soil must be removed and then replaced. Secondly, because of low densities of animals, traps that can be left in place weeks or months are needed.

All trapping devices used are modifications of standard pitfall traps. Laška et al. (2008) used a tube in which a pitfall trap can be set at a given depth. The device requires extensive digging to set up. The most detailed and specialized MSS trap design is that of López and Oromí (2010), the description of which comes from their paper. They noticed that when the same pitfall traps were removed and replaced repeatedly, there was considerable disturbance to the MSS, resulting from the collapse and modification of the environment. Based in part on a design by Owen (1995), they developed a permanent MSS pitfall trap which produces minimal disturbance once it has stabilized. The main component of the trap is a 75 cm-long PVC pipe with an inside diameter of 11 cm. Many small holes (5–7 mm in diameter) are drilled along its surface, except for a 8 cm band at the bottom and a 10 cm band at the top (see Fig. 4.13). The pipe dimensions and the distribution of the holes can vary according to particular requirements of each study, but the diameter of holes must be small enough to avoid the entrance of non-hypogean fauna, particularly small vertebrates (mice, shrews, small reptiles, etc.). A plastic tray 8 cm high and 11 cm in diameter (a width that fits perfectly inside the pipe, see Fig. 4.14A) is fitted with a nylon cord as a handle. The nylon cord is slightly longer than the pipe and is tied to two points on the opposite lips of the tray with small hard plastic tubes (labelled pt in Fig. 4.14A). This tray can be made from the bottom of any plastic bottle with the appropriate diameter. An 8 cm high, 3 cm diameter plastic bait container (labelled bc in Fig. 4.14A) is fixed to the centre of the tray with a nut, bolt and washers

(Fig. 4.14B). The nut, bolt, and washers are covered with silicone to prevent leakage of fluid and rusting. The tray will contain the preservative liquid (propylene glycol, labelled pg in Figure 4.14A), where the specimens will fall, and the bait container holds the bait (e.g. blue cheese, liver, or fresh fish) avoiding its contact with the propylene glycol. This bait in the bait container is covered by a piece of mesh, and the bait container is fastened with a perforated cap allowing the bait odour to exit while preventing animals from reaching the bait. The third component of the trap is a silicone cap closing the top of the pipe, like those used by plumbers as temporary caps to protect pipe threads (Fig. 4.13).

Figure 4.13 Basic features of an MSS trap. **A:** silicone cap sealing the pipe, with lateral cuts for easy removal. **B:** PVC pipe with abundant holes. From López and Oromí (2010).

continued

Box 4.1 *Continued*

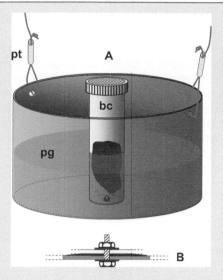

Figure 4.14 Details of an MSS trap (see Fig. 4.13). **A:** tray with propylene glycol (pg), the bait container (bc) and the nylon tied with small plastic tubes (pt). **B:** detail of the union between the bait container (white) and the tray (grey) with a screw, washer and silicone (black). From López and Oromí (2010).

Figure 4.16 Filling stones into the hollow around the trap installed in the MSS (see Fig. 4.15). From López and Oromí (2010).

The PVC pipe is installed inside the hole such that the perforations of the pipe stay at the MSS level, and the top remains a few centimetres below ground surface. The small stones previously removed when digging the hole are used to fill up the empty space left around the pipe, just up to a few cm above the highest perforations of the latter (Fig. 4.16). The rest of the cavity is filled with compacted soil to isolate

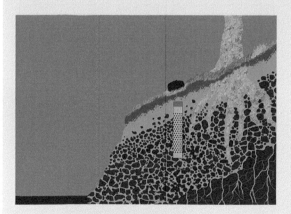

Figure 4.15 An MSS trap installed in the MSS close to a road cut. From López and Oromí (2010).

Suitable MSS is difficult to locate from undisturbed surface terrain, but easier to detect when the terrain is cut for building roads. The traps can be set a few metres upslope from such a location (Fig. 4.15). Once the site is selected, a narrow vertical hole is made by hand from the surface down to reach the MSS, using a 1 m-long, sharp, pointed iron bar.

Figure 4.17 Nesting rings used by V. Růžička to stabilize talus slope to allow placement of pitfall traps. Photo by T. Pipan.

continued

Box 4.1 *Continued*

the system and limit the entrance of surface fauna. The tray, loaded with propylene glycol and the baited bait container, is lowered to the bottom of the pipe, leaving the nylon cord handle protruding beyond the tube. The top of the pipe is closed with the silicone cap trapping the nylon handle, and the whole system is covered with soil and litter to hide and protect the trap. At the end of the sampling interval, the top of the trap is opened, taking care to avoid soil falling inside, and the tray is lifted to the surface with the nylon handle. The contents of the trap are filtered and saved for sorting in the lab, and the tray is deployed again with fresh propylene glycol and bait. The time necessary to obtain animals varies

with site, but in the Canary Islands it has sometimes taken one to two years (!) to get samples.

Working in open talus slopes, Růžička (1982, 1988b) designed a trap to be left for extended periods by drilling a hole in a small board and then placing the pitfall trap in the board. This allowed for stable placement in an unstable debris slope. The most difficult problem with traps on open talus slopes is their initial placement. For every rock removed, one seems to take its place. To counter this problem, V. Růžička (pers. comm.) uses a set of nesting rings that allow the making of any initial hole in the talus slopes (Fig. 4.17).

4.3.3 Overview of the fauna of intermediate-sized terrestrial SSHs

There have been two particular areas of interest with respect to the fauna of terrestrial SSHs—as a location of relict populations of boreal species and as a dispersal corridor and habitat for subterranean species. According to Růžička (2011), talus slopes in Central Europe can be habitats for boreal spiders occurring in stony habitats, boreal forests, and wet habitats (Fig. 4.18). He identifies five such species, including *Bathyphantes eumenis*, whose distribution is shown in Fig. 4.19. The distribution of *B. eumenis eumenis*, which is pigmented, occurs exclusively in sandstone rock complexes of the Czech Republic and Poland (e.g. Růžička et al. 2010). The troglomorphic subspecies *B. eumenis buchari* inhabits talus slopes throughout Central Europe north of the Alps (Růžička 1988a). Species such as *B. eumenis buchari* tend to occupy the colder, lower parts of talus slopes. Every spider and mite species found in open talus slopes has a characteristic horizontal and vertical distribution pattern, many being highly restricted, both vertically and horizontally (Růžička et al. 1995).

The initial focus of interest of French speleobiologists was on the bathyscine beetle *Speonomus hydrophilus* (Fig. 4.20). It is a small (3 mm) beetle common in caves in the Ariège region of France. It is without eyes and with moderately elongated appendages. Its morphology suggests that it may be

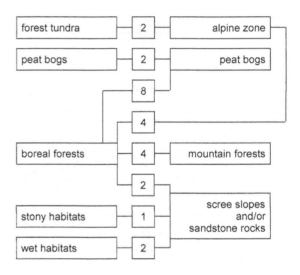

Figure 4.18 Ecological distribution of spiders found in Czech Republic that are also found in European boreal habitats. The lines connecting the boxes indicate a shared species, the number of such species being listed in the small squares. From Růžička (2011).

more of a soil and MSS inhabitant than cave inhabitant. However, it is quite abundant in some caves (Crouau-Roy 1987). As is characteristic of species in Pyreneean MSS sites, populations are low or absent in the winter. Fig. 4.21 shows the contrast between the Tour Laffont MSS site and the cave Grotte de Montagagne. The MSS site is much higher (1350 m asl) than the cave (670 m asl) and most French MSS sites are high on valley sides, as is Tour Laffont. The

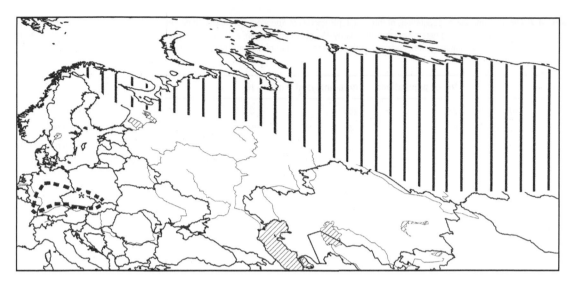

Figure 4.19 Distribution of *Bathyphantes eumenis* in Europe and Asia. The main boreal distribution of *B.e. eumenis* is marked with vertical line patterns, and its central European distribution is marked with an asterisk. The central European distribution of *B.e. buchari* is bounded by the dashed line. From Růžička (2011).

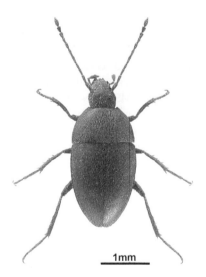

Figure 4.20 Photo of *Speonomus hydrophilus*. Photo © by Prof. Lech Borowiec, used with permission.

disappearance of adults in the winter months is likely explained by their downward movement to escape freezing temperatures. It is also clear from Crouau-Roy's data that reproduction in MSS populations is occurring, as newly emerged adults (imagos) are common in late fall and early spring. It is likely that the winter disappearance of species only occurs in high elevation sites.

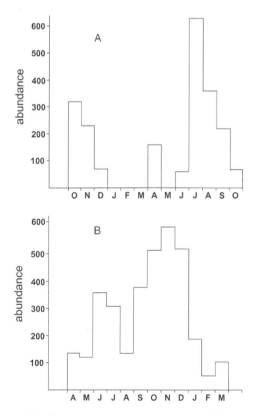

Figure 4.21 Monthly abundance of two populations of *Speonomus hydrophilus* from (A) an MSS site (Tour Laffont) and (B) a cave (Grotte de Montagagne) in France. Adapted from Crouau-Roy (1987).

Table 4.7 Allozyme analysis of populations of three allotopic species of *Speonomus* from Massif de l'Arize, France. H_o is the observed heterozygosity, H_e is the expected heterozygosity under Hardy–Weinberg equilibrium, F_{ST} is a measure of population differentiation under Wright's (1978) island model of population genetics, and Nm is the expected number of migrants per population per generation. Data from Crouau-Roy (1988).

Species	No. of populations	Area analysed (km²)	H_o	H_e	F_{ST}	Nm
Speonomus hydrophilus	23	285	0.124 + 0.055	0.214 + 0.093	0.112 + 0.031	2.0
Speonomus zophosinus	6	24	0.141 + 0.053	0.239 + 0.089	0.081 + 0.069	2.8
Speonomus colluvii	5	40	0.246 + 0.021	0.444 + 0.033	0.050 + 0.024	4.8

Using gel electrophoresis to study allozyme variation, Crouau-Roy (1988) analysed the variation in populations of three species of *Speonomus*, including *S. hydrophilus*, in a small area ranging from 24 to 285 km², depending on the species (Table 4.7). The most remarkable finding of her study was that all populations showed a heterozygote deficiency at all loci, most of which were statistically significant. After considering a series of options, she concluded that the most likely explanation was that all of the populations were inbred, and that heterozygote deficiency resulted. From estimates of F_{ST}, Wright's (1978) measurement of population differentiation among island populations that exchange migrants, it was possible to estimate that the number of migrants per population per generation was between two and five, above the level of individual population differentiation. So there is a puzzle. On the one hand, populations are inbred, suggesting small size and/or isolation. On the other hand F_{ST} values indicate considerable migration. One possibility

is that there is strong selection operating against heterozygotes resulting from mating between migrants and residents, but the entire issue needs more study. What is clear is that the MSS is hardly a dispersal highway. Nearby populations are often differentiated, and in turn differentiated from cave populations.

4.3.4 Species composition and richness in intermediate-sized terrestrial SSHs

Crouau-Roy (1987) provides a tabulation of the taxonomic position and ecological category (troglobiont vs soil dweller vs generalist) for extensive collections from three MSS sites and one cave in the Ariège region of France (Table 4.8). Microarthropods are underrepresented in the samples but there are interesting generalities and differences that emerge. Bathysciinae beetles, especially *Speonomus hydrophilus*, dominate all four sites, comprising 74 and 86 per cent of the fauna. *S. hydrophilus* alone

Table 4.8 Faunal composition of three MSS sites and one cave in the Ariège region of France. Ecological classifications follow those of Crouau-Roy (1987). Frequencies for each site add up to one.

Group	Ecological classification	Tour Laffont	Bellissens	Col de Marrous	Grotte de Montagagne
Bathysciinae Coleoptera	Troglobionts	0.75	0.86	0.77	0.74
Other Coleoptera	Soil	0.01	0.00	0.02	0.01
Diptera	Soil	0.23	0.11	0.09	0.20
Collembola and Diplura	Troglobionts	0.01	0.00	0.10	0.01
Diplopoda—*Typhloblaniulus*	Troglobionts	0.00	0.00	0.00	0.04
Other Diplopoda	Generalists	0.00	0.01	0.00	0.00
Arachnids	Generalists	0.00	0.01	0.02	0.01
Total number of individuals		2893	4602	565	4167

accounted for between 72 and 86 per cent of the fauna. Percentages for the cave were slightly lower than for the MSS sites, but the abundance of Bathysciinae in general and *S. hydrophilus* in particular is impressive. Except for Col de Marrous, Diptera, especially Phoridae, were the second most abundant group. In Col de Marrous, Entomobryidae Collembola were common. Only Grotte de Montagagne had more than 1 troglobiotic *Typhloblaniulus*.

For a nearby area in the Ariège region, Gers and Najt (1983) enumerated the Collembola found in MSS sites as well as describing a new species. Of the ten species of Collembola, two (*Pseudosinella virei* and *Oncopodura tricuspidata*) are common troglobionts in local caves. Of the two, only *P. virei* was common in the MSS. One rare MSS species, *Lipothrix lubbocki* is primarily a deep soil species. The species they described, *Isotoma juberthiei* was primarily an MSS species but a few individuals were found in the soil as well. The other six species, *Ceratophysella denticulata*, *Triacanthella perfecta*, *Protaphorura* gr. *armata*, *Onychiurus* cf. *aguzouensis*, *Folsomia quadrioculata*, and *Lepidocyrtus lanuginosis* are more generalist species and all have been found in caves or cave entrances. Of these six, only *Protaphorura* gr. *armata* was abundant in the MSS. This pattern of a mixture of specialists and generalists, or troglobionts and non-troglobionts, was also the pattern of species composition in hypotelminorheic and epikarst habitats (chapters 2 and 3, see also Pipan and Culver 2012a).

The most extensive species list available for an MSS site is that compiled by H. López and P. Oromí (Table 4.9) for an erosional MSS site in a laurel forest at Teno, Tenerife, Canary Islands, and published in Pipan et al. (2011a). The majority of species (41 of 73) are generalist species, showing no obvious morphological specialization to either soil, MSS, or lava tubes. Another 22 species are listed by López and Oromí as soil specialists (edaphobionts). Given the proximity to the soil, this is not surprising and is similar to the findings of Crouau-Roy (1987) at covered talus MSS sites in the French Pyrenees (Table 4.8) where edaphobionts comprised between 11 and 24 per cent of the individuals found. A total of 10 troglobionts were found at the Teno site, including at least two species, the spider *Dysdera madai* and the beetle *Loboptera tenoensis*, which

Table 4.9 Species list for erosional MSS site in laurel forest in Teno, Tenerife, Canary Islands. Data from Pipan et al. (2011a).

Taxa	Troglobionts	Edaphobionts	Generalists
Arachnida			
Acari		4	
Araneae	3		4
Opiliones			1
Pseudoscorpionida	1	1	2
Hexapoda			
Collembola		7	
Diplura		1	
Insecta			
Blattaria	2		
Coleoptera	3	5	21
Hemiptera			2
Diptera			3
Hymenoptera			6
Chilopoda		1	
Diplopoda	1	3	1
Crustacea: Isopoda			1
Total	10	22	41

are not only endemic to MSS, but found only at the Teno site.

La Gomera, one of the Canary Islands, is interesting because it has no lava tubes, but does have an extensive MSS. P. Oromí and H. López (unpublished) have found a total of 11 obligate subterranean-dwelling species on the islands, all collected in MSS traps (Table 4.10). Four of these species are likely edaphobionts, that is, soil species, but the remaining seven have troglomorphic features. Only one species, the staphylinid beetle *Micranops bifossicapitatus* is found outside of La Gomera. While it is tempting to consider these cases of colonization and evolution in the MSS, the situation is complicated because the island, which is quite old by Canarian standards (12 million years), may have had lava tubes in the past, and so these species may have invaded the MSS from caves! We return to the topic of colonization and evolution of the SSHs in chapter 13.

Data from the Teno site summarized in Table 4.9 are the result of many pitfall traps from 1988

Table 4.10 List of troglobiotic species from La Gomera, Canary Islands, an island without caves. Edaphomorphic morphology is one adapted to life in the soil (see chapter 7). Data from P. Oromí and H. López (unpublished).

Group	Species	Ecological category
Coleoptera: Carabidae	*Pseudoplatyderus amblyops*	troglomorphic
	Lymnastis gaudini gomerae	edaphomorphic
Coleoptera: Staphylinidae	*Domene jonayi*	troglomorphic
	Micranops subterraneus	edaphomorphic
	Micranops bifossicapitatus	troglomorphic
Coleoptera: Scydmaenidae	*Euconnus specusus*	troglomorphic
Coleoptera: Curculionidae	*Laparocerus oromii*	edaphomorphic
	Oromia n.sp.	edaphomorphic
Isopoda: Armadillidae	*Venezillo* n.sp.	troglomorphic
Lithobiomorpha: Lithobiidae	*Lithobius* n.sp.	troglomorphic
Julida: Nemasomatidae	*Thalassisobates emesesensis*	troglomorphic

to 2009, including long-term pitfall trapping (Box 4.1). Because of the relatively low yield of MSS and talus traps, there has been only one report on how many samples must be taken or how long pitfall traps must be left in order to find most of the species. Gers (1992) reported that, based on 12 baited pitfall traps set 1 m apart and left for 21 days, three traps were sufficient to collect all troglobionts and 8 were needed to collect all species. Unfortunately, Gers (1992) did not create species accumulation curves by resampling (see Box 2.2). Since the MSS trap designed by López and Oromí (2010) allows for sample collection without removing the trap, it would be interesting to remove the samples at frequent enough intervals (perhaps a month) so that multiple samples could be analysed.

4.3.5 Adaptations in intermediate-sized terrestrial SSHs

A repeated pattern among the disjunct talus populations of spiders and beetles in the Czech Republic was that their appendages tended to be relatively longer than the appendages of individuals from populations in boreal habitats. The spider *Theonoe minutissima* is found in talus slopes and peat bogs in the Czech Republic, and in peat bogs and forests in Norway and Finland (V. Růžička 1998). Spiders from Czech talus slopes (but not Czech peat bogs) had

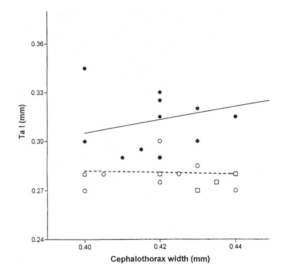

Figure 4.22 Scatter plot of the length of the tarsus of the first appendage against cephalothorax width (a measure of body size) for populations of the spider *Theonoe minutissima*. Lines are least squares fits to the data. Filled circles are individuals from talus slopes in the Czech Republic. Open circles are individuals from peat bogs in the Czech Republic and open squares are individuals from peat bogs and forests in Norway and Finland. From V. Růžička (1998).

proportionally longer appendages (as measured by the length of the tarsus of the first appendage) than peat bog populations (Fig. 4.22). J. Růžička (1998) did a more extensive morphometric analysis of the *Choleva agilis* species group of Leiodidae beetles, including five described and three undescribed species. He measured 21 morphological characters, and on the basis of forward-selection regression analysis, picked the eight most significant variables (including appendage lengths, body size, and eye size) for Canonical Variate Analysis. Růžička distinguished three major habitats: caves, talus, and epigean (surface) habitats. The 95 per cent confidence interval envelopes for the three habitats are quite distinct (Fig. 4.23). On the first CVA axis talus populations are intermediate, and on the second CVA axis talus populations are below cave and epigean samples. Loadings on the first CVA axis are primarily appendage lengths and eye size, and talus populations are intermediate with respect to both groups of variables. The second CVA axis is primarily a measure of size with some body shape features

as well (e.g. the shortest distance between the base of the right antenna and the anterior margin of the right eye). Somewhat surprisingly, the talus slope populations have somewhat larger body sizes than either cave- or surface-dwelling populations. When he compared the eight species, the species clustered according to habitat, with the cave-dwelling species distinct from the talus- and surface-dwelling species. A caveat about this very elegant analysis is that there is no phylogenetic information available so some of the morphological similarities may not be due to similar selection pressures in similar habitats, but to phylogenetic relatedness.

Gers and Najt (1983), in their description of an MSS specialist Collembola species, *Isotoma juberthiei*, reported that there were (1) more antennal sensillae, (2) longer and thinner antennal sensillae, and (3) longer bristles on the manubrium relative to surface-dwelling species of *Isotoma*.

The spider *Dysdera madai*, found exclusively in MSS habitats on Tenerife in the Canary Islands, is an especially interesting species with respect to adaptations. Nine troglobiotic species of the spider genus *Dysdera* are known from Tenerife, seven of which each have their own epigean sister species, and the other two (*D. esquiveli* and *D. hernandezi*) are a pair of troglobiotic sister species (Arnedo et al. 2007). *D. madai* is exclusively found in the MSS of Teno, is the least troglomorphic, and is the only one coexisting with its closest relative, the epigean *D. iguanensis*, which also occurs in the same MSS station. The other eight troglobionts are restricted to lava tubes, with the exception of a unique record of the cave-dwelling *D. esquiveli* in the MSS, and rank from intermediate to highly marked troglomorphism. It might be assumed that cave *Dysdera* species are more troglomorphic than typical MSS inhabitants. However, in the consensus parsimony trees obtained by Arnedo et al. (2007), *D. madai* is not the most basal among subterranean species. Moreover, the phylogenetically distant, mostly cave-dwelling and more troglomorphic *D. esquiveli* was found in the MSS together with *D. madai* in an area many kilometres from any cave. Therefore, we cannot conclude that any species occurring in the MSS or in caves is characterized by a degree of troglomorphism or by a particular phylogenetic position. We return to this topic in chapters 10 and 12.

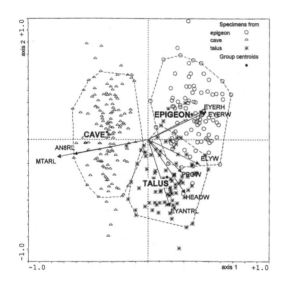

Figure 4.23 Canonical Variate Analysis of eight species in the *Choleva agilis* group from central European sites. Eight morphological variables, chosen by a forward selection procedure, were used. Specimens from caves are marked by an open triangle, specimens from talus by an asterisk, and specimens from surface habitats (labelled epigeon) by open circles. 95 per cent confidence intervals for the three ecological groups are indicated by a dotted line. Vectors indicate variable loadings (PROW—maximum width of pronotum; ELYW—maximum width of the elytra; AN8RL—length of right antennal segment; MTARL—length of the right metatarsus; EYERH—height of the right eye; EYERW—width of the right eye; EYANTRL—the shortest distance between the base of the right antenna and anterior margin of the right eye; and HEADW—maximum width of the head). From J. Růžička (1998), used with permission of Museo Regionale di Scienze Naturali, Torino.

4.4 Summary

Several groups of biologists have been interested in the fauna living in darkness in habitats close to the surface, such as talus slopes, because of the possibility that these habitats harbour troglomorphic species (because the habitat is dark), or relict populations of more northerly distributed species (because the habitat is cold). These habitats can be created by a variety of processes, including rock fall from cliffs and valley walls, erosion of rock surfaces including lava, dissolution of soluble rock, and even spaces created by tree roots. The terminology used to describe these sites is confusing and not consistent. Juberthie et al.'s (1908a, 1908b) phrase 'milieu souterrain superficiel' was first used to describe covered talus in non-calcareous rock, but has been expanded to include most intermediate-sized terrestrial SSHs. Gers (1992) suggests the term macropores, which stresses the connection of terrestrial SSHs with the soil. The lifetime of most terrestrial SSHs (epikarst is a possible exception) is probably on the order of thousands or tens of thousands of years. As they develop, soil clogs the habitat and eventually all the macropores may disappear. A summary of terrestrial SSHs and their position along a continuum of connectivity with the soil and void size is shown in Fig. 4.3.

Intermediate-sized terrestrial SSHs are especially well developed, or at least better studied, in areas of moderate to high relief. Well-studied sites include open and closed talus slopes in the Czech Republic, erosional MSS and epikarst sites in Slovenia, a variety of MSS sites in the French Pyrenees, and volcanic sites in the Canary Islands. Terrestrial SSHs may not occur in the tropics because of rapid filling of macropores by laterite and clay, but this has not been confirmed.

Most of what is known about the physico-chemistry of terrestrial SSHs is about temperature patterns. Open and covered talus slopes differ in their temperature regimes primarily at the surface, where temperature oscillations on open talus can

be extreme. A few tens of centimetres beneath the surface, temperature fluctuations in all measured MSS sites are much less extreme than at the surface, but greater than in deeper subterranean sites such as caves or lava tubes. Terrestrial SSH temperatures have a seasonal pattern, and some even show a daily pattern, albeit of small amplitude. Temperature fluctuations decline with increasing depth. Some talus slopes in the Czech Republic remain much colder throughout the year because they act as a cold air trap. Some terrestrial SSHs, especially those in epikarst, may be aquatic sites during some time periods as water levels in the epikarst fluctuate.

There are two sources of organic carbon to terrestrial SSHs, although quantitative assessment of the two has proven to be elusive. One is dissolved organic carbon in soil water, and the other is the organic carbon flux resulting from the movement of soil organisms, both laterally and vertically.

Most sampling of terrestrial SSH habitats is done by excavating a hole in the habitat, inserting an appropriately designed pitfall trap, and reconstructing the habitat above the trap. Typically traps are left in place for months. López and Oromí (2010) have designed a trap that can be left more or less permanently in place.

The fauna collected in terrestrial SSHs typically includes generalist species, soil specialists (edaphobionts), troglobionts, and, in Czech talus slopes, species otherwise found in boreal habitats. Studies on the bathysciine beetle *Speonomus hydrophilus* and related species indicate that there are permanent, reproducing populations in the MSS, that the populations go deeper into the MSS in the winter, and that there is a high level of population differentiation even in the face of moderate levels of migration. Troglobiotic species and populations found in terrestrial MSS typically show modifications for subterranean life (e.g. eye reduction and appendage lengthening) compared to surface-dwelling species, but are less modified than species from caves.

CHAPTER 5

Calcrete aquifers

5.1 Introduction

Calcrete is a term used to describe both pedogenic and non-pedogenic carbonate deposits in arid terrains (Mann and Horwitz 1979). Also termed caliche or hardpan (Reeves 1976), it occurs throughout arid and semi-arid regions, including the Sonoran Desert of North America, the Kalahari Desert of southern Africa, the High Plains of western USA, and the Yilgarn and Pilbara of Western Australia[1]. From a biological standpoint, all that is known about the fauna of calcrete aquifers is from Western Australia and the Northern Territory, where a diverse, specialized fauna was discovered in 1998 and intensively studied since that time (Humphreys 2001, 2008, Guzik et al. 2010). In addition to the presence of speleobiologists alert to the possibilities of a groundwater fauna in calcrete aquifers, the landscape of the Yilgarn and Pilbara itself offers several advantages for the study of subterranean life in calcrete aquifers.

Firstly, it is an ancient terrestrial landscape, unglaciated since the Permian (Humphreys 2001) with ample time for a specialized fauna to evolve. The long terrestrial history of western Australia has resulted in erosion down to the Precambrian basement, but with the development of groundwater calcrete aquifers along palaeodrainage paths. Secondly, many calcrete habitats in the region are often accessible to sampling because of the presence of pastoral wells and large numbers of boreholes that extend into the aquifer (Eberhard et al. 2009, Guzik et al. 2010). The boreholes were drilled for mineral exploration and to locate water, which is needed for the extensive mining operations that occur in the region (e.g. Johnson et al. 1999).

[1] Western Australia refers to the state and western Australia refers to the region.

The basic morphology of the Western Australia calcrete aquifers is that of extensive (tens of square kilometres) but very shallow deposits, often only 10 m or so in thickness very near the surface. For example in the Sturt Meadows calcrete in the Yilgarn region, calcrete aquifers were accessed by boreholes between 5 and 11 m deep (Bradford et al. 2010). On the other hand, in the Pilbara where calcrete aquifers also have been studied extensively, calcrete deposits are thicker, aquifers are deeper, and often include alluvial and colluvial aquifers as well (Eberhard et al. 2009). For example, borehole depths in the Robe River in the Pilbara, an especially biologically diverse aquifer (see Culver and Pipan 2013), ranged from 13 m to 28 m. By our definition of shallow subterranean habitats, Yilgarn calcrete aquifers are for the most part SSHs, while Pilbara calcrete aquifers are not. Nonetheless, we have chosen to include them with SSHs, both because many of them exactly fit the definition of an SSH, but also because it is worthwhile including them for comparative purposes. Non-Australian authors rarely discuss calcrete aquifers, from either a biological or hydrogeological point of view. Whenever possible, we will emphasize the aspects and types of calcrete aquifers that share features with other SSHs.

5.2 Chemical and physical characteristics of calcrete aquifers

5.2.1 Evolution of calcrete aquifers

Arid parts of Australia contain hundreds of bodies of massive carbonate deposits tens of kilometres long that are associated with present or past drainage channels. These carbonate bodies are often called 'groundwater calcrete aquifers' since their origin is

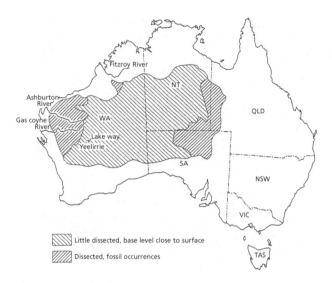

Figure 5.1 Distribution of groundwater calcrete aquifers in Australia. From Mann and Horwitz (1979), used with permission of Taylor and Francis. See also Fig. 1.12.

related to groundwater (Mann and Horwitz 1979). In Australia, groundwater calcrete aquifers extend south to about 30° S (Fig. 5.1). Fossil groundwater calcrete aquifers, such as occur in parts of the Pilbara, tend to be deeper and thus not SSHs. Calcrete distribution in Australia has been explained on the basis of two parameters—rainfall and evaporation. If there is continual moisture, any accumulated carbonate tends to be dissolved and removed rather than precipitated. Mann and Horwitz (1979) also suggest that periods of concentrated but rare rainfall lead to the development of bigger calcrete aquifers because of increased filtration of water. High evapotranspiration is also important for the direct chemical precipitation of calcium carbonate. Therefore, there are many actively precipitating calcrete systems that are very shallow, occurring at depths of less than 5 m below the surface. Overall, precipitation in calcrete-forming areas is less than 200 mm and potential evapotranspiration (PET) is greater than 3000 mm per year.

The landscape of calcrete-forming regions typically has strings of playas (salt lakes) spread along palaeo-valleys incised into the Precambrian basement rocks (Humphreys 2001). Playas are a surface manifestation of palaeodrainage channels, rivers that stopped flowing when the climate became arid. Some of these channels are incredibly old geologically. Humphreys (2001) reported that some of the palaeochannels were incised in a plateau of Precambrian rocks prior to the breakup of

the southern supercontinent Gondwana, perhaps as early as the Permian. The calcrete aquifers then undergo dissolution to create high secondary porosity in the rock, but rarely exhibit surface manifestations of karstification. Given the current aridity in the areas of calcrete aquifers, it is unclear under what precipitation conditions karstification occurs or occurred, but there was great variation in rainfall in western Australia throughout the Pleistocene.

Formation of a calcrete aquifer has four stages (Fig. 5.2, Mann and Horwitz 1979). In the first stage there is a broad drainage channel with alluviated fill and a shallow groundwater system beneath. In the second stage, following periods of recharge, Ca^{2+} and CO_3^{2-} are transported laterally, resulting in precipitation of $CaCO_3$ as the water becomes saturated. Precipitation tends to occur close to the surface and in lower portions of the drainage basin where concentrations are higher. In the third stage, precipitation proceeds and as the precipitate coalesces, it may push upwards, forming surface mounds. In the final stage, a steady-state equilibrium is reached as carbonates lifted above the water table continually erode and new carbonate is produced beneath (Fig. 5.2). Apparently epikarst does not form (see chapter 3) in these arid conditions, and the surface carbonate is often a hard, laminated capping of calcite (Mann and Horwitz 1979). Figure 5.3 shows the relationship between playas and calcrete aquifers in the Lake Way region. The initial formation

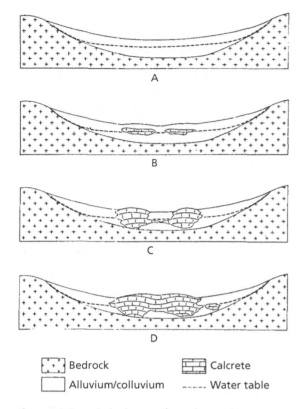

Bedrock

Alluvium/colluvium

Calcrete

----- Water table

Figure 5.2 Stages in development of groundwater calcrete. A. Shallow groundwater system in a broad drainage channel. B. Initial carbonate precipitation. C. Growth of pods and domes. D. Maturation of calcrete and surface reworking. From Mann and Horwitz (1979), used with permission of Taylor and Francis.

of calcrete aquifers may be as old as the Oligocene, following the onset of continental aridity (Morgan 1993). Many calcrete aquifers are continuing to form, making dating difficult. Based on the likely age of isolation of beetles in the family Dytiscidae (Leys et al. 2003), the Yilgarn calcrete aquifers are at least five to eight million years old.

5.2.2 Physico-chemistry of calcrete aquifers

Because of the aridity of calcrete regions and the highly episodic nature of precipitation events, there is considerable spatial and temporal variability in water chemistry. In the area near Lake Way, Yilgarn, the water table is as much as 30 m below the surface at the edge of the catchment area, but near the calcrete zone, the water table is only 2–5 m below the ground surface. Likewise, calcite solubility is higher away from the calcrete zone, and the water is supersaturated with respect to calcite in the calcrete zone (Fig. 5.4). Lake Way itself is an ephemeral lake, and the water table fluctuates from above ground to 2 m below ground surface (Mann and Horwitz 1979, Humphreys et al. 2009).

Perhaps more important biologically are changes in salinity. Because there is no external drainage from these palaeo-channels, salinity tends to increase due to evaporation. The variability of salinity and its spatial pattern led Humphreys et al.

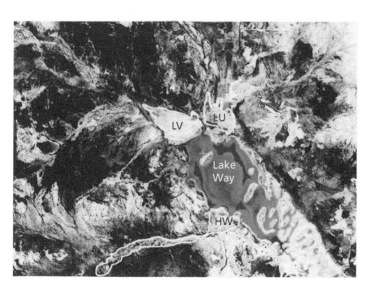

Figure 5.3 Image of the region around Lake Way in Western Australia, showing the juxtaposition of the major calcrete aquifers (areas enclosed by *pale lines*) formed from the several palaeo-drainages entering the playa. Several large open pit mines are in the area, the largest visible north of the Lake Violet calcrete. The three calcrete aquifers (LV, Lake Violet calcrete; LU, Uramurdah Lake calcrete; HW, Hinkler Well calcrete) each support an endemic fauna. The image depicts an area about 40 by 60 km. From Humphreys et al. (2009). (See Plate 13)

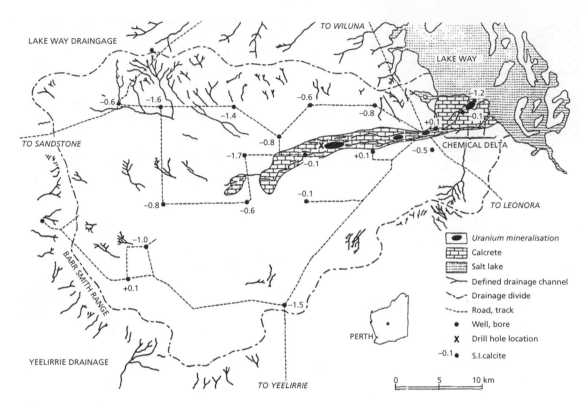

Figure 5.4 Plan of the calcrete unit on the southwestern side of Lake Way, Western Australia. The solubility index for calcite (observed pH − pH at saturation) is shown at each sampling point. A negative index denotes undersaturation and a positive index, supersaturation. From Mann and Horwitz (1979), used with permission of Taylor and Francis.

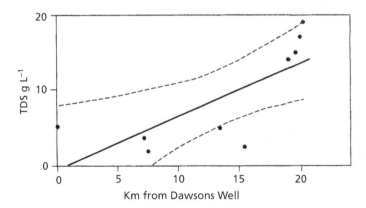

Figure 5.5 Change in salinity (measured as total dissolved solids) along the length of the Hinkler Well calcrete aquifer, from Dawson Well downstream to Lake Way, Western Australia (see Fig. 5.3). From Humphreys et al. (2009).

(2009) to the very useful observation that the subterranean drainages into playas were essentially groundwater estuaries because of salinity changes, even though there is no connection with the sea. As water moves downstream in the palaeo-channels, salinity increases (Fig. 5.5), and CaCO$_3$ is deposited near the mid-line of the drainage. Near the entrance to the playa, a chemical delta forms (see Fig. 5.4), where the calcrete unit broadens. Humphreys et al. (2009) described this system as an estuary with low salinity water near the intake (500–2000 mg L^{-1} TDS) that is alkaline and rich in bicarbonate.

As it moves downstream, it becomes less alkaline and much more saline (20,000 to >300,000 mg L^{-1} TDS)[2]. Morgan (1993) concluded that a separate geochemical system is associated with the formation of each salt lake along a palaeo-river. Each of these groundwater estuaries is connected to a separate calcrete and isolated from other calcrete aquifers, even those in close proximity, by strong salinity gradients.

5.2.3 Analogues with other habitats

In many ways, calcrete aquifers, even if we only consider those at shallow depths, are unique among SSHs. They are associated with arid regions, and salinity is a major environmental parameter that serves to constrain distributions and dispersal. While all SSHs, except the soil, have patchy distributions, calcrete aquifers would seem to represent the extreme in this regard. Like karst regions in general, calcrete aquifers are patchy at large scales (100 km and greater), but there are many more isolated and semi-isolated calcrete aquifers in western Australia than there are karst regions in temperate zones of equivalent area. Figure 1.12 shows this very clearly, and this may be one of the reasons that diversity of stygobionts appears to be very high in calcrete aquifers in Western Australia (Guzik et al. 2010). That is, there is ample opportunity for isolation and speciation. The size of calcrete aquifers is intermediate between karst regions, which tend to be much larger (thousands of km²), and many SSHs (often less than one km²). An extreme of this is the hypotelminorheic, which is patchily distributed on a very fine scale (see Fig. 1.11 for an area in Virginia, USA). Although we could not find data on either secondary porosity or pore size in calcrete aquifers, the fact that the fauna is neither highly elongate nor miniaturized suggests that the pores are relatively large, at least with respect to the invertebrates. Most species seem to be less than 5 mm in length, suggesting perhaps a moderate pore size (perhaps on the order of one cm in diameter or less), larger than the epikarst and interstitial, but certainly smaller than caves. The pores are certainly not large enough to be enterable by humans.

[2] TDS for seawater is 30,000–40,000 mg L^{-1}.

Calcrete aquifers do share some characteristics with other SSHs. Like most other SSHs, their maximum vertical extent is small so that vertical development is not particularly important. Because of the strong redox potentials (a potential source of chemical energy for chemoautotrophic bacteria) found in many calcrete aquifers (Humphreys et al. 2009), which they share with anchialine habitats (subterranean habitats with direct connections to the sea), the presence of chemoautotrophic bacteria would be a potential food supply for invertebrates dwelling in calcrete aquifers (*cf.* Humphreys 1999).

5.3 Biological characteristics of calcrete aquifers

5.3.1 Organic carbon and nutrients in calcrete aquifers

The source of the organic carbon and other nutrients at the base of the calcrete food web is puzzling, especially since there is a diverse, abundant macro-invertebrate community. The obvious source is organic carbon and other nutrients dissolved in surface water percolating into the calcrete aquifers. However, since these are very arid regions, surface productivity is highly limited, and organic carbon in percolating water that derives from this surface productivity is also likely to be very low. This source is all the more problematic because precipitation tends to occur in a few bursts, rather than scattered throughout the year. As is the case for lava tubes (see chapter 8), tree roots may provide an important source of organic carbon, but this has not been demonstrated for calcrete aquifers. Jasinska and Knott (2000) documented roots entering coastal caves in Western Australia, but they did not study calcrete aquifers. Bradford et al. (2013) did demonstrate that the likely source of organic carbon in a Sturt Meadows calcrete aquifer was terrestrial rather than aquatic. The existence of redox environments (Humphreys et al. 2009) allows for the possibility of chemoautotrophy (Engel 2012), an intriguing possibility given the high diversity of these communities. Finally, Humphreys (2001) noted that it is possible that the sampling sites themselves—boreholes—may be a source of organic carbon, given their direct connection with the surface, but surely this does not explain the existence of this fauna.

5.3.2 Overview of the fauna of calcrete aquifers

The Western Shield, containing the Yilgarn and Pilbara regions, has had a continuous freshwater history since the Proterozoic. If there is an extant ancient subterranean fauna anywhere in the world, the Western Shield is a good candidate. And indeed, there is an ancient subterranean freshwater fauna (Karanovic 2007), based primarily on its disjunct Tethyan distribution. Some of the calcrete fauna is younger, but still older than many subterranean faunas elsewhere. Leys et al. (2003), using DNA sequence information, dated the invasion of Dytisicidae beetles to between 4 and 9 mya, and Cooper et al. (2008) suggested dates of up to 11 mya for subterranean isopods. No DNA-based dates are available for the more ancient components of the calcrete fauna.

In the search for an ancient fauna, special attention has been paid to Gondwana and Tethyan distributions of subterranean fauna, disjunct distributions of subterranean species that follow the shoreline of the ancient Tethys Seaway or the ancient southern continent of Gondwana (Poore and Humphreys 1998). A classic example of a Gondwana distribution is the exclusively subterranean crustacean order Spelaeogriphacea, found only in South Africa, Brazil, and Western Australia. In Western Australia it is found in shallow lacustrine and groundwater calcrete aquifers in the Pilbara (Table 5.1.). It is interesting to note that the Spelaeogriphacean habitats are slightly saline and with low oxygen, but with a pH typical of calcrete areas (Poore and Humphreys 2003). The implication is also that calcrete aquifers have been around for tens of millions of years. Poore and Humphreys suggested that colonization by Spelaeogriphacea may have taken place during the late Cretaceous or early Jurassic, about 140 million years ago! Wilson (2008) showed a biogeographical connection between India and Western Australia for phreatoicidean isopods.

Overall, the stygobiotic fauna of the Yilgarn is distinguished by a rich diving beetle fauna (Dytiscidae) with widespread sympatry (Leys et al. 2003, Watts and Humphreys 2004, 2006, 2009, Guzik et al. 2009, 2011). There are a number of crustacean

Table 5.1 Site description and water quality at two collection sites from which the spelaeogriphacean *Mangkurtu kutjarra* was taken at Roy Hill Station, Fortescue River Valley, Western Australia. From Poore and Humphreys (2003).

Parameter	Battle Hill Well	Aerodrome Bore
Site description	Well-maintained, uncovered, cement-lined well on raised stone slab	Overgrown, open borehole, fetid
Conductivity (mS cm^{-1})	6.65	5.67
Salinity (g L^{-1})	3.94	3.36
Temperature (°C)	25.5	29.2
O$_2$ (mg L^{-1})	1.5	0.71
O$_2$ saturation (%)	18.6	9.2
pH	7.13	7.2
Depth to water (m)	3.6	9.0
Depth of water (m)	4.0	2.0

stygobiotic taxa in the Yilgarn (Humphreys 2001), including Bathynellacea (Cho et al. 2006, Abrams et al. 2012), Amphipoda (King et al. 2012), Copepoda (Karanovic and Cooper 2011), Ostracoda (Karanovic 2007), and Isopoda (Cooper et al. 2008). The likely phylogenetic age of these different lineages varies widely, a situation that also occurs in the Pilbara. The Pilbara, generally older and deeper than the Yilgarn, has an even richer stygobiotic fauna (Reeves et al. 2007). Humphreys (2001) identified three assemblages in the Pilbara:

1. Coastal plain of the Robe and Fortescue Rivers with a Tethyan community characterized by *Stygiocaris* (Decapoda), *Haptolana* (Isopoda), and *Halosbaena* (Thermosbaenacea), which reach an elevation of 300 m, the approximate level of the late Eocene sea level.
2. The Western Fortescue Plain characterized by Spelaeogriphacea with Gondwana affinities.
3. Sites upstream of the Western Fortescue Plain and the upper Ashburton catchment characterized by flabelliferan isopods.

Each of these assemblages also includes Amphipoda, Ostracoda, Syncarida (Bathynellacea), and Oligochaeta (Phreodrilidae) (Humphreys 2001). See Box 5.1 for sampling techniques.

Table 5.2 Stygobiotic species in the three calcrete aquifers associated with Lake Way, Western Australia (see Fig. 5.3). From Humphreys et al. (2009).

Order	Family	Species	Hinkler Well calcrete	Lake Violet calcrete	Uramurdah Lake calcrete
Bathynellacea	Bathynellidae		•		
	Parabathynellidae	*Gen.* nov. sp. 1	•		•
		Gen. nov. sp. 2		•	•
		Gen. nov. sp. 3		•	
		Atopobathynella wattsi			•
		Atopobathynella nov. sp. 1	•		
Coleoptera	Dytiscidae	*Limbodessus macrohinkleri*	•		
		L. hinkleri	•		
		L. raeae	•		
		L. wilunaensis		•	
		L. hahni			•
		L. morgani			•
Cyclopoidea	Cyclopidae	*Fierscyclopes fiersi*	•	•	
		Mesocyclops brooksi			•
		Metacyclops laurentiisae	•	•	
		Halicyclops ambiguus			•
		H. kieferi		•	•
Harpacticoida	Ameiridae	*Haifameira pori*	•	•	
		Nitocrella trajani			•
		Parapseudoleptomesochra karamani	•		•
		P. rouchi			•
	Diosaccidae	*Schizopera austindownsi*		•	
		S. uramurdahi			•
Oniescidea	Philosciidae	*Andricophiloscia pedisetosa*			•
	Scyphacidae	*Haloniscus longiantennatus*			•
		H. stilifer			•
		Haloniscus sp. nov.			•
Podocopida	Candonidae	*Candonopsis dani*		•	
		Gomphodella sp.			•
Totals			10	9	17

Not only are calcrete aquifers rich in species, species ranges are small, and almost all are endemic to a single calcrete aquifer. Species lists for the three nearby aquifers of Lake Way indicate little overlap of species (Table 5.2) in spite of the aquifers being within 20 km of each other (Fig. 5.3). Of the 28 stygobionts known from the three calcrete aquifers (Lake Violet, Uramurdah Lake, and Hinkler Well), 14 are known from a single calcrete. Jaccard indices (see section 3.3.6) were very low, ranging from 0.04 to 0.188.

In general, calcrete stygobionts originate from interstitial freshwater lineages (e.g. Bathynellacea), near-marine lineages (e.g. *Halicyclops* in the Cyclopidae [Copepoda]), and ancient freshwater lineages (e.g. crangonyctid Amphipoda). Only one taxon with stygobiotic representatives in calcrete aquifers has a clear affinity with the salt lakes themselves— oniscoidean isopods of the genus *Haloniscus* (Humphreys et al. 2009, Taiti and Humphreys 2001). There is an eyed, surface-dwelling species, *H. searli*, that is widespread in playas across southern Australia, a distribution that is itself remarkable given the isolated nature of playas and the absence of desiccation-resistant eggs in the species. Immediately to the north, numerous stygobiotic *Haloniscus* occur in calcrete aquifers, each limited to a single aquifer (Taiti and Humphreys 2001, Cooper et al. 2008).

5.3.3 Colonization of calcrete aquifers

Based on environmental conditions in the Pilbara, where rivers are still seasonally active, Humphreys (2001) hypothesized that headwater calcrete communities would become isolated as a result of reduced water flow and the downstream progression of salinity. If this is the case, then phylogenies should reflect the drainage patterns. The phylogeny of amphipods in isolated headwater aquifers in the Pilbara matches the drainage pattern, with different lineages within a river clustered together in the phylogeny (Finston et al. 2007). As a consequence, most of the molecular variance was partitioned among tributaries across the Pilbara (Table 5.3).

By contrast, there was no such association between phylogeny and drainage for dytiscid beetles in the

Yilgarn, where there are no seasonally active surface rivers. Molecular phylogenies of dytiscid beetles do not map the drainage pattern but rather support multiple independent invasions of calcrete aquifers across a wide region (Leys et al. 2003, Leijs et al. 2012). This pattern of long isolation of each calcrete aquifer, each with different species, has been called the archipelago model (Cooper et al. 2007). Cooper et al. (2007) found evidence for at least 22 species in two amphipod families, each restricted to a single aquifer. Mitochondrial sequence divergence suggested that the calcrete aquifers had been isolated at least since the Pliocene, coinciding with a major aridity phase. They point out that this observed pattern could be the result of multiple invasions by surface species or by subsurface migration following colonization, in general two hypotheses that are especially difficult to separate (Culver et al. 2009). However, their data and the data of Guzik et al. (2008), Leys et al. (2003), Leijs et al. (2012), and Abrams et al. (2012) suggest no dispersal, either surface or subsurface, since the Pliocene 5–10 million years ago.

The best estimates of age are from dates of divergence of sympatric sister species of dytiscid beetles (calculated from mtDNA divergence), and place isolation between nine and three million years ago, with the age of isolation increasing towards the north (Fig. 5.6). This corresponds to the direction of

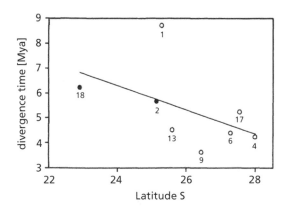

Figure 5.6 Latitudinal variation in divergence times of eight sympatric sister pairs of stygobiotic dytiscid beetles in Western Australia. The open circles show species pairs belonging to the Bidessini; the black circles show species pairs belong to the Hydroporini. From Leys et al. (2003), used with permission of Blackwell Publishing.

Table 5.3 Analysis of molecular variance partitioning of total genetic variance, and Φ statistics, roughly equivalent to the F_{ST} statistics of Wright for the amphipod genus *Chydaekata* from Pilbara calcrete aquifers. Φ_{ST} was significant at p<0.001. Nearly identical results were obtained for the amphipod genus *Pilbarus*. From Finston et al. (2007).

Comparison	% variance	Φ_{ST}
Among tributaries	98.7	0.99
Within tributaries	1.3	

Box 5.1 Biological sampling of calcrete aquifers

Unlike other SSHs, with the partial exception of some hyporheic sites, the sampling points for calcrete aquifers were designed for another purpose—usually the location and assessment of water resources (e.g. Johnson et al. 1999). For example, the Pilbara contains over 3700 bores and wells, less than 500 of which have been sampled biologically (Eberhard et al. 2009). In the Sturt Meadows calcrete, boreholes were approximately 10 cm in diameter and reached a depth of 10.3 m (Allford et al. 2008). After mineral prospecting was completed, they were sealed with a concrete plug. Biologists replaced the concrete plug with a 1.5 m long PVC sleeve to stabilize the borehole. Nevertheless, the boreholes were uncased for most of their length. Water levels were typically 0.8–4.0 m beneath the surface (Allford et al. 2008).

Allford et al. (2008) tested several sampling schemes involving:

1. drawing a haul net (weighted plankton net, basically a phreatobiological Cvetkov net [Malard 2003]) through the water column (Fig. 5.7) to the bottom of the bore and retrieving slowly, a procedure repeated 5–10 times;
2. collecting bore water from a discrete-interval sampler, a 2.2 L Kemmerer SS Water Sampler consisting of a steel cylinder 75 mm in diameter and 50 cm long; and
3. filtering water pumped from the borehole using an impeller pump capable of pumping rates of 5 L min⁻¹.

The haul net turned out to be the most efficient in terms of number of species, number of animals, and time spent per sample. Of course the discrete-interval sampler would be useful to investigate vertical habitat separation of the stygobionts. In the Pilbara, where many of the calcrete aquifers are deeper than 10 m, haul nets are used to sample wells. These nets are designed to be dropped down to a substrate and take a substrate sample.

Although nearly all of the interest has been in the aquatic fauna, there is a terrestrial fauna, although it has been little studied (Humphreys 2001). Halse and Pearson (2012) pointed out that haul nets also collect terrestrial fauna. They suggested that scraping the net along the borehole is a more efficient means of collecting the terrestrial fauna than by suspending traps in the boreholes. All orders, except Diptera, Spirobolida (Class Diplopoda), and Spirostreptida

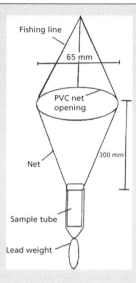

Figure 5.7 Line drawing showing the main components of a haul net. The opening consists of a rigid ring made of PVC piping or cable to ensure the mouth remains open. The conical net (mesh size 250 µm to prevent clogging) concentrates the sampling into a plastic collection vial. A lead weight is connected to the bottom of the net. From Allford et al. (2008). Courtesy CSIRO Publishing (http//www.csiro.au/nid/120/paperIS07058.htm).

(Class Diplopoda) were better represented in haul samples. The difficulty with this approach is that the soil fauna as well as the SSH fauna is collected, and traps may be more selective than haul nets for the terrestrial fauna of calcrete aquifers, as opposed to the soil fauna.

Humphreys (2001) pointed out several artefacts of sampling the fauna through boreholes:

1. Boreholes are conduits for organic carbon, and this might allow displacement of stygobionts by epigean taxa.
2. Sampling mixes samples from physical-chemically stratified water.
3. The borehole provides artificial substrate and gaseous exchange interface.
4. It introduces iron as a substrate for chemoautotrophy.

However, there is no other sampling option aside from boreholes.

aridification during the Pliocene. The colonization of calcrete aquifers by dytiscid beetles is perhaps the best example of colonization of subterranean habitats by climatic forcing, the so-called Climate Relict Hypothesis (Barr 1968), and it is better documented than the many likely cases of colonization of caves as the result of Pleistocene climate changes.

5.3.4 Speciation in calcrete aquifers

In the Yilgarn, the stygobiotic group that has been best studied—the diving beetles (Dytiscidae)—has several interesting features. All stygobiotic species in these groups occur in single calcrete aquifers. Watts and Humphreys (2004, 2006, 2009) listed a total of 99 styobiotic dytiscids from 52 calcrete aquifers in the Yilgarn (Western Australia) and the Ngalia Basin (Northern Territory). In an earlier analysis of 35 stygobiotic species, Leys et al. (2003) found that there were 28 independently evolved species in two tribes—Bidessini and Hydroporini—based on mtDNA phylogenies. Leys et al. (2003) found only one case of related species (*Nirridessus cueensis* and *N*. cf. *cueensis*) in adjacent calcrete aquifers, indicating that either these calcrete aquifers were connected in the past, or that there was dispersal from one calcrete to the other. With this one exception, Leys et al. (2003) proposed that each stygobiotic species (or sympatric pair) evolved independently from surface-dwelling ancestors. For example, species in the *N. hinkleri* clade are found in five isolated calcrete aquifers belong to three entirely separate palaeo-drainage systems including both sides of the divide between oceanic and interior drainages. Cooper et al. (2007) reported a similar pattern for stygobiotic amphipods in the Yilgarn. This pattern corresponds in some ways to distribution of troglobionts and stygobionts in highly dissected limestones, such as the Appalachians in Virginia, USA (Culver and Pipan 2009). However, in this and other similar cases, geographically adjacent species are often phylogenetically related. This makes the Australian beetle and amphipods cases especially clear examples of multiple invasions with little or no possibility of subsequent subterranean dispersal. Cooper et al.'s (2007) description of this system as a subterranean archipelago in the Australian arid zone is especially apt.

What is even more remarkable is the pattern found within individual calcrete aquifers. In the Yilgarn, Leijs et al. (2012) found 11 aquifers with three sister species of dytiscids, 16 with two sister species, and 18 with one. They point out that this sympatric distribution of sister species could be the result of speciation within the aquifer (sympatric, parapatric, or micro-allopatric) or the result of multiple colonizations from the same surface ancestral species. Because there were multiple cases of sister species sympatry at the calcrete aquifer scale, Leijs et al. (2012) were able to devise a statistical test to separate the two hypotheses (Fig. 5.8). The multiple invasion model fails to account for the observed fractions of aquifers with pairs or triplets, except when the number of species in the ancestral species pool was very low (Fig. 5.8A, B). On the other hand, a within-aquifer speciation model could account for the observed fractions for a wide range of niche colonization probabilities (Fig. 5.8C), the main parameter of the within-aquifer model, based on the pattern of limiting similarities of dytiscids (Vergnon et al. 2013). This system provided a unique opportunity for a statistical test of modes of speciation, but as Leijs et al. (2013) pointed out, speciation within the aquifer could be sympatric, parapatric, or micro-allopatric.

Guzik et al. (2009) studied the fine-scale phylogeography of diving beetles in the genus *Paroster* in the Sturt Meadows Calcrete in the Yilgarn. The total area of the aquifer is 43 km^2, and they sampled beetles from a grid of 112 mineral exploration bores spaced 100 and 200 m apart, in an area of 3.5 km^2. Three *Paroster* species are present in this calcrete aquifer—*P. macrosturtensis*, *P. mesosturtensis*, and *P. microsturtensis*, a monophyletic clade. Based on sampling within the grid of boreholes, Guzik et al. (2009) found that two of the species (*P. macrosturtensis* and *P. mesosturtensis*) showed geographical differentiation and coalescence of mtDNA haplotypes dating back one million years. To estimate the spatial groups, they used a technique, SAMOVA 1.0, a simulated haplotype annealing in which individuals were grouped based on geographical closeness and genetic similarity (Dupanloup et al. 2002). Their results are shown in Table 5.4. The three 'SAMOVA' groups for *P. mesosturtensis* occurred

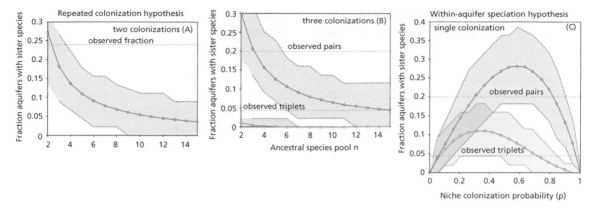

Figure 5.8 Results of colonization modelling. (A and B) Repeated colonization model. (A) The relationship of the size of the ancestral species pool and the fraction of the aquifers containing sister species after two colonization events. An initial niche colonization probability (p1) of 0.5 was used as this maximizes the probability of sister pairs. The last colonization probability (p2) was set to 1. The observed fraction of aquifers with sympatric sister species (11/45) is also indicated. (B) The predicted fraction of aquifers containing sympatric sister pairs and triplets (formula 3) calculated based on three colonization periods and a niche colonization probability (p1 = p2 = 0.4) that maximizes the probability of pairs and triplets. Horizontal lines indicate the observed fraction of aquifers with sister pairs and triplets. The last colonization probability (p3) was set to 1. (C) Within-aquifer speciation model. The relationship between the initial niche colonization probability and the predicted fraction of aquifers containing sympatric sister pairs and triplets calculated with single colonizations and subsequent divergence within aquifers. Horizontal lines indicate the observed fraction of aquifers with sister pairs and triplets. The models are calculated using the observed number of aquifers with one (18 aquifers), two (16 aquifers), or three (11 aquifers) species. The shaded areas in (A)–(C) represent the 5% and 95% percentiles as confidence limits from 10,000 randomizations. The model assumed that all open niches were filled by speciation (q = 1). From Leijs et al. (2012).

Table 5.4 Analysis of molecular variance (AMOVA) results estimated with ARLEQUIN (Excoffier et al. 2005) using SAMOVA groups in three stygobiotic *Paroster* species from the Sturt Meadows calcrete aquifer, Western Australia. From Guzik et al. (2009).

Species	Source of variation	Sum of squares	Variance components	% of variation
P. mesosturtensis	Among SAMOVA bore groups	228.0	2.8 Va	40.1
	Among bores within SAMOVA bore groups	190.1	0.03 Vb	0.5
	Within individual bores	814.9	4.2 Vc	59.4
	TOTAL	1233.1	7.0	100.0
P. macrosturtensis	Among SAMOVA bore groups	9.9	3.1 Va	44.8
	Among bores within SAMOVA bore groups	169.3	0.7 Vb	10.9
	Within individual bores	233.3	3.0 Vc	44.4
	TOTAL	412.5	6.8	100.0
P. microsturtensis	Among bores	11.4	0.02 Va	0.7
	Within individual bores	149.2	2.6 Vb	99.3

in (1) the southwest edge of the borehole field, (2) the northwest quarter of the borehole field, and (3) throughout the rest of the field. For *P. macrosturtensis*, there was one isolate on the northern border of the borehole field. While these groups are not at the level of differentiation of cryptic species, they do account for roughly 40 per cent of the molecular variance (Table 5.4). This geographically structured variation in these two species suggests that there is very great potential for cryptic species in a calcrete, given that the area studied was a small fraction of the Sturt Meadows calcrete (Guzik et al. 2009).

Leijs et al. (2013) caution against assuming that such micro-allopatric fragmentation is the basis for reproductive isolation among all sympatric sister species. If this is universally important, then larger calcrete aquifers should have more species than smaller ones. This is not the case (Leijs et al. 2013). Sympatric sister species were found in very large, linear aquifers and in very tiny aquifers, with areas of less than 3 km².

5.3.5 Interspecific competition in calcrete aquifers

Vergnon et al. (2013) found very strong support for the theory of limiting similarity, which states that, due to interspecific competition, species must be sufficiently different in order to coexist. In 34 calcrete aquifers with either two or three species of dytiscid water beetles, the ratio of the body sizes was remarkably constant, with an approximate ratio of 1.6 (Fig. 5.9). This was all the more impressive because the absolute sizes of the beetles varied greatly from calcrete aquifer to calcrete aquifer. Fišer et al. (2012) also found interspecific competition to be important in communities of *Niphargus*

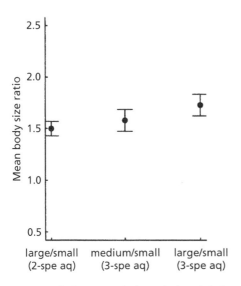

Figure 5.9 Mean body size ratios (with standard species) of coexisting species of dytiscid beetles in calcrete aquifers in the Yilgarn, Western Australia. Mean body size ratios between coexisting species belonging to consecutive size ranks did not differ significantly (F = 1.5416, p = 0.23). From Vergnon et al. (2013).

amphipods in caves and interstitial habitats in the Dinaric Mountains (see chapter 6). In their study, actual assemblages were morphologically more variable than communities assembled at random.

Whether speciation is micro-allopatric or sympatric, and they are very similar in scale and likely in dynamics, it would seem that interspecific competition is a major constraint on species richness. Otherwise it is hard to explain why there aren't even more species in calcrete aquifers.

5.3.6 How species-rich is the calcrete aquifer fauna of the Yilgarn?

Calcrete aquifers can harbour a relatively large number of subterranean species. Eberhard et al. (2009) estimated that the Pilbara region has over 400 stygobiotic species. However, the Pilbara calcrete aquifers are deeper and the Pilbara subterranean fauna is not strictly limited to calcrete aquifers (Eberhard et al. 2009, Guzik et al. 2010). Guzik et al. (2010) estimated that there are 4140 stygobionts and troglobionts in Australia, mostly from western Australia. While there are several important technical problems with their analysis (Culver et al. 2012c), there is no doubt that the western Australian subterranean fauna in general, and the calcrete fauna in particular, is very far from depauperate, as previously thought, and ranks among the richest subterranean sites globally. Among SSHs only lava tubes (see chapter 8) have a similar ranking.

The potential regional stygobiotic diversity in the Yilgarn is large. There have been 47 calcrete aquifers sampled out of more than 200 (Guzik et al. 2011). Based on species accumulation curves, Allford et al. (2008) estimated that there are eight stygobiotic species in the Sturt Meadows aquifer. Given that each calcrete aquifer has a separate set of species, it is tempting to just multiply this number by the number of calcrete aquifers to yield an estimate of more than 1600 species, which is more than all stygobiotic and troglobiotic species known from North America. Several caveats are in order. Firstly, all caves and all calcrete aquifers are not equally diverse. For example, only a handful of the 47 sampled calcrete aquifers have 3 dytiscid beetle species, as does Sturt Meadows. This problem was called the fallacy of isotropy by Culver et al. (2012c). Secondly, undescribed species, when closely studied, do not all become new species; some are variants of existing species (Culver et al. 2012c). Nevertheless, the richness of the calcrete aquifer fauna is remarkable, and its discovery must rank as the most significant discovery of previously unknown subterranean biodiversity in the past several decades.

5.4 Summary

Calcrete aquifers are carbonate deposits in desert terrains. While occurring in desert areas throughout the world, the fauna has only been studied in the calcrete aquifers of the Pilbara and Yilgarn of Western Australia, and to a lesser extent in the Northern Territory. Calcrete aquifers are typically tens of km long but only 10 m or less thick. Those in the Yilgarn are typically within 10 m of the surface, but those of the Pilbara are not. They form during drying periods with high evaporation but also when there are periods of rare but concentrated rainfall. Formation is associated with salt lakes (playas), and results from evaporation of carbonate-rich water. Water in calcrete aquifers can have very high salinity, which may be a barrier to dispersal within the calcrete. Overall, the system is like an estuary, but unconnected to any ocean. The habitat is highly patchy and archipelago-like, but quite distinct from other SSHs, not only in its chemical characteristics, but also in its geographical location. Based on the size of the invertebrates inhabiting calcrete aquifers (1–5 mm), pore size is small to moderate.

Virtually nothing is known about the sources of organic carbon in calcrete aquifers. Some may come from percolation of rainwater but rain is uncommon and very sporadic. Roots may provide a source but this has not been demonstrated. Finally, there are redox environments in which chemoautotrophy may occur, but this is just speculation.

Because of its long geological history, some of the fauna of calcrete aquifers is likely very ancient, part of the Gondwana and Tethys fauna. A classic example of a Gondwana component is the crustacean order Spelaeogriphacea. All or nearly all stygobionts, whether ancient or modern in origin, are limited to a single calcrete aquifer or palaeodrainage. One group is specialized in the subterranean saline and hypersaline water near the playas—the isopod genus *Haloniscus*. A particularly rich and well-studied component of the fauna are the diving beetles in the family Dytiscidae. Colonization of calcrete aquifers was the result of increased aridity, especially during the Pliocene. The Climate Refuge Hypothesis, originally invoked to explain the colonization of north temperate zone caves as a result of glaciation, seems to explain the colonization of calcrete aquifers, albeit with aridity as the forcing agent.

The restriction of nearly all species to a single calcrete aquifer, and the lack of relationship of species in adjacent aquifers, suggest that dispersal is rare. The occurrence of pairs and triplets of sister dytiscid species is quite common, and evidence points to speciation within the aquifer, either microallopatric, parapatric, or sympatric.

Many calcrete aquifers have more than one dytiscid beetle species, and these species invariably differ in size, suggesting interspecific competition was an important selective agent, resulting in niche separation. Even within an aquifer there is some geographical differentiation in a species, suggesting the possibility of cryptic speciation. Overall, species richness is high, and calcrete aquifers are globally important sites of subterranean biodiversity.

CHAPTER 6

Interstitial habitats along rivers and streams

6.1 Introduction

Interstitial habitats can be either relatively shallow with regular interchanges with the surface, or deep with no interchange with the surface. Shallow aquatic interstitial habitats are comprised of water-filled spaces between grains of unconsolidated sediments. These habitats occur in littoral sea bottoms and beaches, freshwater lake bottoms, and the beds and margins of rivers and streams. Of these aphotic habitats, the one that is consistently shallow is the underflow and bed of rivers and streams. Relative to the strict sense SSHs, the underflow of rivers and streams is a small-pore-sized habitat, part of the 'milieu perméables en petit', one of the two types of subterranean habitats proposed by Botosaneanu (1986). The terrestrial analogy to these aquatic interstitial SSHs is the soil and to a certain extent the MSS, the subject of chapters 4 and 7.

The hyporheic zone, the best-studied by far of all interstitial habitats, is the surface–subsurface hydrological exchange zone beneath and alongside the channels of rivers and streams. A more technical definition is provided by Krause et al. (2011):

A temporally and spatially dynamic saturated transition zone between surface and ground water bodies that derives its specific physical (e.g. water temperature) and biogeochemical (e.g. steep chemical gradients) characteristics from mixing of surface- and ground water to provide a dynamic habitat and potential refugia for obligate and facultative species.

The hyporheic of rivers is an ecotone between surface and groundwater. Figure 6.1 illustrates the transitional nature of the hyporheic.

Except for the soil, there is no better-studied SSH than the hyporheic. The number of review papers on various aspects of the hyporheic probably outnumbers the total number of papers published on any of the other SSHs. The reason for this is not the subterranean aspects of the habitat, but rather its importance for streams and rivers. The hyporheic is an integral part of all riverine ecosystems, and Boulton and Hancock (2006) demonstrated that rivers and streams are groundwater-dependent ecosystems. Because of the vulnerability of stream and river beds to physical disruption, accumulation of pollutants, and clogging by fine sediment, the hyporheic is a focus of river managers as well (see Buss et al. 2009). These approaches from the point of view of the river and the riverine landscape are literally a top-down perspective. Historically, some of the earliest studies of the hyporheic were bottom-up, ones from the perspective of the subterranean fauna in general and stygobionts in particular (Hancock et al. 2005). The name hyporheic was coined in fact by a Romanian speleobiologist, Traian Orghidan (1959), but it was not until the hyporheic captured the attention of stream ecologists such as Stanford and Gaufin (1974) that research on the hyporheic took off.

Our approach is also a bottom-up perspective, with an emphasis on the subterranean (aphotic) aspects of the habitat. While river biologists make a distinction between the hyporheic zone, with its steep biogeochemical gradients, from groundwater, a subterranean perspective is that both are aphotic habitats, and the boundary between the hyporheic and groundwater is not always clear. In practice, most of the sampling of shallow interstitial habitats

Shallow Subterranean Habitats. David C. Culver and Tanja Pipan.
© David C. Culver and Tanja Pipan 2014. Published 2014 by Oxford University Press.

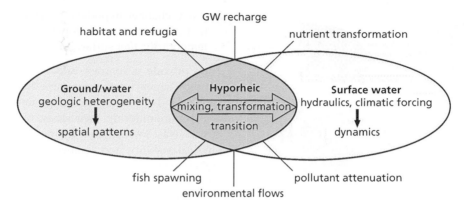

Figure 6.1 Hydroecological and biogeochemical functions of the hyporheic as a mixing and transition zone with environmental fluxes (energy and matter) between groundwater and surface water environments. Modified from Krause et al. (2011), used with permission of John Wiley & Sons.

Table 6.1 Summary of comparative physical and biological characteristics of groundwater, hyporheic, and surface water environments. Modified from Krause et al. (2011).

	Descriptive characteristics of environment		
	Groundwater	Hyporheos	Surface water
Physical characteristic			
Light	Constant darkness	Constant darkness	Daylight fluctuations
Current velocity	Low	Intermediate	High
Annual and daily temperature range	Very low	Low	High
Substrate stability	High	Intermediate	Low
Gradient of physico-chemical parameters	Low	Steep	Moderate
Biological characteristic			
Habitat diversity	Low	Intermediate	High
Food webs	Simple and short	Intermediate	Complex and long
Primary productivity	None	None	High

is in the hyporheic because it is the most accessible. Krause et al. (2011) provide a useful comparison of the physical and biological characteristics of surface, hyporheic, and groundwater (Table 6.1).

The connection between the hyporheic and permanent groundwater (phreatic water) can be very direct or without any direct connection at all (Fig. 6.2). In rare cases, hyporheic and groundwater are absent, and the stream flows on impermeable rock (Fig. 6.2a). In all other cases, there is water flow vertically and laterally from channel water and/or groundwater. When there are unconsolidated sediments along the stream bank, the hyporheic can extend tens of metres from the stream bank. Working on Sycamore Creek, a desert stream in Arizona, USA, Boulton et al. (1992) distinguish three kinds of hyporheic habitats[1]:

[1] Boulton *et al.* (1992) actually refer to the groundwater (phreatic) habitat as a hyporheic habitat where it occurs beneath the stream. This is certainly in keeping with the definition of hyporheic (under flowing water), but is not accepted usage.

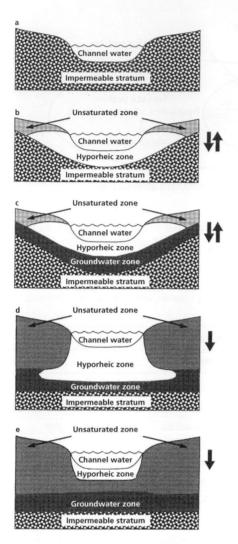

Figure 6.2 Conceptual cross-sectional models of surface channels and beds showing relationships of channel to hyporheic, groundwater, and impermeable zones. Thick arrows indicate direction of water fluxes between surface stream and underlying reservoirs. (a) No hyporheic zone. (b) Hyporheic zone created only by advected channel water. (c) Hyporheic zone created by advection by both channel water and groundwater. (d) Hyporheic zone created only by infiltration of channel water beneath the stream bed (no parafluvial flow). (e) A perched hyporheic zone created only by infiltration of channel water beneath the stream bed. From Malard et al. (2002), used with permission of John Wiley & Sons.

1. a shallow hyporheic directly beneath the stream, and extending to a depth of 50 cm;
2. a parafluvial habitat that extends laterally from the stream, similar to the riparian zone of Chestnut and McDowell (2000); and

3. a dry channel hyporheic, a temporary habitat that fills with water following rains, and exists temporarily after surface water disappears.

They provide a contrast in some basic physico-chemical characteristics for these three hyporheic habitats and contrast them with groundwater (phreatic) communities. In their study, oxygen concentration was lowest, and hydraulic conductivity, physical stability, and temperature stability were highest in phreatic habitats. The final hyporheic type (Fig. 6.2e) is a rather curious one, completely isolated from groundwater. On a small scale, it closely resembles the hypotelminorheic (see chapter 2).

Gibert et al. (1990) and Vervier et al. (1992) championed the ecotone perspective for the hyporheic, and the subterranean–surface contact in general. They argued that because of the strong contrast between surface and subsurface (see Table 6.1), the boundary between the two was especially important, as it was a transfer point of energy and matter between two contrasting habitats. The hyporheic is what they call a lotic ecotone, where there are large-scale transfers of energy, materials, and animals. The ecotone is also a filter and Vervier et al. (1992) described three filters:

1. the mechanical filter, which is determined by the size of the interstices between sand, rock, etc. and their connectivity, i.e. porosity[2];
2. the photic filter, the filter imposed by light (or lack of light); and
3. the biochemical filter which is the result of constraints on biological and chemical processes that are linked to the activity of living organisms.

The direction of transfer is two-way, and varies both spatially and temporally. In a typical stony-bottomed stream, there are areas of upwelling and downwelling (Fig. 6.3) that are the main areas of transfer. Water and nutrients thus spiral between surface and subsurface flows. While the dependency of the river on groundwater has been emphasized by Boulton and Hancock (2006) and others who utilize the top-down (river ecosystem) approach, the hyporheic and the groundwater components are

[2] The ratio of openings to total volume.

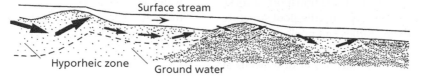

Figure 6.3 Surface–subsurface hydrological exchanges in the hyporheic zone induced by spatial variation in stream bed topography and sediment permeability. Dark areas represent fine sediments. Arrows indicate direction of flow and their width corresponds to flux rate of advected channel water with the sediments. Downwellings bring both organic matter and oxygen to the hyporheic zone. Modified from Malard et al. (2002), used with permission of John Wiley & Sons.

surface-water dependent ecosystems. It is the interaction zone that is the transfer point of oxygen, organic carbon, and nutrients to the hyporheic and groundwater.

Transfers of matter and organisms through and into the hyporheic depend on the permeability of the substrate. If sediment load increases, fine particles can settle in the gravel bed reducing permeability, or slightly larger particles can form a subsurface seal (Fig. 6.4). In urban streams, such as Rock Creek in Washington, DC (USA), sediment load from stormwater runoff is so heavy that nearly all of the hyporheic of some tributary streams is completely clogged and sealed (Culver and Šereg 2004). This not only has degraded the hyporheic, it has also truncated connections of upstream springs and seeps.

Other SSHs (e.g. epikarst) can be viewed as ecotones (Gibert et al. 1990), as can caves themselves (Moseley 2009). However, it is only with the hyporheos that the ecotone concept has proved useful beyond a few generalities.

6.2 Chemical and physical characteristics of the hyporheic

6.2.1 Geological setting of the hyporheic

For the most part, studies of the hyporheic are of the hyporheic in stony-bottomed streams, although there is an analogous boundary (ecotone) in sand- and mud-bottomed streams (e.g. Ward and Palmer 1994). Gravels and sands typically come from

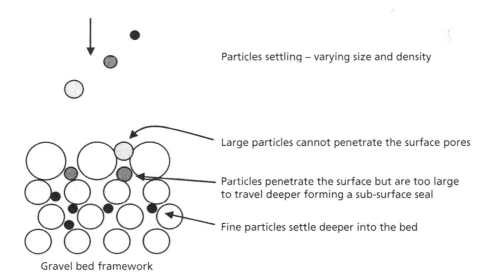

Figure 6.4 Sediment infiltration into river beds, showing passive infiltrations into a gravel bed by different-sized particles. Modified from Buss et al. (2009).

weathering in higher elevation areas with the formation of slope debris, with the transport of rocks and stones to create the typical riffle-pool structure of streams and rivers. Riffles and pools correspond to upwellings and downwellings in the hyporheic zone (Fig. 6.3). Typically, water downwells at the upstream end of the riffle and upwells at the downstream end. From this perspective the river is a mosaic of surface–subsurface exchange patches. These exchanges have a major effect on nutrient dynamics (see section 6.3.1) as well as the movement of animals, especially from groundwater to surface water.

The position of riffles and pools is dynamic and the position of riffles is likely on the order of tens of years (Buss et al. 2009). The path of a stream or river also changes course, leaving meander cutoffs and oxbow lakes, probably on the order of hundreds of years. During floods, individual rocks and gravels can move when velocities are high enough, but the position of riffles usually does not change that rapidly. The type of the underlying and upstream bedrock has an impact on the connectivity between the surface channel, hyporheos, and groundwater, and in the extreme case of low primary permeability and few fractures, the hyporheic may not be present at all (Fig. 6.2). Even the detail of the arrangement of patches is biologically important. Palmer et al. (2000) found that the geometry of the hyporheic exchange patches has an important effect on densities of copepods and larval chironomids in a sand-bed stream that could not be explained by non-spatial information.

6.2.2 Hydrology of the hyporheic

The detailed study by Rouch (1988, 1992) of a 15 m reach of the small 5 m wide Lachein Stream (see Fig. 1.5), near Saint Giron in the French Pyrenees, illustrates the relationships between geomorphical units and the spatial heterogeneity of hyporheic flow (Malard et al. 2002). Rouch identified three geomorphological units:

1. two gravel bars, one on the left bank and one on the right bank (Fig. 6.5);

2. an erosion zone characterized by high surface-water velocities on the right bank upstream of the gravel bar; and

3. a pool on the downstream left bank.

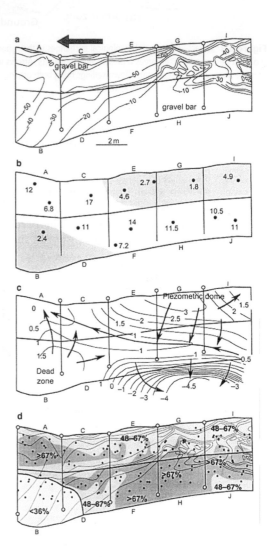

Figure 6.5 Hydraulic conductivity, interstitial flow, and dissolved oxygen concentration in a 75 m² area of Lachein Stream, Pyrenees, France. There are two gravel bars (A,C on the right bank and D,F,H,J on the left bank), an erosion zone (I,G,E, and C), and a pool (B). (a) bathymetric map (water depth in cm); (b) hydraulic conductivity (in units of 10^{-4} m s^{-1}); (c) piezometric map (relative height in cm) and flow lines, and (d) categorical map of dissolved oxygen concentration (% saturation). Black dots indicate sampling points. From Rouch (1992).

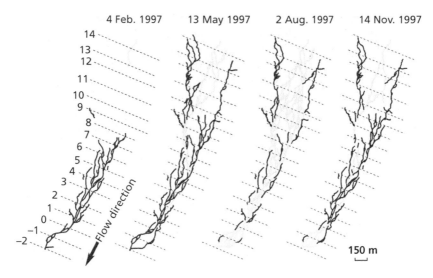

Figure 6.6 Distribution of patches fed by subsurface (turbidity < 9 NTU) and glacier water in the flood plain of the Roseg River, Switzerland, during a period of base flow (February, Q = 0.2 m³ s⁻¹), the expansion phase (May, Q = 1.5 m³ s⁻¹), the maximum extension of the flood plain channel network (August, Q = 5.7 m³ s⁻¹), and the contraction phase (November, Q = 0.7 m³ s⁻¹). Dark channels are fed by subsurface water and light channels are fed by glacial water. From Malard et al. (2002), used with permission of John Wiley & Sons.

The two gravel bars, with high hydraulic conductivity[3] (10^{-3} m s⁻¹) formed an area of preferred interstitial flow in which dissolved oxygen averaged 74 per cent of saturation. The distribution of invertebrates (dominated by harpacticoid copepods) also had fine-scale differentiation (see section 6.3.5).

The hyporheic is not only spatially variable, it is temporally variable as well. In a study of the groundwater-fed channels of the Roseg River, Switzerland, Malard et al. (2002) showed that the size, distribution, and connectivity of exchange patches varied seasonally (Fig. 6.6). The total length of the floodplain network varied from about 7 km in the winter to 24 km during summer melting of the glacier. Defining an exchange patch as a continuous length of channel with a turbidity of less than nine NTUs (nephelometric turbidity units), patchiness (the number of patches divided by the total length of patches fed by subsurface water) was higher in May and August, and turnover rate of patches (disappearance and appearance) was highest in November.

This may represent an extreme case due to the proximity to a glacier, but it occurs everywhere.

6.2.3 Geography of the hyporheic

Hyporheic habitats are widespread, occurring in the tropics (e.g. Mary and Marmonier 2000, Chestnut and McDowell 2000) and throughout the temperate zone. Stony-bottomed streams with a gravel hyporheic require a source of rocks, so flat areas tend to have sand or mud hyporheic zones, such as in much of the Midwest of the USA. Until very recently, the bulk of studies of the hyporheic, especially from a subterranean perspective, were of European rivers, especially the Rhône and its tributaries near Lyon, France and the Danube wetlands near Vienna, Austria, which remain the two most thoroughly investigated rivers. Other well-studied sites include the Never Never River in Australia, Sycamore Creek in Arizona, USA, and the Flathead River in Montana, USA.

6.2.4 Physico-chemistry of hyporheic habitats

Detailed temperature records are available for two sites in the Rhône River basin in France:

[3] Hydraulic conductivity, K, is Q/A (dl/dh) where Q is the quantity of water per unit time, A the cross-sectional area, and dh/dl the hydraulic gradient. The units of hydraulic conductivity are m s⁻¹, and range over more than 12 orders of magnitude (Heath 1983).

- Morcille, a small tributary stream of the Rhône River in the Beaujolais Hills at an elevation of 550 m asl in a wooded area (Piscart et al. 2011).
- Méant wetland, a cutoff meander of the Rhône River in an agricultural area at an elevation of 200 m asl (Dole 1984, Dole and Mathieu 1984).

Temperature dataloggers were placed for approximately 440 days at depths of 10 and 50 cm in the substrate. Both sites were upwellings of groundwater, and both sites had stygobionts, including species in the amphipod genus *Niphargus*. The patterns are summarized in Table 6.2 and Figure 6.7. At both sites

there was a strong seasonal component to temperature, at both depths. The annual pattern for Morcille was straightforward, with a maximum in July and August, and a minimum in February. In the summer of 2012, temperatures at the 50 cm-deep site were consistently lower than the upper site, presumably the result of groundwater upwelling. The effect was present in 2011 although not as strongly expressed. The annual pattern for Méant was complicated in the summer of 2011 by several dips in June and July, likely the result of increased groundwater pumping in the area (M.J. Dole-Olivier, pers. comm.).

Table 6.2 Basic statistics for temperature (ºC) for the Méant and Morcille hyporheic sites in the Rhône River basin, France. See Figure 6.7.

Description	n[1]	Mean	Standard deviation	Coefficient of variation	10% Quantile	90% Quantile	Minimum	Maximum	Range
Morcille −10 cm	10,617	10.9	3.6	32.6	6.3	15.1	1.4	18.9	17.5
Morcille −50 cm	10,617	10.9	2.8	25.8	7.1	14.3	4.1	16.3	12.3
Méant −10 cm	10,637	13.1	3.1	23.7	9.7	18.2	7.2	23.8	16.6
Méant −50 cm	10,637	13.0	2.8	21.4	9.8	17.5	8.3	20.5	12.2

[1]8760 hours in a year.

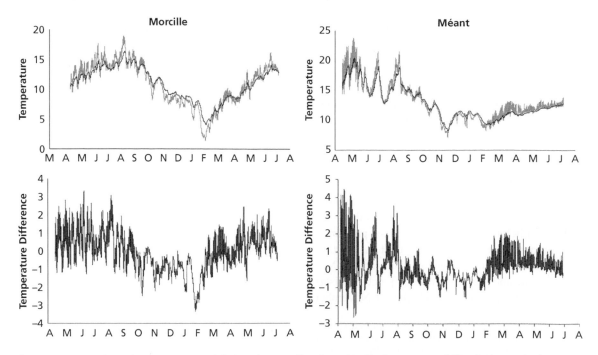

Figure 6.7 Top panels: Hourly temperature records for hyporheic upwelling sites at Morcilles Stream, France (left) and Méant wetland, France (right) from April 2011 to July 2012. The grey line is temperature at a depth of 10 cm and the black line is temperature at a depth of 50 cm. Bottom panels: Hourly differences in temperature between the two depths.

The difference in temperature between the 10 cm and 50 cm sites indicates that there is significant damping of daily fluctuations, hence the spiky nature of the graphs of the differences in temperature. This is especially pronounced during the summer (Fig. 6.7). There is little seasonal damping, and with the exception of Morcille in January and February, the difference curves fluctuate around zero. The damping of temperature with depth is largely the result of the damping of extremes of temperature, rather than damping of longer-term trends. At both Morcille and Méant, the range of temperature at 50 cm depth is reduced by more than 4 °C compared to 10 cm depth (Table 6.2) while the 10–90 per cent quantile range is reduced by 1.6 °C in Morcille and only 0.8 °C in Méant. Likewise, coefficients of variation show relatively small reductions. Thus, the hyporheic may be a refuge from temperature extremes, especially in the summer.

Among the chemical parameters, one of the most biologically relevant ones is that of oxygen. In his study of Lachein Creek, France, Rouch (1992, see Fig. 6.5) showed that oxygen varied from less than one per cent to more than 84 per cent saturation. While low oxygen levels are obviously a stress factor for hyporheic invertebrates (Malard and Hervant 1999), some species are typically found only in areas of reduced oxygen, and show adaptations to reduced oxygen, including reduced metabolic rate, increased energy stores (glycogen and arginine phosphate), and more rapid glycogen resynthesis following anoxia (Hervant and Malard 2012). In Rouch's (1988) study, two stygobiotic species were found only in the hypoxic hyporheos of a pool—the harpacticoid copepod *Elaphoidella bouilloni* and the bathynellan *Vandelibathynella vandeli*. Boulton et al. (1992) reported that Oligochaeta abundance was negatively correlated with dissolved oxygen.

Oxygen also plays a critical role in controlling several important biogeochemical processes in streams, especially nitrification, denitrification, and a variety of redox reactions such as denitrification, sulphate reduction, and methanogenesis (Malard et al. 2002). The balance between the various processes, especially those in the nitrogen cycle, depends on the amount of oxygen in the upwelling groundwater. In general, the hyporheic zone is an area of rapid nitrogen transport. For example, high nitrification occurs when NH_4^+-rich groundwater meets O_2-rich stream water (Chestnut and McDowell 2000). Whether a subsurface–surface patch acts as a source or sink of nitrate or phosphate depends to a large extent on dissolved oxygen concentrations (Malard et al. 2002). In turn, oxygen concentrations are related both to residence time within groundwater and hydraulic conductivity. Hydraulic conductivity can have a major impact on concentrations of nitrates, ammonia, and organic nitrogen, as Chestnut and McDowell (2000) showed for a tropical stream in the Luquillo Mountains of Puerto Rico. Boulton et al. (1992) provided comparative data on O_2, NO_3^-, NH_4^+, and soluble reactive phosphorus (SRP) for Sycamore Creek, Arizona. Among the three hyporheic sites and the phreatic zone, O_2 and NH_4^+ concentrations were highest, and NO_3^- concentrations were the lowest in the shallow hyporheic (Table 6.3). SRP was highest in the phreatic and parafluvial.

Malard et al. (2002) provided a useful, albeit hypothetical, overview of nitrogen dynamics along a hyporheic flow path (Fig. 6.8). Firstly, dissolved oxygen is consumed by aerobic respiration, and nitrification of ammonium produces increased nitrate concentration along the hyporheic flow path. As dissolved oxygen is depleted, denitrification may remove nitrate, if the flow path is long enough. If it is even longer, chemoautotrophy in the form of methanogenesis may occur (Jones et al. 1995). Thus short-length hyporheic flows may be a source of nitrate to the stream but longer flows would be a sink.

Table 6.3 Comparison of physico-chemical variables among three hyporheic habitats and the phreatic zone in Sycamore Creek, Arizona, USA. Means with the same superscript are not significantly different (p>0.05) by Scheffé's multiple comparison test used with the Kruskal–Wallis non-parametric test. From Boulton et al. (1992). With permission of E. Schweizerbart'sche Verlagsbuchandlung OHG (www.schweizerbart.de).

Variable	Shallow hyporheic	Parafluvial	Dry channel hyporheic	Phreatic
DO (mg L^{-1})	6.30[b]	2.82[a]	2.37[a]	3.42[a]
NO$_3^-$ (µg L^{-1})	50[b]	252[a]	67[b]	135[a,b]
NH$_4^+$ (µg L^{-1})	281[a]	40[a]	25[a]	46[a]
SRP (µg L^{-1})	47[a,b]	100[a]	35[b]	103[a]

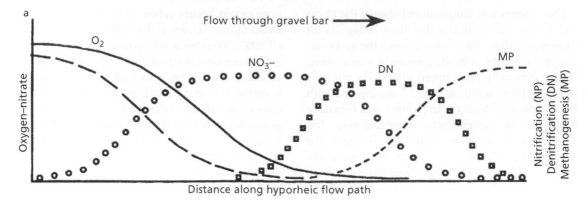

Figure 6.8 Hypothetical trends of dissolved oxygen (O_2), nitrate (NO_3^-), and methanogenesis (MP) along a hyporheic flow path within a gravel bar. From Malard et al. (2002), used with permission of John Wiley & Sons.

The organic content of the sediment is also crucial for nitrate dynamics because it controls the availability of oxygen (Jones and Holmes 1996). The details and complexities of nutrient dynamics in the hyporheic are beyond the scope of this book, and are mainly of interest from the point of view of the river, rather than the subterranean realm (see Jones and Mulholland 2000).

6.2.5 Analogues to the hyporheic

Hypotelminorheic habitats and their associated seeps are in a sense the end member of hyporheic habitats. Hypotelminorheic habitats are the headwaters of a small, probably intermittent surface stream, and are similar to the system in Fig. 6.2b where the hyporheic (i.e. hypotelminorheic in this context) is perched above the groundwater zone. The size of the hypotelminorheic is also much smaller. Perhaps most importantly, the hypotelminorheic cannot be sampled with the standard hyporheic sampler, the Bou–Rouch pump (see Box 6.1).

There are other aquatic SSHs with small habitat dimensions in addition to those associated with rivers and streams. Lake bottoms also have an interstitial zone that can also harbour many stygobionts (Wang and Holsinger 2001), but typically the depth of lakes exceeds 10 m. For example, Lake Tahoe, Nevada and California, USA has two endemic stygobiotic *Stygobromus* amphipods (*S. lacicolus* and *S. tahoensis*) that are found at depths between 60 and 495 m (Wang and Holsinger 2001). There are also

aquifers that extend to less than 10 m from the surface but they also typically extend to greater than 10 m beneath the surface.

Finally, calcrete aquifers (chapter 5) are another type of aquatic SSH. Calcretes, whose fauna has only been studied in western Australia, are a feature of arid karst areas, and are much less continuous than the hyporheic (Fig. 1.12), and perhaps calcretes are more analogous to epikarst than the hyporheic.

6.3 Biological characteristics of hyporheic habitats

6.3.1 Organic carbon and nutrients in the hyporheic

The amount and flux of organic carbon has been of special interest to biologists studying the hyporheic because (1) it is the main source of carbon for groundwater communities, (2) it is the site of much of the processing of organic carbon in the stream, and (3) may be the 'bottom-up' control of the hyporheic community. Because of the close proximity to the surface, differences in productivity in different streams and rivers, and differences in hydraulic conductivity, amounts of organic carbon vary considerably from place to place. Marmonier et al. (2000) provided values of dissolved organic carbon (DOC) for two hyporheic sites, one on the Rhône River in France (Grand Gravier section) and one on Vanoise brook, in the Rhône River basin (Fig. 6.9). Mean values for six different sampling sites ranged

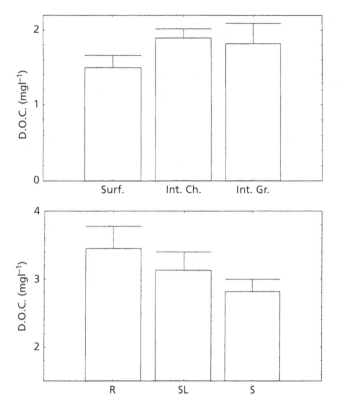

Figure 6.9 Dissolved Organic Carbon (DOC) component of interstitial water at two hyporheic sties in the Rhône River in France. A. Surface (Surf.), 40 cm deep in the underflow of the stream (Int. Ch.), and 40 cm deep in a gravel bar (Int. Gr.) in Vanoise brook. Standard errors are based on six samples. B. Underflow 1.5 m from bank (R), at the shoreline (SL), and 1.5 m from the shore (S) of the Grand Gravier sector of the Rhône. At each site, 12 samples were taken from depths of 20, 50, and 100 cm. Modified from Marmonier et al. (2000). Reproduced with permission of Elsevier.

from 1.8 mg L^{-1} at Vanoise brook at a depth of 40 cm in a gravel bar to 3.4 mg L^{-1} at the Rhône at depths of 20, 50, and 100 cm in the underflow of the river. In general, dissolved organic carbon declines with depth (Datry et al. 2005). Vervier et al. (1992) reported DOC values for several sites at depths from the uppermost layer of the soil, to the B soil horizon at a depth of 30 cm to a depth of 50 cm in the hyporheic (Table 6.4). While these values came from several different sites, the trend of decline of DOC with depth is clear.

DOC values have been reported from several other hyporheic sites and generally fall in the range of 1.0–10 mg L^{-1}. Another intensively studied hyporheic habitat is the meander arm (oxbow) of the Danube River near Vienna, Austria—the Lobau wetlands that comprise the Danube Flood Plain National Park. Danielopol et al. (2000) reported a mean value of DOC of 3.71 mg L^{-1}. In their study of a tropical stream in Luquillo Experimental Forest in Puerto Rico, Chestnut and McDowell (2000) found the following values for different hyporheic sites:

• parafluvial (riparian) 10 m from the stream: 5.1 ± 2.5 mg L^{-1};
• parafluvial (riparian) 1 m from the stream: 2.6 ± 2.4 mg L^{-1}; and
• hyporheic 1.40 ± 1.70 mg L^{-1}.

Table 6.4 Dissolved Organic Carbon (DOC) concentration in subsurface waters. A—alluvial interstitial water, B—interstitial soil water. From Vervier et al. (1992).

Site	Water type	Depth	Mean DOC (mg L^{-1})
Second-order creek, Quebec, Canada	A	50 cm	11.10
Mountain catchment, Alberta, Canada	A	saturated zone	5.90
	B	50 cm	21.20
Brook Valley, New Hampshire, USA	B	E horizon	31.50
	B	Upper B horizon	5.91
	B	B horizon (30 cm)	2.96

There was a lateral gradient of DOC, and all values from the hyporheic were higher than in the stream itself, with DOC concentrations of only 0.80 ± 0.30 mg L⁻¹. As Chestnut and McDowell (2000) pointed out, the hyporheic acts as a trap for organic carbon and nutrients primarily because DOC values are strongly negatively correlated with hydraulic conductivity at their site.

DOC is only one component of total organic carbon, but it is typically the dominant component at least in subterranean streams (Simon and Benfield 2001). Vervier et al. (1992) provided information about particulate organic matter (POM) for a gravel bar in a bypassed section of the Rhône River (Fig. 6.10). The quantity of POM declined from the upstream infiltration zone in the gravel bar to the downstream part of the bank as a result of processing in the hyporheic zone. Values of POM, at least at the site of sinking water, were much higher than for DOC, but also variable. Vervier et al. (1992) pointed out that in a sense invertebrates are also part of the flux of particulate organic matter if the

current is strong enough that they cannot move upstream.

Because hyporheic habitats typically have no primary production except for the possibility of chemoautotrophy in near-anoxic parts (Jones et al. 1995), it seems likely that the biotic community, both at the microbial and invertebrate level, is controlled by the flux of organic carbon, i.e. bottom-up regulation. Foulquier et al. (2011) tested the hypotheses that both microbial communities and invertebrate communities were bottom-up regulated. The system they used was a series of wells connected to a shallow (<5 m from the surface) aquifer in the eastern Lyon metropolitan area, France, consisting of three highly permeable glaciofluvial aquifers that drain into the Rhône. This study area was, in a broad sense, a parafluvial hyporheos, but it is much larger and extends deeper than typical parafluvial sites. Access to the sites was through wells, some of which are recharged with surface storm water. Storm water increases the concentration of organic carbon and nutrients, and they were able to compare

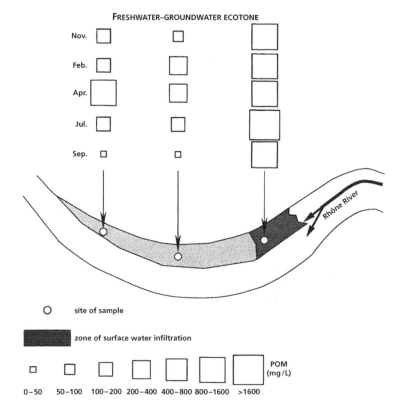

Figure 6.10 Quantity of particulate organic matter (POM, ash-free dry weight) in a gravel bar in a bypass section of the Rhône River, France. The channel is viewed from above and the gravel bar is situated on the inside of the meander. Sampling stations were located at the head, middle, and downstream end of the gravel bar. Samples were collected five times from November to September. From Vervier et al. (1992).

such artificially recharged sites with control sites that were not recharged. Not only were DOC and phosphate elevated in the recharge sites, but O_2 levels declined and temperature increased. Microbial biomass and activity were significantly higher in the artificially recharged sites, but invertebrate communities were not significantly more abundant. Invertebrates did not show any relationship with microbial biomass. Foulquier et al. (2011) suggested that the invertebrate community was controlled by environmental stresses, particularly oxygen depletion and groundwater warming. Strayer et al. (1997) pointed out that the invertebrate density in hyporheic habitats should increase with DOC until dissolved oxygen becomes limiting. Apparently in Foulquier et al.'s study, this point was reached. In a parallel study, Foulquier et al. (2010) showed experimentally, using slow filtration columns, that the stygobiotic hyporheic amphipod *Niphargus rhenorhodanensis* had little impact on microbial abundance and biomass. They suggested that one of the reasons for this was the slow metabolic rate of *N. rhenorhodanensis*, which is probably low as a result of adaptation to low food or low oxygen (see chapters 11 and 12).

6.3.2 History of biological study of the hyporheic

Interest in the hyporheic began with subterranean biologists, and the Romanian biologist Orghidan (1959) first invented the word hyporheic. Working at the Emil Racoviţă Institute of Speleology in Bucharest, Romania, he was well acquainted with the European perspective that subterranean biology was much more than the biology of caves. At the time he was working, Motas (1958) developed the idea of phreatobiology, the general idea of the study of the biology of porous aquifers. Early work was limited by the ability to collect organisms. For the hyporheic, the best technique available was to dig a hole along the side of a stream, wait for it to fill with hyporheic water, then extract and filter the water for organisms (sometimes called Karaman–Chappuis holes after the two biologists who first developed it [Karaman 1935, Chappuis 1942]). It was highly limited, not only in efficiency, but also in the restricted kinds

of habitats where it could be used, and because it was non-quantitative.

All this changed with the development of the Bou–Rouch pump (Bou and Rouch 1967, Box 6.1). For the first time, there was a sampling device that could be used in a wide variety of shallow subterranean habitats, including the middle of rivers. It was limited in its application only by where shallow subsurface water occurs, where one could pound in a steel pipe, and where sediments were not so fine that the pump clogged. Since the amount of water pumped can be measured, samples of the same volume can be compared. Boulton et al. (2003) suggested a standard sampling regime of three to five replicates of 5 L samples. Large-scale sampling of several European rivers commenced. Especially active was the Subterranean Hydrobiology and Ecology Research Team at University Claude Bernard Lyon I, France. Interest by stream ecologists was first awakened by the discovery of Stanford and Gaufin (1974) that the stonefly *Paraperla frontalis* spends its entire nymphal life cycle in the hyporheic. By the 1980s, the importance of the hyporheic as essential components of river and stream ecosystems was widely recognized (Hancock et al. 2005)

6.3.3 Overview of the fauna of hyporheic habitats

In a very influential paper, Gibert et al. (1994) proposed the categorization of the interstitial fauna in general and the hyporheic fauna in particular into three categories, which were named according to the widely used Schiner–Racoviţă system of ecological classification of subterranean organisms—stygobite (stygobiont), stygophile, and stygoxen (stygoxene). However, the categories were heterogeneous and did not necessarily correspond to the original definition (Fig. 6.12). The classification reflected the authors' background in speleobiology and made a basic habitat distinction between surface waters and aphotic sediments. Stygoxenes were occasional visitors to the hyporheos and the example provided is the dipteran family Simuliidae. Stygophiles were a highly heterogeneous group consisting of: (1) 'occasional hyporheos'—species that could spend part of their life cycle (typically nymphal stages) in sediments, the example being the mayfly *Caenis*;

Box 6.1 Collecting devices for hyporheic habitats

The Bou–Rouch pump (Fig. 6.11) allows the sampling of relatively shallow interstitial water, especially in the beds of streams and rivers, and along their banks (Bou and Rouch 1967). The device consists of a mobile stainless steel pipe that is driven to various depths in the bed sediment with a pump fixed on top. The principle of the method is to create a disturbance and maintain an interstitial flow around the pipe that is sufficient to dislodge interstitial organisms. Because of its high discharge rate, the pump probably samples both swimming organisms and species intimately linked to sand particles (Malard 2003). The pumped water is filtered through a net typically with a mesh size of 150 µm. Practical experience with the pump indicates that most of the species are collected in the first 5 L of water pumped. There have been various modifications of the pump, including battery-operated versions, but the basic principle of the pump has

been used by all its variants. Since the publication of the original paper by Bou and Rouch there have been over 100 papers citing it, an indication of its impact.

M.J. Dole-Olivier (pers. comm.) points out a limitation of the Bou–Rouch pump. It cannot be used to quantitatively sample the benthos at the top of the streambed because stream water would be pumped directly through the device, inflating the actual volume. She proposes using a Hess sampler in conjunction with the Bou–Rouch pump. A Hess sample is for a given area but this can be converted to a volume by estimating porosity and sampling at a given thickness, e.g. 10 cm. A Hess sampler is more efficient and so if it is used after a Bou–Rouch sample, then a correction can be made for the efficiency differences of the two devices. While this is a cumbersome procedure, it is much simpler than the only other alternative—a frozen core sample.

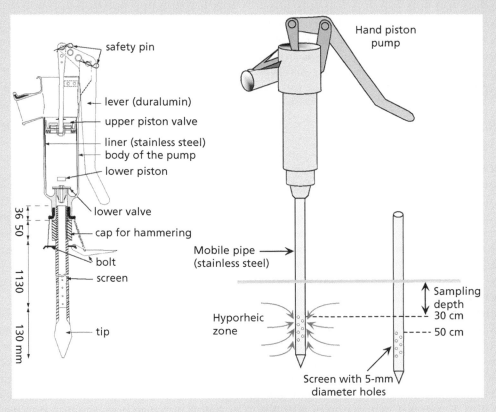

Figure 6.11 Design of Bou–Rouch pump used to sample hyporheic invertebrates. From Bou (1974).

(2) 'amphibites'—species with an obligate dependence on aphotic sediments that completed part of their life cycle (typically nymphal stage) in the sediments, the example being the stonefly *Isocapnia* (see Stanford and Gaufin 1974); and (3) 'permanent hyporheos'—species that can spend their entire life cycle in the sediments or in surface waters, the example being the copepod *Acanthocyclops viridis*. Included in the stygobites are (1) 'ubiquitous'—species that live in sediments or caves, the example being the amphipod *Niphargus rhenorhodanensis*, and (2) 'phreatobite'—species that live only in sediments and not in caves, the example being the amphipod *Salentinella delamarei* (Fig. 6.12).

A more hyporheic-specific example is provided by Gibert et al. (1990). Rather than focusing on a subterranean–surface distinction, they focused on the hyporheic itself (Fig. 6.13). Some species (the crustaceans *Salentinella*, *Microcharon*, and *Bathynella*) are found only beneath the hyporheic, in permanent groundwater; some (the amphipod *Niphargus*) are found in both groundwater and the hyporheic; some are hyporheic specialists (the amphipod *Niphargopsis*); some are found in both surface waters and the hyporheic (the amphipod *Gammarus*), and some are found only in the stream (the mayfly *Heptagenia* and the caddis fly *Hydropsyche*). Perhaps fortunately, no new terminology was

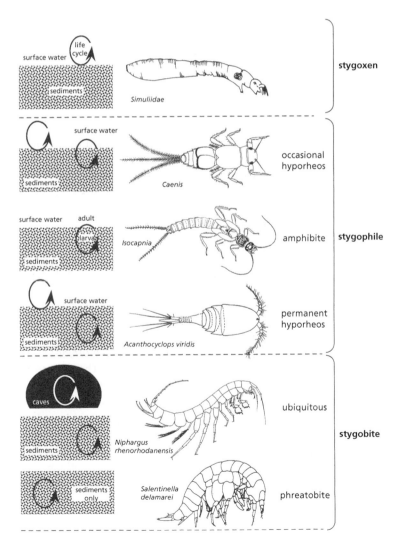

Figure 6.12 A classification of groundwater fauna based on its life cycle and its presence in surface and subterranean environments. From Gibert et al. (1994). Reproduced with permission from Elsevier.

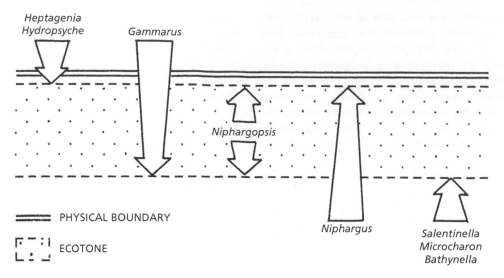

Figure 6.13 Diagram of distribution of invertebrates in relation to the hyporheic, in this case the underflow of the Rhône River, France. The dashed line indicates the extent of the hyporheic. From Gibert et al. (1990).

added by Gibert et al. (1990) to describe these categories. The vertical and horizontal distribution of these species changes during and after a flood (Fig. 6.14). At the time of flooding, *Gammarus*, an

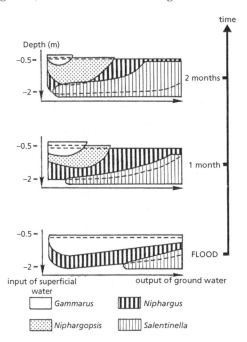

Figure 6.14 Dynamics of spatial distribution of amphipods after a flood in the underflow of the Rhône River, France. From Gibert et al. (1990).

indicator of surface water, sinks or actively migrates into the surface sediments. Because of the physical conditions present during flooding, the boundary of the hyporheic is pushed deeper into the sediment. As flood waters recede, *Gammarus* becomes more restricted to the upper layers of the sediment.

Malard et al. (2003b) separated taxa found in the hyporheos of the Roseg River in Switzerland by their primary habitat—permanent groundwater, permanent hyporheos, and occasional (temporary) hyporheos. The Roseg River is a glacial-fed river with a strong annual hydrological cycle of high water due to glacial melt occurring from May to September. The temporary hyporheos are those sites that are dry part of the year, typically winter. An idea of the diversity of taxa occurring in the hyporheic can be obtained from their data (Table 6.5). They identified 30 taxa although the number of species was much greater. Chironomid larvae, facultative inhabitants of the hyporheos, were the most common taxa and occurred in nearly all samples. Aquatic insects dominated the temporary hyporheos and the meiofauna dominated the permanent hyporheos. The freshwater polychaete worm *Troglochaetus* was the most common stygobiont. As a rule, stygobionts were not common, but were widespread in the samples (Table 6.5).

Table 6.5 Relative density and frequency of occurrence (percentages) of taxa collected in the hyporheic zone in 108 10 L Bou–Rouch samples in the Roseg River flood plain, Switzerland. Modified from Malard et al. (2003a).

Taxonomic group	Mean density + S.D.	Frequency of occurrence
Stygobionts		
Troglochaetus cf. *beranecki*	28.3 ± 80.0	68.5
Parastenocaris glacialis	10.7 ± 37.1	50.9
Bathynellidae	2.3 ± 3.7	55.6
Niphargus cf. *tatrensis*	0.2 ± 0.8	12.0
Permanent hyporheos inhabitants		
Nematoda	30.0 ± 60.9	91.7
Oligochaeta	14.4 ± 21.2	83.3
Ostracoda	11.5 ± 51.5	38.9
Harpacticoida	9.4 ± 43.7	51.9
Cyclopoida	4.0 ± 11.7	36.1
Hydracarina	0.8 ± 1.5	38.9
Temporary hyporheos inhabitants		
Chironomidae	58.3 ± 120.8	98.1
Plecoptera	5.7 ± 11.1	80.6
Heptageniidae	2.9 ± 10.3	34.3
Baetidae	1.3 ± 7.0	25.9
Limoniidae	1.1 ± 2.3	46.3
Limnephilidae	0.7 ± 2.8	25.0
Ceratopogonidae	0.6 ± 1.8	32.4
Crenobia alpina	0.4 ± 1.8	14.8
Empididae	0.2 ± 0.8	12.0
Simuliidae	0.1 ± 0.5	6.5

6.3.4 Species richness in the hyporheos

Of all the sampling devices available for SSHs, the Bou–Rouch pump is the most quantitative, even more so than epikarst drip samples unless total water volume is measured for those. Yet, there have been no attempts to compare species richness patterns on a continental or global scale[4]. Perhaps the main reason for this is that the hyporheic as a whole is both extremely species-rich and highly structured.

[4] A partial exception to this statement is the PASCALIS (Protocols for the Assessment and Conservation of Aquatic Life In the Subsurface), a large-scale European study of aquatic subterranean biodiversity (Gibert 2005). Hyporheic samples were included in this study but were combined with samples from other habitats.

For example, in sites at the glacial-fed Roseg River (Switzerland) that were part of the permanent hyporheos, Malard et al. (2003b) report more than 65 species. The fauna also had species that were stygobionts and species that were temporary inhabitants of the hyporheic zone (see Table 6.5). Samples taken at downwellings are much less rich than samples taken at upwellings, yet downwellings and upwellings are not necessarily easily recognizable in the field. There is the additional problem of which species to include, i.e. stygobionts and/or stygophiles and/or stygoxenes in Gibert et al.'s (1994) classification (see Fig. 6.12). In spite of these problems, a comparative analysis of species richness of the hyporheic in different regions should

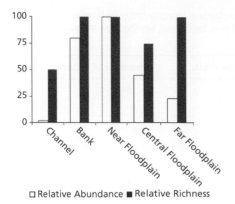

Figure 6.15 Bar graph of relative abundance and species numbers of macroscopic stygobiotic crustaceans in sites in the Flathead River floodplain in Montana, USA. Maximum abundance was 220 individuals per sample and the maximum number of species was four. Data from Ward et al. (1994).

be a research priority. Rouch and Danielopol (1997) listed the number of ostracod, cyclopoid copepod, and harpacticoid copepod species reported from different hyporheic sites, mostly in Europe. The Lobau wetlands in Austria had the richest ostracod species assemblage while Lachein Brook in France had the richest cyclopoid and harpacticoid assemblage, with one exception. The exception was not a hyporheic site but a water-saturated soil site in Brazil (Reid 1984, see chapter 7). Sket and Velkovrh (1981) enumerated the interstitial fauna of the Sava River and the floodplain on Ljubljansko polje (Slovenia) and found some of the most diverse sites in terms of macroscopic crustaceans, with up to eight *Niphargus* amphipods at one site. These studies are

all qualitative comparisons because no correction for sampling intensity was made.

Comparative species richness data, by hyporheic habitat type, are available for two sites—the Flathead River in Montana, USA (Ward et al. 1994) and Sycamore Creek in Arizona, USA (Boulton et al. 1992). Ward et al. (1994) sampled an extensive (greater than 2 km wide) alluvial flood plain, using 10 m-deep sampling wells at the Flathead River, and split the hyporheic into five subdivisions—the underflow of the Flathead River, its banks, and near, middle, and far floodplain (Fig. 6.15). Species richness and abundance of macroscopic stygobiotic crustaceans, including amphipods, isopods, and bathynellans, were relatively constant across sites, typically four species, except in the underflow of the Flathead River, where only two stygobionts were found. In contrast, abundance varied considerably among habitats, with a maximum at the river bank and near floodplains sites. Some species showed different patterns to the overall abundance trend. For example, the isopod *Microcharon* was most common in the underflow of the Flathead River and in far floodplain sites (Ward et al. 1994). Because only macroscopic crustaceans were included, this is likely only a small fraction of total crustacean species richness and abundance.

Boulton et al. (1992) split the hyporheic of Sycamore Creek into four categories—hyporheic (underflow of Sycamore Creek), dry channel hyporheic (roughly the equivalent of Malard et al.'s [2003a, 2003b] occasional hyporheic), parafluvial (the equivalent of Ward et al.'s [1994] floodplain sites), and phreatic. In this desert stream, they found that the

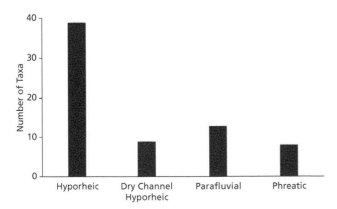

Figure 6.16 Bar graph of number of taxa, which typically correspond to taxa higher than species, for invertebrates in Sycamore Creek, Arizona, USA. Data from Boulton et al. (1992).

Plate 1 The authors at a seepage spring at Scott's Run Park, near Washington, DC, USA. Photo by W.K. Jones, with permission. (See Figure 1.1)

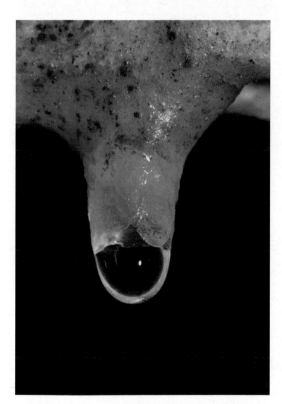

Plate 2 Water drop from epikarst at the end of a small stalactite in Postojna Planina cave system, Slovenia. Photo by J. Hajna, with permission. (See Figure 1.2)

Plate 3 Photo of exposed MSS in a *Fagus* forest at Mašun, Slovenia. Photo by T. Pipan. (See Figure 1.3)

Plate 4 Calcrete aquifers in southwestern Australia. **A**. Disused and uncapped water extraction bores (wells) in Lake Austen calcrete aquifer formerly supplying water to Big Bell gold mine near Cue, Western Australia. **B**. Deliberate exposure of groundwater in calcrete aquifer to provide watering point for stock. Ngalia Basin, Northern Territory. Photos by W. Humphreys, with permission of the photographer and the Western Australian Museum. (See Figure 1.4)

Plate 5 Photo of Lachein Creek, France, one of the first sites where the hyporheic was studied (Rouch 1991). Pipes into the stream are sites for sampling chemical and physical parameters as well as pump sites for the fauna. In the photo are Charles Gers, Raymond Rouch, and Thomas C. Kane. Photo by D. Culver. (See Figure 1.5)

Plate 6 Photo of a matrix of soil, rocks, and roots in the Pyrenees in Ariege, France, one of the sites studied by Gers (1992). The predominant habitat is soil, but the MSS and MSS fauna are present. Photo by T. Pipan. (See Figure 1.6)

Plate 7 Photo of typical passage in La Cueva del Viento on the island of Tenerife, Canary Islands, Spain. Photo by J. S. Socorro Hernández, with permission. (See Figure 1.7)

Plate 8 Field of a'a lava near Emesine Cave in Hawaii, USA. Photo by T. Pipan. (See Figure 1.13)

Plate 9 Photographs of the amphipod *Stygobromus tenuis potomacus* (top) and the isopod *Caecidotea kenki* (bottom), both inhabitants of the seepage spring at Pimmit Run, George Washington Memorial Parkway, Virginia, USA. Photos by W.K. Jones, used with permission. (See Figure 2.11)

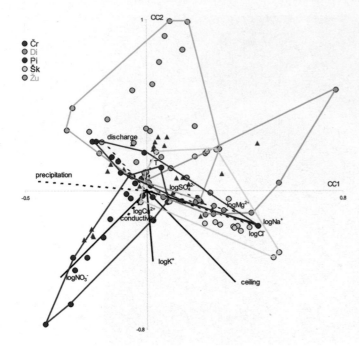

Plate 10 Ordination diagram based on species composition and abundance data in drips in five Slovenian caves. Lines indicate the environmental variables and their orientation on the canonical axis. Triangles indicate different species and dots represent individual drips. Convex hulls enclose the drips for individual caves. From Pipan 2005a, used with permission of ZRC SAZU, Založba ZRC. (See Figure 3.17)

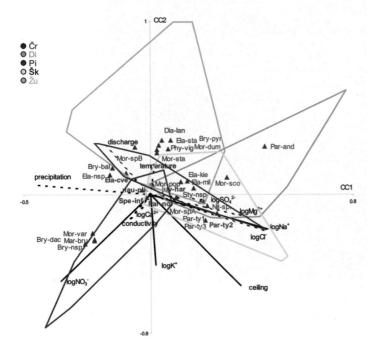

Plate 11 Ordination diagram shown in Fig. 3.17 with species indicated by abbreviations as follows: CYCLOPOIDA, *Diacyclops languidoides* Dia-lan, *Speocyclops infernus* Spe-inf; HARPACTICOIDA, *Bryocamptus balcanicus* Bry-bal, *B. dacicus* Bry-dac, *B. pyrenaicus* Bry-pyr, *Bryocamptus* sp. Bry-nsp, *Elaphoidella cvetkae* Ela-cve, *E. kieferi* Ela-kie, *E. stammeri* Ela-sta, *E. millennii* Ela-mil, *Elaphoidella* sp. Ela-nsp, *Maraenobiotus* cf. *brucei* Mar-bru, *Moraria poppei* Mor-pop, *M. stankovitchi* Mor-sta, *M. varica* Mor-var, *Moraria* sp. A Mor-spA, *Moraria* sp. B Mor-spB, *Morariopsis dumonti* Mor-dum, *M. scotenophila* Mor-sco, *Nitocrella* sp. Nit-sp., *Parastenocaris nolli alpina* Par-nol, *P.* cf. *andreji* Par-and, *Parastenocaris* sp. 1 Par-ty1, *Parastenocaris* sp. 2 Par-ty2, *Parastenocaris* sp. 3 Par-ty3, *Phyllognathopus viguieri* Phy-vig, cf. *Stygepactophanes* sp. Sty-nsp, From Pipan et al. 2006b, used with permission of ZRC SAZU, Založba ZRC. (See Figure 3.18)

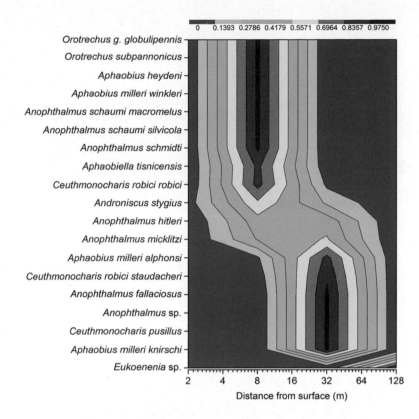

Plate 12 Comparative normalized spatial density map of 19 troglomorphic taxa in small caves in north-central Slovenia. From Novak et al. (2012). (See Figure 3.20)

Plate 13 Image of the region around Lake Way in Western Australia, showing the juxtaposition of the major calcrete aquifers (areas enclosed by *pale lines*) formed from the several palaeo-drainages entering the playa. Several large open pit mines are in the area, the largest visible north of the Lake Violet calcrete. The three calcrete aquifers (LV, Lake Violet calcrete; LU, Uramurdah Lake calcrete; HW, Hinkler Well calcrete) each support an endemic fauna. The image depicts an area about 40 by 60 km. From Humphreys et al. (2009). (See Figure 5.3)

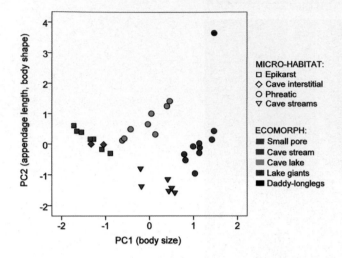

MICRO-HABITAT:
□ Epikarst
◇ Cave interstitial
○ Phreatic
▽ Cave streams

ECOMORPH:
■ Small pore
■ Cave stream
■ Cave lake
■ Lake giants
■ Daddy-longlegs

Plate 14 Principal Components Analysis on morphometric traits (mean values) of 33 *Niphargus* species and populations from seven cave communities in the Dinaric karst. The first two axes (PC1 and PC2) together explain 97.5 per cent of the total variation. Cave microhabitats (symbols) and proposed ecomorphs (grey shading, or colours in Plate 14) are only partly in agreement. There is no morphological distinction between inhabitants of the epikarst and the cave interstitial, and there are three distinct morphological groups within the phreatic habitat. From Trontelj et al. (2012) with permission from John Wiley & Sons. (See Figure 12.1)

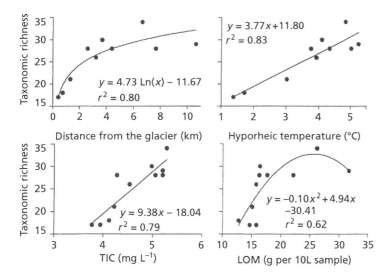

Figure 6.17 Relationships between taxonomic richness of hyporheic invertebrate assemblages and distance to the glacier, hyporheic water temperature, concentration of total inorganic carbon (TIC), and ash-free dry mass of loosely associated organic matter (LOM) for the floodplain of the Roseg River, Switzerland. From Malard et al. (2003b), used with permission from John Wiley & Sons.

greatest diversity of taxa, by more than three-fold, was in the underflow of Sycamore Creek (Fig. 6.16).

In their extensive study of the glacier-fed Roseg River in Switzerland, Malard et al. (2003b) looked for correlates of hyporheic invertebrate diversity, and found four environmental variables that were strongly correlated with species richness (Fig. 6.17). Species richness increased linearly with the log of distance from the glacier, hyporheic temperature, and total inorganic carbon (TIC). Species richness reached a maximum at an intermediate level of loosely associated organic matter (LOM), particulate organic matter easily separated from the sediment. The only taxon to decline with distance from the glacier was the harpacticoid *Maraenobiotus insignipes*, a species typically found at the margin of glaciers. All three of the other variables also increased with downstream distance, and TIC and temperature were also higher in areas of hyporheic upwelling. Upwelling may stimulate microbial production and in turn explain the correlation of species richness with these variables. The decline of species richness with high values of LOM is perplexing, and likely due to correlations among predictor variables.

In what is perhaps the only site in the world to be protected primarily for its hyporheic fauna, Danielopol and Pospisil (2001) reported on the crustaceans, molluscs, and rotifers sampled by Bou–Rouch pumps in Danube Flood Plain National Park, near Vienna, Austria. In a small 0.8 km² area, known as Lobau C, they found 11 stygobionts and at least 13 other species (see also Culver and Pipan 2009).

6.3.5 Ecology of hyporheic organisms

One of the themes to emerge from the now several decades-long study of hyporheic habitats is that of heterogeneity, both spatial and temporal. Dole-Olivier et al. (1993) pointed out the four-dimensional nature of the hyporheic, with longitudinal, lateral, vertical, and temporal variation. We have encountered examples of variation in all these dimensions, including upwelling and downwelling in the longitudinal dimension (Fig. 6.3), differences in species richness and composition between underflows and the parafluvial (Fig. 6.15), temporal changes as a result of flooding (Fig. 6.14), and changes with depth (Fig. 6.13). It is important to recognize that these changes can occur at very small geographical scales. While most recent studies of hyporheic ecology are at the scale of at least kilometres if not tens of kilometres, variation, just as is the case with epikarst (see chapter 3), occurs on the scale of metres. The best example of this is Rouch's now classic study of a 75 m² section of Lachein Brook in France (Rouch 1988, 1992). Not only do physico-chemical parameters vary over distances of one or two metres, they do so in a way that is understandable in terms of upwellings and downwellings, and differences in hydraulic conductivity (Fig. 6.5), and so do the distributions of hyporheic invertebrates (Fig. 6.18). Two

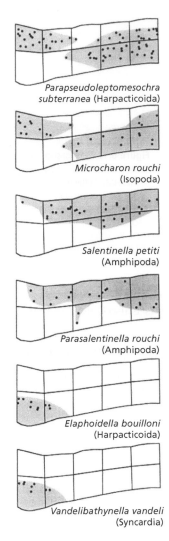

Figure 6.18 Distribution of six stygobiotic species living in the hyporheic of Lachein Brook, France. See Fig. 6.5 for physico-chemical data. From Rouch (1988). Copyright © 1988, Masson. Reprinted with permission of Cambridge University Press.

species—the bathynellan *Vandelibathynella vandeli* and the harpacticoid copepod *Elaphoidella bouilloni*—are limited to a hypoxic dead zone of low hydraulic conductivity. Two others—the harpacticoid copepod *Parapseudoleptomesochra subterranea* and the isopod *Microcharon rouchi*—are limited to gravel bars, and the other two—the amphipods *Salentinella petiti* and *Parasalentinella rouchi*—are found in the erosion zone characterized by high surface velocity.

Fišer et al. (2012) investigated the impact, if any, interspecific competition had on communities of

Niphargus amphipods living in interstitial aquifers in rivers and shallow wells in the Balkans. Up to six species of *Niphargus* co-occur in the same habitat. They used convex hulls (Cornwell et al. 2006), to represent the morphometrics of each species, which included size and relative length/width of appendages. Using all *Niphargus* species in the region as a source of randomly drawn samples of species, they showed that the morphology of co-occurring *Niphargus* was consistently more disparate than species drawn at random (Fig. 6.19).

Orghidan (1959), who coined the term hyporheic, also pointed out that the hyporheic could serve as a refuge for the invertebrate inhabitants of surface streams from environmental stresses, including flooding and drying. Called the hyporheic refuge hypothesis (HRH), it has been the subject of much speculation and experimentation (Dole-Olivier 2011). The hyporheic zone can offer several advantages for surface invertebrates:

1. further space for colonization;
2. higher temporal stability (dampened physico-chemical variations and greater water permanence); and
3. generally lower levels of competition and predation.

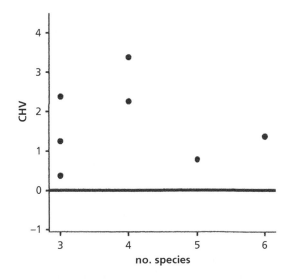

Figure 6.19 Convex hull volumes (CHV, y-axis) as measures of functional diversity of *Niphargus* communities in interstitial sites in the Balkans. Positive values indicate higher morphological diversity than expected if an equal-sized community were assembled at random. From Fišer et al. (2012).

It has been increased temporal stability that has been the focus of most of the work on the HRH. Naturally occurring environmental disturbances can be either 'pulse' disturbances, i.e. floods, or 'ramp' disturbances whose strength steadily increases over time, i.e. drying (Dole-Olivier 2011). These disturbances are important for benthic stream communities because these communities are characterized by low resistance (ability to withstand a disturbance) but high resilience (capacity to recover from disturbance).

There are three phases of the HRH. The first is the infiltration phase, in which organisms enter the hyporheic to escape from disturbance. This may be especially apparent in unstable streams such as the glacier-fed Roseg River in Switzerland, where hyporheic densities are often greater than benthic densities (Malard et al. 2001). The second phase is the survival phase. Survival depends not only on the biology of the species involved, but also the hydrological details, such as permeability and the presence of groundwater discharge. The third phase is emigration back to the surface stream. The downwelling zone is the primary refuge from flooding, since organisms will be moved by the current into the hyporheic, and the upwelling zone is the primary refuge from drying since it is fed by groundwater.

While the hypothesis itself is very straightforward, and it would seem hard to imagine that it is not true at least in some circumstances, demonstration of the hypothesis has proven to be very elusive. Of the 39 studies that test the hypothesis, roughly half find no evidence to support it (Dole-Olivier 2011). Dole-Olivier (2011) pointed out that this is due to a variety of reasons, especially the failure to take into account the spatial hydrological dynamics, e.g. upwelling and downwelling zones. There are also severe problems associated with equivalent sampling techniques of both benthic and hyporheic samples, the best available technique being the freezing core technique (Stocker and Williams 1972).

While HRH is posed as an ecological hypothesis, it also has evolutionary implications. In particular, if the stresses of flooding and drying are important factors promoting dispersal (infiltration in the terminology of the HRH), and the organisms survive, then emigration may not occur, and

this would result in permanent colonization of the hyporheic. In some case, emigration may be delayed, as is the case for the stoneflies such as *Isoperla* that only emerge (emigrate) at the short-lived adult stage. Many crustaceans, especially stygobionts, do not emigrate to surface habitats at all. It is an open question whether stygobionts, typically in the groundwater and lower parts of the hyporheic, disperse upward from deeper groundwater to take advantage of the relatively abundant resources in the hyporheic, also part of the aphotic zone, or whether the hyporheic is the site of initial colonization of the groundwater. The topic of dispersal and colonization of SSHs is the subject of chapter 13.

In contrast to the widely accepted hypothesis, based on extensive empirical evidence, that stygobionts and troglobionts have very restricted ranges due to limited dispersal potential, Ward and Palmer (1994) suggested that the hyporheic and associated phreatic habitats form a dispersal corridor along the river course, and that other habitats can be colonized from this 'interstitial highway'. Lefébure et al. (2007) tested these competing hypotheses with the widespread stygobiont *Niphargus rhenorhodanensis*. It is distributed throughout the Rhône River basin, and has colonized multiple subterranean habitats: karst and alluvial, SSHs and deep subterranean. Based on sequences of the COI mitochondrial gene, and the 16S and 28S rRNA genes, they concluded that *N. rhenorhodanensis* was polyphyletic and comprised at least six separate cryptic species. However, based on the analysis of hyporheic populations along the alluvial corridor of the Ain River, there appeared to be dispersal among the populations, which were separated by a maximum linear distance of 275 km. Lefébure et al. (2007) also suggested that dispersal was not post-Pleistocene, but earlier, and that populations of *N. rhenorhodanensis* survived underneath ice sheets (see Holsinger et al. 1983).

In contrast, Eme et al. (2013) found that five hyporheic species of *Proasellus* (*P. cavaticus*, *P. slavus*, *P. strouhali*, *P. synaselloides*, and *P. walteri*) had extensive ranges (up to 1300 km in linear extent) that were not the result of lumping together of cryptic species. There were a number of cryptic species found as a result of mitochondrial and nuclear DNA sequencing, but these newly found species all had very small ranges, except in the case of *P. walteri*. *Proasellus* species all had ranges which were in part covered by Pleistocene glaciation. Thus there does seem to be an interstitial highway for *Proasellus* (see chapter 13 for details).

6.3.6 Adaptations to hyporheic habitats

Species from several different groups of stygobionts that are found in interstitial aquifers show some of the features of troglomorphy shared by nearly all stygobiotic species, especially the loss of eyes and pigment (Coineau 2000). But because of the small pore size of interstitial habitats, they have unique features of miniaturization, segment reduction, and fewer spines and bristles, that contrast with troglomorphy. Coineau (2000) suggested that this is the result of progenesis, precocious sexual maturation. Miniaturization can be extreme, as in the case of the amphipod *Ingolfiella* which is less than 1 mm in length. Ingolfiellids from other habitats are often more than 10 mm in length. It is also clear from comparative morphological studies that pore size is the dominant selective factor in the evolution of miniaturization in these habitats (see chapter 12).

The question is whether the hyporheic is a small-pore subterranean habitat? It is typically listed as so, in common with alluvial aquifers. However, there is reason to doubt that the answer is that simple. A relatively large number of large-sized stygobiotic species occur in hyporheic habitats, especially amphipods. Many species in the genus *Niphargus* in Europe and the genus *Stygobromus* in North America are found in these habitats (e.g. Wang and Holsinger 2001). Some of these species are quite large. For example, *N. rhenorhodanensis* from hyporheic habitats is greater than 10 mm but Ginet (1985) indicated that the body is thin and the appendages relatively short. Thinness and appendage shortening may actually be more important modifications for life in a small pore-size environment than body size itself (see chapter 12). The presence of amphipods this size and without apparent reduction in segmentation or spination suggests that there are some large-pore habitats in the hyporheic. On the other hand, its thinness may be the result of selection, not for small size but for small diameter (Trontelj et al. 2012). It is very dif-

ficult to test this directly. Because of particle size sorting, lower gravels, those in the aphotic hyporheic zone, will have fewer large spaces than upper parts of the stream bed. This does not tell us if they exist, however, and it is difficult if not impossible to determine this directly from the substrate. For example, porosity and hydraulic conductivity can be measured but this gives us no information on the distribution of particle size and the spaces between particles.

6.4 Summary

The interstitial habitats along rivers and streams are generally small-pore SSHs, ones where the dissolution of bedrock plays little if any role. The best-studied and most ubiquitous of these habitats is the hyporheic, the underflow of rivers and streams. In addition to being of interest as a subterranean habitat, the hyporheic also has a critical role in river and stream ecosystems. This results in two views of the hyporheic, one focusing on its subterranean connections ('bottom up') and the other on its connection to the river above ('top down'). Hyporheic habitats can be subdivided in various ways, including lateral extensions (riparian and parafluvial), and temporary hyporheic habitats. Although best-studied in Europe, hyporheic habitats occur globally, including the tropics and arid zones.

The hyporheic is an ecotone between surface and subsurface habitats, and is a filter in three ways—a photic filter, a mechanical filter, and a biochemical filter. Transfers occur in both directions, and a characteristic feature of the hyporheic is that there are series of upwellings of groundwater and downwellings of surface water. Physical and chemical differences can occur at very small spatial scales, as Rouch (1988, 1992) demonstrated for Lachein Creek, France. In an area of 75 m², a wide variety of hydrological and chemical conditions were present, including hypoxic zones. Oxygen concentrations are among the most critical parameters of the hyporheic (Malard and Hervant 1999) and some species are only found in areas of low oxygen concentrations. The hyporheos is also the site of nitrification, denitrification, etc., and may be either a nitrate source or sink, depending on local conditions.

Organic carbon is typically higher than in other SSHs, and particulate organic matter (POM), in addition to dissolved organic matter (DOM), may be an important organic carbon source. POM usually declines along a hyporheic flow path and DOM typically declines with depth. Although there is no photoautotrophy in the hyporheic because there is no light, chemoautotrophy in the form of methanogenesis is likely widespread but quantitatively unimportant (Jones et al. 1995). Control of the microbial hyporheic community is likely through the amount of organic carbon, but the invertebrate community is likely controlled by environmental factors such as oxygen depletion (Foulquier et al. 2011).

Interest in the hyporheic dates back to at least the 1950s (Orghidan 1959), but sampling, especially quantitative sampling, was not possible until the development of a device to pump water from the hyporheic (Bou and Rouch 1967). The hyporheic fauna is usually subdivided into three major groups—(1) stygobionts that are obligate subterranean species, (2) stygophiles, a heterogeneous assemblage of species that spend part or all of their life cycle in the hyporheic, but are also found in surface habitats, and (3) stygoxenes, occasional inhabitants of the hyporheos. Some species are specialists for the hyporheic itself (e.g. *Niphargopsis*), found neither in surface waters nor in phreatic waters.

Little is known about large-scale patterns of species richness in spite of the possibility of quantitative, comparative sampling. At the local level, species richness can be quite high, typically with up to 100 species and perhaps 10–20 stygobionts. In a comparative study of different habitats of the hyporheos, Boulton et al. (1992) found that diversity was highest in the direct underflow of the stream.

From a short-term ecological perspective, the hyporheos may serve as a refuge from flooding, from drying, and from competition and predation. In spite of its obvious appeal, the hyporheic refuge hypothesis (HRH) has been difficult to demonstrate, in part because it depends on hydrological details such as the location of upwellings and downwellings, details that are often not taken into account (Dole-Olivier 2011). The HRH may also be important in evolutionary time, since it provides a context and an explanation for the dispersal of surface-dwelling invertebrates into subterranean habitats.

Based on detailed studies of the phylogeography of the amphipod *Niphargus rhenorhodanensis*, Lefébure et al. (2007) concluded that dispersal of hyporheic species along river corridors did occur, but it was limited in extent, in the order of 200–300 km.

Finally, although the hyporheos is typically considered a small-pore habitat, many of the species found in the hyporheic are rather large, more than 10 mm in the case of some amphipods. This combined with the possibility of some large spaces due to the presence of rocks and gravels suggests that the assumption of miniaturization and appendage reduction should be re-examined in this context.

CHAPTER 7

Soil

7.1 Introduction

Soil belongs on the list of SSHs if we define them as aphotic shallow subterranean habitats. Historically, however, the study of the biology of soil has had a very different trajectory than the study of caves and other subterranean habitats. Of course, soil is vastly more important economically than all other subterranean habitats, because of the importance of soil and soil fertility in the human food supply. At an even more basic level, there has been a deliberate exclusion of soil in the study of subterranean biology. This goes back at least to Racoviţă's famous 1907 essay 'Sur les Problèmes Biospéologiques' where he distinguishes between endogean[1] and subterranean habitats. However, he largely excludes any discussion of the soil fauna, even though he emphasizes the fauna of cracks and crevices in caves, which would include epikarst in present terminology (see chapter 3). He excluded the soil fauna because he argues (largely correctly) that it is distinct and that 'the majority of hypogean [endogean] animals do not all find the necessary conditions for their survival in the subterranean environment' (Moldovan 2006), even though he points out that there is overlap of the fauna, especially those that are 'searching only darkness, humidity, and protection against temperature variation'. Even by Racoviţă's time, the study of biology of soil was a separate discipline, and one senses that he deferred to this separation.

Nearly a century later, the prominent Slovenian biologist Sket (2004) explicitly excluded soil from the list of subterranean habitats because it 'contains rich and varied food resources.' By this logic, any

subterranean habitat with rich and varied food resources, including caves with bat guano and most SSHs, especially MSS (chapter 4) and lava tubes (see chapter 8), would be excluded, a nonsensical conclusion.

The exclusion of soil and its fauna from subterranean biology, whether due to tradition (according to Racoviţă) or the presence of rich food sources (according to Sket), is unfortunate. As we discuss in section 7.3.4, the soil fauna shares important morphological features, not only with a deep subterranean cave fauna (e.g. loss of eyes and pigment), but even more so with other SSHs (e.g. reduced body size). If interstitial aquatic habitats (see chapter 6) with their small pore sizes are part of the domain of subterranean biology, then soils, also with small pore size, should also be a part of subterranean biology. Beyond the shared environmental features (lack of light, and small pore size in the case of some SSHs [see Fig. 1.14]), soils are an important medium for the transport of organic carbon and nutrients. Gers (1992, 1998) holds that view that the MSS can be best understood in the context of the soil, and the movement of organic matter both by gravity and the action of animals through the soil. Not only does soil surround nearly all MSS habitats, with the possible exception of some recent lava flows, but also it is an integral part of the hypotelminorheic, also a habitat that exists within a soil (and leaf litter) matrix (see chapter 2). Soil also overlies epikarst, and the amount and type of organic carbon reaching epikarst is result of microbial interactions in the soil.

In this chapter, we focus on those aspects of soils and soil biology that are connected with subterranean biology, and the biology of SSHs in particular. These include:

[1] Racoviţă (1907) uses the term hypogean in the sense that of 'animals buried deep in the earth', thus a synonym of endogean.

Shallow Subterranean Habitats. David C. Culver and Tanja Pipan.
© David C. Culver and Tanja Pipan 2014. Published 2014 by Oxford University Press.

- characterization of soils as a subterranean habitat, with special attention given to porosity, in keeping with its importance in other SSHs;
- soil as a source of, and a transport medium of, organic carbon and nutrients to other SSHs;
- soil components of the matrix of SSHs;
- the frequency and position of larger cavities (MSS habitats) in the soil;
- an overview of the soil fauna, especially the obligate soil-dwelling fauna;
- patterns of diversity and abundance of soil animals; and
- convergent morphological adaptations of soil animals.

We are not the first to suggest that the study of soil and subterranean animals can mutually inform both disciplines. Coiffait (1958) did an extensive study of soil Coleoptera, especially in the Pyrenees. His work deserves a much wider readership because he emphasizes the parallels (and differences) between the soil fauna and the cave fauna. It is clear from his work that a failure to understand the differences and similarities of morphology of troglobionts and edaphobionts[2] can lead to major errors in interpretation of the primary habitats of many subterranean-dwelling species.

The list of topics about soil biology listed above does not span the range of soil biology, and there are many important topics, including soil fertility, soil–surface interactions, soil types and evolution, and carbon sequestration in the soil. For these and other topics in soil biology, more general treatments of soil biology (e.g. Bardgett 2005) should be consulted.

7.2 Chemical and physical characteristics of the soil

7.2.1 Relationship of soil to other SSHs

Writing prior to the discovery of the MSS or epikarst, Coiffait (1958) diagrammed the relationship between some major epigean and subterranean

(hypogean) habitats, and the likely dispersal of animals between them (Fig. 7.1). This clearly shows the connections between epigean and soil habitats, and among subterranean habitats, including the soil. As did Gers (1992, 1998), he assigned a central role to the soil. It is possible to tentatively assign a position in his diagram for not only the MSS and terrestrial epikarst, but also the hypotelminorheic. The latter habitat may be what he thought was a connect between phreatic water and the soil, which he based on the presence of *Niphargus* amphipods in rocks and soil, a likely hypotelminorheic site. He also pointed out some other SSHs that are worthy of further investigation. They include animal burrows, ant nests, and the humus layer of the soil. At least occasionally, troglomorphic species have been found in these habitats. Culver et al. (2012a) reported a record of the stygobiotic amphipod *Stygobromus tenuis* from an ant mound nest; Heads (2010) argued that troglomorphy in crickets evolves directly in leaf litter and humic layers; and Hoch et al. (2006) reported an eyeless, reduced pigment planthopper, *Notuchus kaori* from a *Paratrechina* ant nest in New Caledonia.

7.2.2 Soil structure

In vertical profile, soils are divided into layers (Fig. 7.2), which are often quite distinct (Bardgett 2005). The uppermost layer (L) is leaf litter, which in forested areas, is often deep enough to have an aphotic zone, enough so that troglomorphy may evolve there (Heads 2010). Below litter are the F and H layers, litter in intermediate stages of decomposition; the F layer is composed of partly decomposed litter from earlier years; the H layer is well-decomposed litter mixed with mineral material from below. Together they comprise the O-horizon, which is rich in organic matter. The A horizon is the uppermost mineral horizon, characterized by organic matter in the form of humic acids. The B horizon is the mineral subsurface horizon without rock structure. It is characterized by the accumulation of clays, iron, and aluminium. The final soil layer is the C horizon, or regolith, an unconsolidated mineral horizon that retains rock structure. In karst regions, this is the start of the epikarst (chapter 3). Tree roots do not penetrate into this layer, including

---

[2] Edaphobionts are species that spend their entire life cycle in the soil, and are not found anywhere else. In parallel with the terminology for cave animals, edaphophiles and edaphoxenes can also be defined.

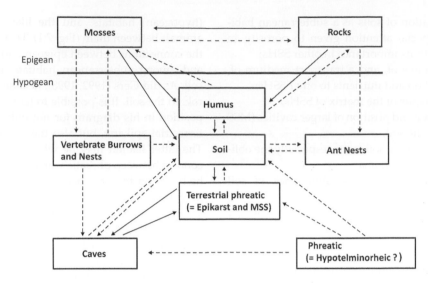

Figure 7.1 Relationships between various shallow subterranean habitats and surface habitats, with a focus on the soil. Modified from Coiffait (1958).

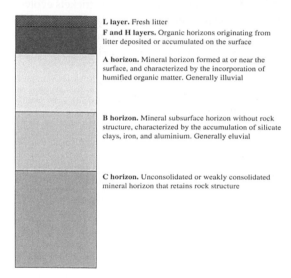

L layer. Fresh litter

F and H layers. Organic horizons originating from litter deposited or accumulated on the surface

A horizon. Mineral horizon formed at or near the surface, and characterized by the incorporation of humified organic matter. Generally illuvial

B horizon. Mineral subsurface horizon without rock structure, characterized by the accumulation of silicate clays, iron, and aluminium. Generally eluvial

C horizon. Unconsolidated or weakly consolidated mineral horizon that retains rock structure

Figure 7.2 Schematic representation of a soil profile showing major surface and subsurface horizons. From Bardgett (2005), used with permission of Oxford University Press.

the epikarst (Schwartz et al. 2013); and organic content is low.

A variety of soil types (termed orders) exist, such as spodosols (podsols). Soil formation is affected by a number of factors, especially the underlying parent rock (e.g. limestone compared to granite), climate, topography, and time (Bardgett 2005).

From the point of view of the distribution and occurrence of the soil fauna, soil texture and structure are very important. The typical soil texture ternary (triangle) diagram is shown in Fig. 7.3, with the three vertices representing the proportion of particles of different sizes (clay, silt, and sand). The relative proportions are important because they determine the ability of the soil to retain water and nutrients, and because they determine (in part) pore size and the connectedness of the habitat (Bardgett 2005). Although the higher the proportion of sand, the greater the pore size and the potential for movement through the soil, it does not indicate the presence of larger cavities in the soil. Large cavities, including the space around roots (see Fig. 4.3) and the cavities created by burrowing animals, are part of the interemediate-sized terrestrial subterranean habitats, discussed in chapter 4. These large cavities, or macropores, are typically destroyed during the sampling of soil sediments, and are analogous to secondary porosity in carbonate aquifers (Heath 1983). Macropores are also important for water storage, and some macropores become filled with water during precipitation events.

Pore space in soils is a major constraint on the size of the organisms present. For example, nematode diameters are approximately 30 μm and require a space at least this large in order to move about,

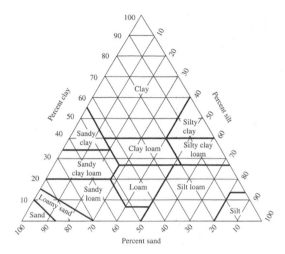

Figure 7.3 Ternary diagram of the composition of three soil texture classes (sand, clay, and silt). From Bardgett (2005), used with permission of Oxford University Press.

the surface. Coiffait (1958) did a series of 24 measurements of soil temperatures sites in the Pyrenees in July, at an elevation of 2200 m asl. At all three sites, temperature variation at depths of 5 cm and greater varied less than 3 °C. Temperatures at the soil surface varied between 24 °C and 38 °C. The extreme surface temperature variation is no doubt the result of the high elevation of the sites, but the damping of temperature within 5 cm of the surface is remarkable. Coiffait also reported annual variation in soil temperature at a soil site near Toulouse, France (elevation approximately 150 m) of 9 °C at a depth of 30 cm. If these daily and annual profiles are typical of soils, soils are at least no more variable than caves (e.g. Cigna 2002), and less variable than other SSHs (Culver and Pipan 2008). The buffering capacity of the soil is obviously great.

7.3 Biological characteristics of the soil

7.3.1 Organic matter and nutrients in soil

Jones and Willett (2006) discussed the methodological problems with measuring soil carbon and nitrogen, and provided suggestions on standardization with some representative values for organic carbon and nitrogen in some Welsh soils (Table 7.1). In general, the values or organic carbon are 10–100 times higher than those in other SSHs. However, all of these samples were taken near the top of the soil, either in the O-horizon or A-horizon, at a maximum depth of 20 cm. The amount of organic carbon declines dramatically with depth. There are two

although they can possibly burrow to a limited extent. This in turn requires particles of 150–250 µm, approximately the size of sand. Thus, soils with higher sand-sized particles are likely to have more invertebrates. Larger invertebrates, such as earthworms and some beetles, are able to burrow through the soil. Most microarthropods—mites and springtails—are not able to burrow.

7.2.3 Soil temperature

As is the case with other SSHs, temperature fluctuations decline rapidly with depth or isolation from

Table 7.1 Nitrogen and organic carbon in Welsh soils. Data from Jones and Willett (2006).

Site	Soil type	Dominant vegetation	Depth (cm)	Horizon	NO_3^- (mg N L^{-1})	NH_4^+ (mg N L-1)	DON (mg N L^{-1})	DOC (mg C L^{-1})	DOC/DON
A	Dystric cambisol	Grazed grassland	2–15	A	0.9 ± 0.3	0.2 ± 0.0	3.2 ± 0.2	47 ± 2	15
B	Humic cambisol	Deciduous woodland	0–5	O	21.4 ± 0.1	3.4 ± 0.1	22.9 ± 1.7	343 ± 15	15
C	Leptic podzol	Coniferous forest	0–5	O	130.5 ± 0.6	12.8 ± 0.1	168.5 ± 6.1	2925 ± 67	17
D	Leptic podzol	Decidous woodland	5–15	A	49.3 ± 9.7	0.6 ± 0.5	10.9 ± 3.8	74 ± 16	7
E	Eutric cambisol	Grazed grassland	0–20	A	65.6 ± 6.7	0.8 ± 0.1	16.3 ± 5.9	174 ± 1	11
F	Calcic cambisol	Deciduous woodland	0–15	A	3.8 ± 0.3	0.2 ± 0.0	4.5 ± 0.7	66 ± 9	15
G	Eutric cambisol	Grazed grassland	0–10	A	30.4 ± 0.9	0.7 ± 0.1	12.2 ± 0.3	146 ± 3	12
H	Dystric gleysol	Grazed grassland	0–15	A	20.5 ± 0.2	0.7 ± 0.1	8.1 ± 0.5	97 ± 6	12

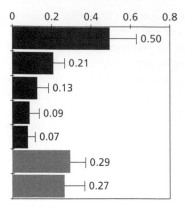

Figure 7.4 Profile of soil organic carbon from a forested site in Wales. Black bars indicate the proportional distribution of total organic carbon in the first soil metre in 20 cm intervals. The grey bars indicate the proportion of additional carbon in the 2 m and 3 m layers. Lines indicate standard errors. Modified from Jobbágy and Jackson (2000).

sources of organic carbon in the soil—organic carbon from the surface and from roots. In a large-scale survey of more than 2700 soil profiles, Jobbágy and Jackson (2000) found that overall, organic carbon content declined by roughly 50 per cent in the second metre and in the third metre of soil. The rate of decline was higher in grasslands (57 per cent) than in forests (44 per cent) because roots extend deeper in forested areas. A detailed distribution of organic carbon with depth is shown in Figure 7.4. Also note that the amount of organic carbon in the O- and A-horizons can vary enormously—from 47 to 2925 mg C L^{-1} (Table 7.1). The amount of organic carbon depends on a variety of interacting factors, such as vegetation type, soil drainage, and microbial activity.

Soil organic matter has a dual role with respect to SSHs. Firstly, of course, it is the carbon source for the soil biota; and secondly, the organic matter leaching from the soil is the carbon source for the epikarst, MSS, and to a certain degree, the hypotelminorheic. According to Bardgett (2005), most components of soil organic matter occur as large molecules (e.g. cellulose), which are, over time, broken down to simpler units by extracellular microbial enzymes which can be more readily assimilated by microbes. This may explain the decrease in complex organic molecules observed in epikarst drips relative to soil water (Simon et al. 2010). In the case of epikarst, this is the only source of organic carbon.

Gers (1998) pointed out that in the case of the MSS, organic carbon enters both from the vertical movement of water, and from the lateral and vertical movement of organisms through the soil. For the hypotelminorheic, the microbial processes occurring within the habitat are likely similar to those occurring in the soil. Dissolved organic carbon is also added to the soil from root exudates. This source of organic carbon is the base of the food web of many if not all lava tube communities (see chapter 8). Novak et al. (2012) also suggested that roots, or more specifically, the space around roots, is an important intermediate-sized terrestrial subterranean habitat.

Nitrogen, frequently a limiting nutrient in terrestrial systems (Menge et al. 2012), occurs in the soil both in mineralized form (NO$_3^-$ and NH$_4^+$) and in organic molecules. In three of the eight soils tested by Jones and Willett (2006), dissolved organic nitrogen was the major component of biologically usable nitrogen. This in turn may have important implications for other SSHs. Much of the nitrogen cycle takes place in the soil, including mineralization (conversion of N to NO$_3^-$ and NH$_4^+$); this rate is strongly correlated with the nitrogen supply to plants (Nadelhoffer et al. 1985), and is likely also to be strongly correlated with the nitrogen supply in other SSHs. Available nitrogen in SSHs results from leaching, as shown in Figure 7.5. This nitro-

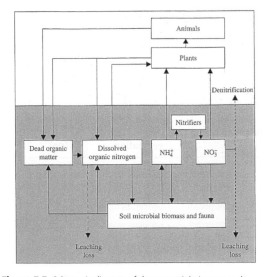

Figure 7.5 Schematic diagram of the terrestrial nitrogen cycle. From Bardgett (2005), used with permission of Oxford University Press.

gen is originally in the form of complex insoluble polymers which are broken down by extracellular microbial enzymes.

7.3.2 Overview of the soil fauna

In terms of numbers of species and numbers of individuals, the soil invertebrate community is among the richest of all terrestrial communities. In his comprehensive study of the soil fauna of the Pyrenees, Coiffait (1958) took a Berlese sample with a volume of 800 cm^3 (8 cL) from 100 sites in the French and Spanish Pyrenees, taken near the top of the soil column (O- and A-horizons). The summary of what he found is shown in Figure 7.6. Numerically, the fauna is dominated by mites and springtails, each from a number of different families with a variety of functional roles. Together with the Protura, they seem to have no adaptations for burrowing or moving particles, and thus are at the mercy of pore size in the soil. Some of the larger invertebrates, especially Coleoptera, are able to burrow at least to a limited extent. The body shape of some of the groups, especially Symphyla and Chilopoda, is elongate with long appendages, suggesting they are found in the uppermost part of the soil, the humic layer and the O-horizon (Fig. 7.2). Symphyla that are in deep soil have shorter legs (L. Deharveng pers. comm.). Three of the groups (Pauropoda, Protura, and Symphyla) are of special interest because of their morphology—all species

lack eyes and pigment (e.g. Scheller 2009), yet are not found in caves or other large cavity subterranean habitats[3]. All of the other groups have at least some representatives without eyes or pigment (even ants [Roncin and Deharveng 2003]). Although not included by Bardgett (2005), earthworms are probably the best-known soil denizen, having been studied by no less a figure than Charles Darwin (1881).

Coiffait (1958) also provided an ecological categorization of all the beetle species he collected (Table 7.2). He used a parallel classification to the Shiner–Racoviţă system of troglobiont–troglophile–trogloxene for the soil with edaphobionts being obligate inhabitants; edaphophiles being regular inhabitants but occurring elsewhere; and edaphoxenes being occasional residents. He also included a separate category (myrmecophiles) for species regularly associated with ant mounds. This ecological classification was probably conflated with a morphological classification because it was not until 1962 that Christiansen distinguished distributions indicating obligate habitat occurrence from morphological convergence, especially with respect to reduced eyes and pigment. At a maximum, 40 per cent of the 194 species and 30 per cent of the 88 genera showed signs of eye and pigment reduction and/or were only found in the soil.

Overall, Coiffait found 20,908 micro- and macroarthropods in a total volume of 0.8 m^3 of soil. Densities of soil microarthropods can reach 300,000 m^{-2} in

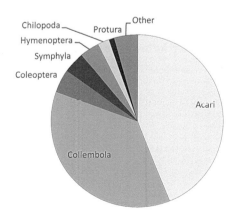

Figure 7.6 Pie diagram of species composition of soil samples from 100 sites in the Pyrenees Mountains. Compiled from Coiffait (1958).

Table 7.2 Pyrenees beetle species categorized according to ecological category. Mymecophiles are associated with ant nests, edaphobionts are obligate soil dwellers, edaphophiles are regular soil dwellers but occur elsewhere, and edaphoxenes are occasional soil dwellers. Data from Coffait (1958).

Ecological category	Number of genera	Number of species
Myrmecophiles	6	6
Edaphoxenes	6	8
Edaphophiles	50	102
Edaphobionts	26	78
Total	88	194

[3] Symphylans are certainly found in deep leaf litter, which may be a large pore soil habitat.

permanent grasslands (Bardgett and Griffiths 1997). The oligochaete family Enchytraeidae can dominate in acid soils in boreal forests, and reach densities of up to 400,000 m^{-2} (Didden 1993).

Species numbers can likewise be very high. Siepel and van de Bund (1988) found 108 species of micro-arthropods in 500 cm^2 of soil in a Dutch grassland. Laneyrie (1960) reported a total of 318 edaphobiotic beetle species from French soils.

Reid (1984) found a highly diverse and abundant aquatic meiofauna in wet campo soils in the cerrado region of central Brazil. The habitat she studied is basically a wetland formed in highly organic, spongy soils. The water table was within 1–2 m of the surface during the dry season, the soil was saturated with moisture, and moisture content was never less than 60 per cent of wet weight of the soil. The community was dominated by nematodes, rotifers, and copepods. Most of the organisms were in the upper 6 cm of the wet campo, and densities of nematodes and rotifers averaged 250 per 10 cm^2 (2.5 × 10^6 m^{-2}).

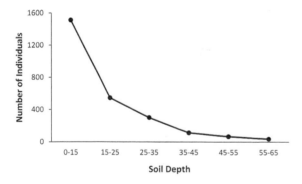

Figure 7.7 Decline of arthropod abundance with depth (up to 65 cm) for a soil site in Sainte-Cécile, in Barcelona Province, Spain, at an elevation of 600 m. Data from Coiffait (1958).

Species richness of copepods—29 harpacticoid and 4 cyclopoid species—was the highest reported for a freshwater system (Reid 1984), including ten new undescribed species of *Forficatocaris* and nine undescribed species of *Parastenocaris*. This can be compared to the 14 copepod species in epikarst drips in Županova jama, Slovenia (see Table 3.8). These rather unusual wet campo habitats are not the only soil habitat where copepods can be found. They are also routinely present in the litter and humic layer in forests (Reid 2001).

While species richness and abundance is high, it rapidly declines with depth. At a site near Sainte-Cécile, Barcelona Province, Spain, Coiffait (1958) took a vertical profile to a depth of 65 cm (Fig. 7.7). There was a rapid fall off of numbers (volume was not reported), with a reduction from 1512 in the first 15 cm to 36 at a depth of 55–65 cm. At this layer, the rate of decline with depth will vary with conditions, especially the thickness of the O- and A-layers. Laška et al. (2011) investigated the distribution of spiders in soil and MSS SSHs in the Czech Republic, and also found a sharp decline with depth. The mean depth for all soil spiders was 10–15 cm. However, not all spider species showed a decline with depth, and some, such as *Centromerus cavernarum*, occur throughout the soil column (Fig. 7.8). One species, *Porrhomma microps*, was only found in deeper layers, and occurred at a mean depth of 61 cm.

The bases of the soil food web are roots and detritus (Fig. 7.9). Feeding on them are fungi and bacteria in the case of detritus and mycorrhizal fungi, and phytophagous nematodes in the case of roots and root exudates. Two groups stand out in the food web as being important and diverse—

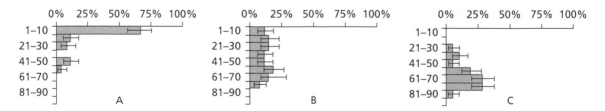

Figure 7.8 Distribution of three spider species with soil depth at a several sites near Hranice na Moravě in the Czech Republic. A is *Centromerus silvicola*, a species found mostly in the most superficial layers; B is *Centromerus cavernarum*, a species found through the soil depths sampled; C is *Porrhomma microps*, a species found mostly in deeper layers. Modified from Laška et al. (2011).

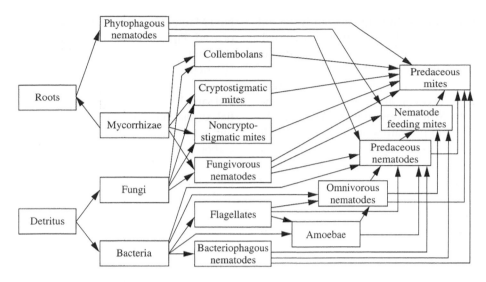

Figure 7.9 Structure of soil food web, excluding macro-invertebrate fauna. From Bardgett (2005), used with permission of Oxford University Press.

Box 7.1 Collecting the soil fauna

A simple device for the extraction of soil arthropods was first developed by Berlese (1905). The device, which can easily be constructed, consists of a funnel on a container containing a preserving fluid (typically alcohol) at the bottom (Fig. 7.10). A wire mesh screen is placed in the funnel, and a soil or litter sample added on top of it. Usually either a fixed volume or soil/litter area is added. A light is placed over the funnel to dry out the sample and force the animals down the funnel. Typically the samples are left for 48 hours or more, by which time drying is complete. There are a number of variants on the procedure. For example, Bedos et al. (2011) do not use a light source but rather let the sample slowly dry for four to eight days, a soft extraction technique. They find that this is a more effective way of extracting small, soft bodied arthropods, particularly springtails and mites.

Nematodes are much more difficult to extract and a series of techniques have been developed to do this (Goodey 1963). They include sieving, decanting, followed by centrifugation in a sucrose solution. Perhaps the most common technique is that of the Baermann funnel. A wire mesh basket, lined with tissue, and containing the sample, is nested on top of a glass funnel. A short rubber tube is attached to the stem of the funnel and sealed with a clamp. Water is then added to cover the sample. Nematodes migrate down

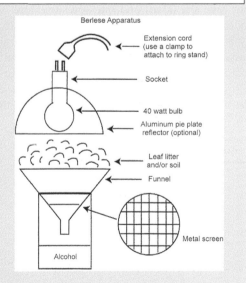

Figure 7.10 Diagram of Berlese funnel, used for the extraction of arthropods from a soil sample.

the funnel and rubber tube and congregate near the clamp. After a few hours or days, a small amount of water is extracted, which contains the nematodes.

nematodes and mites. In Coiffait's study, he did not collect nematodes because they require collecting and preservation techniques different from those of arthropods. Nonetheless, they play a key role as root herbivores, bacteriophages, fungivores, omnivores, and predators (Wall and Virginia 1999). Mites likewise play a diverse role, as fungivores and predators, including nematode predators. While most Collembola are fungivores, they are very diverse in their morphology. The food web in Figure 7.9 is incomplete because it does not include the many macro-invertebrates (e.g. Coleoptera), but numerically the microarthropods dominate.

7.3.3 Species richness in the soil

Ducarme et al. (2004a) surveyed soil microarthropod density and species richness in three soil sites in the B-horizon and two cave sites near Rochefort, Belgium, over several seasons. The granulometry of the sites is shown on a ternary diagram (Fig. 7.11). One cave site (Nou Maulin) had sandy soil, and the other cave site (Han) and one soil site (Han) had silty soil. The other two soil sites had a mixture of clay and silt. Overall densities of Collembola and Acari, and species richness of Acari are shown in Table 7.3. Mites were generally more abundant in the soil samples, while springtails were more variable, but very common in one cave (Nou Maulin).

Nou Maulin Cave is the site with a high proportion of sand (Fig. 7.11), and it had not only more springtails, but also more mites than Han Cave. Mite species richness was also higher, and in fact within the range of values for the soil sites (Table 7.3). For the soil samples, there was a significant overall and partial correlation of mite density with organic content, but not in the case of caves. Interestingly, organic carbon content of cave soil and the B-horizon was roughly the same, averaging between 1.0 and 6.9 per cent, depending on site and season. This makes Sket's (2004) suggested exclusion of the soil fauna from the study of subterranean biology, because of high food resources levels, problematic at best. Porosity showed a significant overall correlation but a non-significant partial correlation with mite density, and porosity and organic carbon were strongly correlated. Collembola showed a similar pattern, but with more non-linearities apparent in the data (Ducarme et al. 2004a).

Mite species richness did not consistently differ between soil and cave sites, whether observed number or Chao estimates of total numbers were compared (Table 7.3). The high number of mite species observed in cave soils is perhaps surprising, but this habitat is little sampled. The observed number of species was between 58 and 93 per cent of the Chao estimates of species, and the two numbers were closer together at the soil sites than at the cave sites. This is probably a reflection of the lower densities

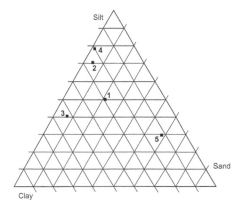

Figure 7.11 Relative proportions of sand, silt, and clay for three soil samples taken in the B-horizon (1 – Epraves; 2 – Han; 3 – Nou Maulin) and two caves (4 – Han; 5 – Nou Maulin) in Belgium. Data from Ducarme et al. (2004a).

Table 7.3 Mean density (individuals dm^{-3}) with standard errors of Collembola and Acari, and number of species and Chao 2 estimates (see Box 2.2) of number of species of mites for five Belgian sites. From Ducarme et al. (2004a), used with permission of Elsevier.

Location	Collembola abundance	Acari abundance	Acari species	Acari Chao 2
1 – Epraves soil	48 ± 4.3	151 ± 11.2	68	87[*]
2 – Han soil	25 ± 3.2	77 ± 7.4	39	42
3 – Nou Maulin soil	92 ± 8.1	225 ± 22.8	74	92
4 – Han cave	19 ± 2.1	9 ± 1.2	32	55
5 – Nou Maulin cave	206 ± 32.5	43 ± 5.2	55	75

[*]Chao 1, not Chao 2.

SOIL 137

Table 7.4 Jaccard similarity index values for mites for five Belgian study sites. From Ducarme et al. (2004a), used with permission of Elsevier.

	Epraves soil	Han soil	Nou Maulin soil	Han cave	Nou Maulin cave
Epraves soil	X	0.37	0.33	0.11	0.22
Han soil		X	0.28	0.15	0.19
Nou Maulin soil			X	0.10	0.17
Han cave				X	0.23
Nou Maulin cave					X

Table 7.5 Number of soil nematode sampling sites, by latitude. Data from Boag and Yeates (1998).

Latitude (°N or S)	Number of single sample sites	Number of repeated sample sites	Number of intensively sampled sites
0–10	0	1	1
10–20	8	0	0
20–30	2	0	1
30–40	2	1	1
40–50	25	12	22
50–60	7	16	25
60–70	0	0	1
70–80	5	1	3
80–90	0	0	0

of mites in caves. Finally, Ducarme et al. (2004a) calculated Jaccard indices (see chapter 3) and found that similarities among soil sites were greater than similarity among cave sites (Table 7.4). This is in accordance with the fact that the soil is more or less continuous, while caves (and most SSHs for that matter) are patchily distributed. Similarities between cave and soil sites were several times lower than within habitat comparisons. And this is consistent with the specialization of species for different subterranean habitats (see section 7.3.4).

Boag and Yeates (1998) assembled a large amount of data on nematode species richness on a global scale, and provide a tantalizing look at global diversity patterns in the soil. They assembled data for 134 sites (Table 7.5), and, recognizing that they were not all sampled with the same effort, divided them into three groups:

- single, unreplicated samples taken from a site at any one time;
- repeated samples, typically as a series of samples taken from the same site at different times; and
- intensive sampling that involved one site on a frequent basis, from various soil horizons, or collating of single from a variety of vegetation variants in the area.

They summarized the number of species found in 10 degree bands of latitude, and also corrected the mean number of species to take into account the relative proportion of the three kinds of samples. We added a summary of the pattern of intensively sampled sites (Fig 7.12).

For all three variables (mean number of nematode species, corrected mean number of nematode species, and mean number of nematode species in intensively sampled sites), the same pattern emerged. Going from the poles, species richness increases with latitude to 40°, then declined precipitously with latitude to 10°. From 10° to 0° species richness regained the levels of 40° to 30°. The amount of data available for tropical and subtropical sites is less than that for temperate sites (Table 7.5), and it is tempting to conclude that more data are needed. However, even in the tropics and subtropics, a significant amount of data are available, and it is hard to imagine how these data can ever be consistent with a tropical–temperate cline in species richness. Termites, which are important components of the humic and O-layer of the soil, do show a tropical–temperate gradient (Eggleton et al. 1994). What the pattern of nematode diversity shows is a pronounced hump at mid-latitude with a second increase in the tropics, or alternatively, a pronounced decline at subtropical latitudes, the so-called horse latitudes.

What is fascinating is that the pattern of obligate terrestrial cave dwellers (troglobionts) shows a similar pattern of high diversity at mid-latitudes, at least in Europe and North America (Culver et al. 2006a). They suggested that this mid-latitude ridge of high subterranean diversity is likely due to higher net primary productivity, which declines in the semi-arid horse latitudes. Alternatively, habitat

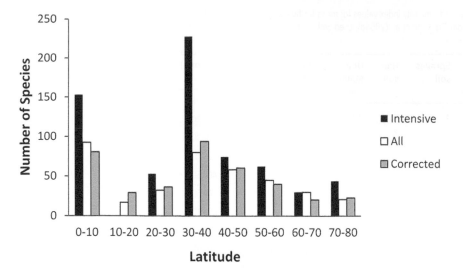

Figure 7.12 Latitudinal trends in soil nematode species richness, based on data compiled by Boag and Yeates (1998). Intensive samples are only those sites with repeated samples in several different habitats. Corrected samples are values scaled to the same proportion of single, repeated, and intensive samples. All samples are averaging for all samples, regardless of their thoroughness.

availability (in their study, cave density) may be higher. Either or both of these explanations may account for the nematode pattern of diversity.

A site of remarkable soil biodiversity is the Hòn Chông hills (Kiên Giang) in southern Vietnam, which Deharveng et al. (2009a) suggested is perhaps the highest known. The small caves in the karst hills are also remarkably diverse, but with a fauna separate from the soil. Together, more than 235 species of arthropods have been reported from soil and caves, with high levels of endemism. Unfortunately, the area is being destroyed by quarrying for cement. Deharveng et al. (2009a) stated:

what we observed and documented in Hòn Chông from 1993 is probably one of the worst current case[s] on earth of multi-species extinction under direct human action.

7.3.4 Morphology of the soil fauna

Studies of species richness of the soil fauna discussed in the previous section did not distinguish between species that are restricted to the soil (edaphobionts) and generalist species, nor between species that show morphological specialization for life in the soil and those that show no apparent

modification for subterranean life. In this section we consider the morphological characteristics of obligate soil dwellers, and compare to species found in other subterranean habitats. While this topic is central to speleobiologists, it doesn't even receive mention in many studies of ecology and biodiversity of soil organisms, and Bardgett's (2005) text did not mention the topic. As far as we know, the first major treatment of the patterns of morphology of soil-dwelling organisms was that of Coiffait (1958) who studied soil-dwelling beetles, followed shortly afterwards by Laneyrie (1960). Coiffait's (1958) monography remains the most comprehensive study from this perspective. However, his work is rather dated because he relied on the evolutionary theory, called organicism (see Vandel 1965), which holds that cave and other subterranean animals are in senescent phyletic lines. Therefore, Coiffait was at pains to show losses of structure, and with a reduced emphasis on adaptation.

Given the pattern of specialized morphology seen for the inhabitants of other SSHs, it is not surprising that soil animals show a similar pattern. In fact, Zacharda (1979) used the phrase edaphomorphy (actually euedaphomorphy) in parallel with troglomorphy.

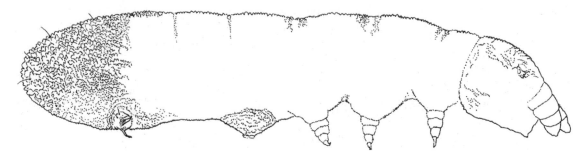

Figure 7.13 External morphology of the soil Collembolan *Stachia minuta*. Overall length is approximately 400 µm. From Bernard (2008). Copyright © Proceedings of the Biological Society of Washington Allen Press Publishing Services, used with permission.

One pattern, carried to an extreme in some Collembola, is miniaturization to a grub-like morphology, with reduced antennae, legs, and furcula (Bernard 2008). These minute Collembola are about 0.4 mm in length (Fig. 7.13), with reduced appendages. This pattern is known from several genera (Bernard 2008, Najt and Weiner 2002). All species in the family Odontellidae, which includes both deep soil and litter species, have small and integrated mouthparts, but what is remarkable about the soil species is their miniaturization.

A second extreme pattern of morphological change in soil organisms is epitomized by the mite *Gordialycus tuzetae* (Fig. 7.14). It occurs in dry sands both along coasts and inland in Africa, Asia, Europe, and North and South America (Coineau et al. 1967, Thibaud and Coineau 1998). It is clearly adapted for moving through small pores. Another elongated species has the highly appropriate name *Nematalycus nematoides* (Strenzke 1954). *G. tuzetae* has entirely lost one pair of legs and another is modified to act as antennae.

In his survey of soil-dwelling beetles of the Pyrenees, Coiffait (1958) listed seven shared morphological features:

1. Reduction in body size.
2. Flattened or thin body shape.
3. Reduction in length of appendages and elytra.
4. Disappearance of the scapular articulation in appendages in staphylinid beetles.
5. Disappearance of wing membrane.
6. Eye loss and sensory compensation.
7. Disappearance of pigment.

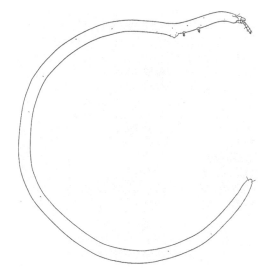

Figure 7.14 External morphology of the soil mite *Gordialycus tuzetae*. Overall body length is 2.8 mm and width is 0.05 mm. From Thibaud and Coineau (1998).

The strong emphasis on losses (all but #2 and part of #6) is typical of French evolutionary thinking at the time, as is the relegation of sensory compensation to a minor role. Nevertheless, the morphological pattern is clear. Coiffait (1958) attributed most of these changes to the evolution of flightlessness and to adaptation (although he would not call it that) to subterranean life. The unique features on this list were the size and appendage reduction as well as shape changes (recall Figs. 7.13 and 7.14); modifications for life in small spaces.

Loss of pigment and eyes is a universal feature of the subterranean fauna, and cannot be used to separate, for example, the cave fauna from the soil fauna. This is in fact a practical question because many samples of SSHs and deep subterranean habitats may have components of both. Examples include many MSS sites where the soil fauna is an important component (Table 4.8, Pipan et al. 2011). The aquatic fauna of the MSS and epikarst both show miniaturization (chapters 3 and 4, Coineau 2000) as does the soil fauna. In all these habitats, the morphology of the specialized fauna has shortened appendages as well as miniaturization.

More insight into the evolution of the morphology of soil organisms comes from Růžička et al. (2011), who compared the morphology of soil and leaf litter populations of the spider *Porrhomma microps*, whose vertical distribution in the soil is shown in Fig. 7.8. Compared to a leaf litter population, the soil population had a shorter metatarsus, but not a thinner cephalothorax. The soil population had shorter appendages but not a thinner body relative to the leaf litter population (Fig. 7.15). Eye size was an approximately 15 per cent reduction in the size of the posterior median eye. Cephalothorax thinning and appendage length reduction were evident when a scree (MSS) population of *P. myops* was compared to an undescribed *Porrhomma* species with affinities to *P. myops* (Fig. 7.16). Individuals in

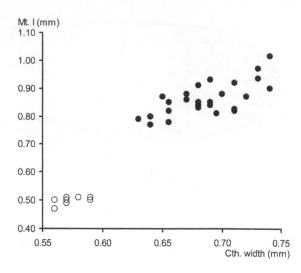

Figure 7.16 Relationship between cephalothorax width and length of metatarsus I in the spider *Porrhomma myops* (specimens from scree slopes and caves, filled circles) and *P.* aff. *myops* from soil (open circles) in the Czech Republic. From Růžička et al. (2011).

the soil population were thinner and had shorter appendages.

7.3.5 Community differences in morphology

In their study of Belgian cave and soil mites, Ducarme et al. (2004b) compared the body sizes of mites in the two habitats. Overall, they found 113 species in soil sites and 72 in caves. They then analysed 41 species (21 from soil and 20 from caves) that were the best indicator species. Indicator species are ones that are both common and occur in most samples of a habitat type. They then compared the morphology of these two sets of indicator species (Fig. 7.17), and found that body lengths of soil indicator species were smaller but with a great deal of overlap. When they compared species by family, the soil species were always smaller than the cave species in the same family. Thus, they were able to detect a signal of selection for decreased size, but there was still a great deal of overlap, even between indicator species groups. They also looked for similar patterns in other morphological features, especially appendage lengths, but found no consistent differences that could not be explained by size alone. This is similar to the pattern found by

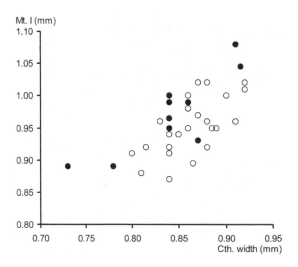

Figure 7.15 Relationship between cephalothorax width and length of metatarsus I in the spider *Porrhomma microps* from the Czech Republic. Specimens from leaf litter (filled circles) and soil (open circles). From Růžička et al. (2011).

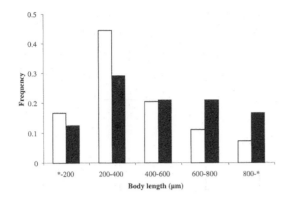

Figure 7.17 Frequency distribution of body length (μm) of indicator mite species of caves (black bars) and deep soil (white bars) in Belgium. From Ducarme et al. (2004b).

Culver et al. (2010) when they compared body size and antennal size for *Stygobromus* species in different aquatic SSHs. *Stygobromus* from different subterranean habitats differed in size but not in shape.

7.4 Summary

Traditionally, speleobiologists have not studied soils or soil fauna, but there are several features of soil habitats and other SSHs that make this unfortunate. All are aphotic habitats, and the fauna shows a number of shared morphological features. In addition, several SSHs are embedded in the soil (MSS, epikarst, and the hypotelminorheic) and at a minimum soil is the matrix through which organic carbon and nutrients are transferred to SSHs. There are other aphotic terrestrial habitats, such as animal burrows and ant nests, that share some features with the soil habitat and fauna.

In vertical profile, soils have several more or less discrete layers, from the O-horizon to the C-layer, or regolith. Organic carbon and nutrients decline with depth (although may be augmented by roots) and abundance of animals declines as well. Any important aspect of soil structure is the frequency of clay, silt, and sand, which is also closely tied to soil porosity. In this context, MSS sites are macropores. In general, pore size imposes a constraint on the size and shape of soil animals. Temperature variation declines with depth, and often strikingly so.

In the upper layers of the soil, organic carbon content is 10–100 times that of other SSHs, but organic content sharply declines with depth, approximately 50 per cent per metre in one study. Extracellular microbial enzymes break down organic molecules making them more easily assimilated both by the soil fauna and by the fauna of other SSHs that depend on dissolved organic carbon from the soil. Much of the nitrogen cycle, including mineralization, takes place in the soil, and nitrogen is often a limiting nutrient in terrestrial ecosystems.

Both in terms of abundance and species numbers, soil faunas are very rich. Tens to hundreds of thousands of arthropods can be found in a cubic metre of soil, and tens to hundreds of species. Numbers of nematodes and oligochaetes are even higher. The aquatic meiofauna of waterlogged soils (wet campo) in Brazil is likewise the most diverse freshwater meiofauna known. The number of species and their abundance fall off with depth, but some species are specialized for deeper soils. Both nematodes and mites play a diverse and key role in soil food webs, and are herbivores, bacteriophages, omnivores, and predators. Berlese funnels are the standard collecting device for soil arthropods; nematodes require more specialized techniques, such as the Baermann funnel.

In a study of Belgian soil habitats, Ducarme et al. (2004a) found that, compared to the mite fauna of caves, the mite fauna of deep soils has approximately the same number of species and the same organic content, but higher densities and higher similarities among sites. Organic matter and granulometry were important determinants of abundance and diversity. Overall, the global pattern of soil nematode diversity shows a maximum at mid-latitudes, a pattern also found in the terrestrial cave fauna.

The typical morphology of soil organisms is miniaturization, body thinning (except for some mites), eye and pigment loss, and appendage shortening. Extreme examples include grub-like springtails and nematode-like mites. They share eye and pigment loss with cave animals, and miniaturization with the interstitial aquatic fauna, and some elements of the MSS. As is the case for other MSS habitats, not all soil-dwelling species show obvious morphological specialization.

CHAPTER 8

Lava tubes

8.1 Introduction

Unlike caves in limestone and other water-soluble rocks, caves in lava are nearly always shallow, typically with ceilings of less than 10 m in thickness, and sometimes only 1 m thick (Palmer 2007). Even the world's longest cave in lava, Kazumura Cave in Hawaii, over 60 km long with a vertical extent of more than 1 km, is never more than 20 m from the surface (Allred 2012). Roots typically penetrate the ceiling of lava tubes, and their superficial nature clearly puts them in the category of SSHs.

The discovery by Howarth (1972) of troglobiotic species living in lava tubes on the island of Hawaii[1] initiated a flurry of interest in this terrestrial fauna, for at least two reasons. Firstly, the taxonomic composition of the Hawaiian troglobionts was very different from that found in North Temperate caves. Most striking were the eyeless, depigmented planthoppers in the genus *Oliarus*, a common inhabitant of Hawaiian lava tubes. These faunal differences pointed to subterranean communities that were different from North Temperate cave communities both in their organization and in their origin. Secondly, there was no good theory at the time to explain the presence of any troglobionts in Hawaiian lava tubes, let alone species from groups with few other subterranean representatives. The prevailing theories about the isolation of animals in caves posited that climate change, especially during the Pleistocene, forced animals into caves and isolated them there. According to this hypothesis, troglobionts and stygobionts should be rare in the tropics, because of the minimal effects of the Pleistocene on Hawaii. At the time of

Howarth's publication, relatively few stygobionts or troglobionts were known from the tropics. The Hawaiian lava tube fauna formed the basis of Howarth's subsequent challenges to existing theories of the origin of cave animals (Howarth 1980, 1987a).

Howarth also organized a highly successful series of 14 publications, *The cavernicolous fauna of Hawaiian lava tubes*, many of which were published in the journal *Pacific Insects*. The Hawaiian lava tube fauna continues to play an important role in discussions about evolution of the subterranean fauna, and the remarkable planthoppers found in lava tubes are model systems both for the study of the evolution of albinism (Bilandžija et al. 2012) and rapid speciation (Wessel et al. 2013).

Shortly after the discovery of the lava tube fauna in the Hawaiian Islands, an even more diverse fauna was found in lava tubes of the Canary Islands. Oromí and his colleagues described a number of new, troglobiotic species, including cockroaches (Martín and Oromí 1987), spiders in the genus *Dysdera* (Ribera et al. 1985), isopods (Dalens 1984), reduviid bugs (Español and Ribes 1983), and staphylinid beetles (Oromí and Martín 1984), as well as planthoppers (Remane and Hoch 1988). Several lava tubes on the island of Tenerife are among the most diverse caves in the world in terms of the terrestrial fauna (Culver and Pipan 2013).

The existence of lava tubes has been known for centuries, but the detailed study of lava tubes and other subterranean features of the volcanic landscape is rather recent. Much of this interest has been fuelled by discoveries on the island of Hawaii, where it is possible to routinely observe lava flows and the formation of new lava tubes (Palmer 2007). Since 1972, there have been periodic meetings devoted to vulcanospeleology (Halliday 1976, www.vulcanospeleology.org). In addition to the

---

[1] We follow convention and use the symbol ` in the word Hawai`i when referring to the island and use Hawaii when referring to the state.

discovery and mapping of many lava tubes, theories for the development of lava tubes and other subterranean features have been developed (Hon et al. 1994, Palmer 2007, Kempe 2012).

8.2 Chemical and physical characteristics of lava tubes

8.2.1 Geological setting of lava tubes

Caves in lava are formed in an entirely different manner than caves in limestone, and have an entirely different geographical distribution. Volcanic activity is concentrated mainly along moving plates of the Earth's crust (Fig. 8.1). Lava tubes have been explored in Iceland, Australia, Italy, South Korea, eastern and southern Africa, Argentina, Canada, western US, NewZealand, and many oceanic islands, especially the Canary Islands and the Azores (Palmer 2007). Most lava caves and other subterranean features are by-products of volcanic processes themselves, and are essentially the same age as the rock. Because Hawaii has easily accessible recent lava fields and an active volcano (Kilauea), much of the exploration and geological research is centred there.

There are two major types of lava—pahoehoe and a`a. Pahoehoe has a smooth surface with ropy wrinkles; a`a is rough and blocky and more viscous than pahoehoe. Most lava tubes are in pahoehoe but the MSS in a`a is more extensive (see chapter 4). Lava tubes[2] may form as the result of the surface cooling of a lava flow, or perhaps more commonly as the result of sequential lava flows, rather than a single flow, with later flows going underneath the older ones and forming cavities by a process called inflation (Hon et al. 1994). This results in stratification of the ceiling, observed in many lava tubes, including Kazumura Cave. Whatever the process of tube formation, the result is often a relatively simple, long tube (Fig. 8.2), but occasionally more or less labyrinthic tubes can form from lava flowing down a slight slope. In addition to the long, sinuous, partly braided passage, nearly all lava tubes have a radiating

pattern of small distributary tubes at the flow front, and many have complexes of sub-parallel tubes, and higher tubes that were vacated when lava flow was captured by underlying tubes (Palmer 2007). In spite of these and other complexities, currently the subject of much attention by vulcanospeleologists, lava tubes are remarkably linear objects (Table 8.1), especially when compared to limestone caves which often have maze-like sections. Furthermore, their passage orientations within a lava flow are generally parallel and downslope.

The density of caves and the density of cave passages can be quite high, as demonstrated by some comparative data for high density cave regions in the USA and Slovenia (Table 8.2). Both cave density and cave passage density are higher than in the karst areas but this is somewhat misleading because the sizes of the lava flows are much smaller (except for one small cave system in Slovenia), at least ten times smaller. Nevertheless, it is clear that lava tubes are not rare habitats, and that the density of passages can be very high (Table 8.2).

Other kinds of subterranean habitats are present, in addition to the MSS habitats discussed in chapter 4. There are some pit craters, fissures, and open volcanic conduits that are deep subterranean habitats, a few of which reach depths of over 100 m. For example, the volcanic crater Na One on Hawaii has a depth of 263 m, and may be the deepest known natural free-fall pit in the USA (Palmer 2007). Shallow subterranean habitats include tumuli (small hills formed by upwelling lava that then recedes), lava rises (injection of lava beneath a hardened crust with later deflation), and lava moulds, which retain the shape of the object covered by lava. Most lava moulds are from trees but the most exotic is Blue Rhino Cave in Washington state, USA, which preserves the outline of a rhinoceros.

Lava tubes are much younger than karst caves. According to Medville (2009), most of the caves formed around Hualalai volcano in Hawaii are between 1500 and 5000 years old. Kazumura Cave was formed between 300 and 500 years ago. Hawaii itself is no older than 500,000 years (MacDonald 1983) and most of the lava flows are much more recent than that. At the other end of the age spectrum, the volcanic Canary Islands range in age from 0.8 to

---

[2] Kempe (2012) suggests the term pyroducts be used instead of lava tubes. Pyroduct was originally coined by Coan in 1843 for 'covered aqueducts . . . covered with mineral fusion.'

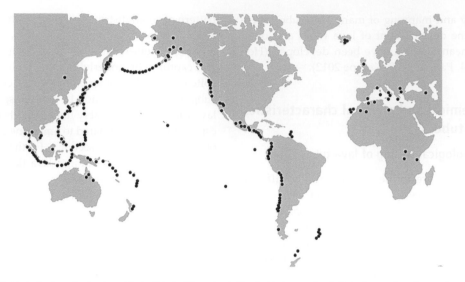

Figure 8.1 World distribution of volcanic activity in historical times (excluding that below sea level). Not all are active at present.

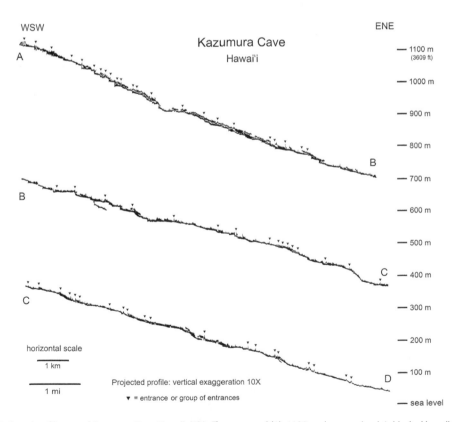

Figure 8.2 Projected profile map of Kazumura Cave, Hawaii, USA. The upper end (A), 1128 m above sea level, is blocked by collapse from road construction. The lower end (D) is a lava plug only 29 m above sea level. The mean slope is 1.9°, conforming to the topography which pre-dated the lava. Profile by K. Allred, C. Allred, and B. Richards. From Palmer (2007), used with the author's permission.

Table 8.1 Comparison of some morphological features of Hawaiian lava tubes. Sinuosity is the ratio of maximum linear extent to main trunk length. Modified from Kempe (2012).

Cave	Total length (km)	Main trunk length (km)	Maximum linear extent (km)	Sinuosity
Kazumura Cave	65.50	41.86	32.10	1.30
Ke'ala Cave	8.60	7.07	5.59	1.25
Epperson's Cave	1.93	1.13	0.80	1.41
Thurston Lava Tube	0.49	0.49	0.43	1.13
Ainahou Ranch System	7.11	4.82	4.27	1.13
Ke'auhou Trail System	3.00	2.27	1.99	1.13
Hu'ehu'e Tube	10.80	6.17	5.13	1.20
Clague's Cave	2.73	1.39	1.18	1.15
Pa'auhau Civil Defense Cave	1.00	0.58	0.50	1.14
Whitney's Cave	0.65	0.51	0.44	1.15

22 million years (Table 8.3). Lava tubes are abundant in islands up to 16 million years in age (Lanzarote), while troglobiotic species richness is highest on Tenerife, 12 million years old, although most lava tubes on these islands were formed from flows that were much younger. The two youngest islands, El Hierro and La Palma, as well as Tenerife, still have volcanic activity. The longest cave in the Canary Islands is Cueva del Viento on Tenerife, with a surveyed length of more than 20 km (Oromí 1995, Martín Esquivel and Izquierdo Zamora 1999). Because of the old age of the islands, it is quite possible that some islands,

Table 8.2 Comparison of cave density in lava flows and karst areas. Lava flow data from Medville (2009). The two US counties are those with the highest cave and passage density (Culver unpublished), and the data for Slovenia are courtesy of J. Hajna.

	Area (km²)	Caves/km²	Cave passage (m) per area (km²)
Hawaiian lava tubes			
Huehue	9.09	1.54	1337
Hualalai Ranch	5.40	7.04	5604
Lama Lua	17.30	3.35	1108
Kiholo Bay	6.92	14.16	1539
Rte. 190	2.23	3.14	4561
Limestone caves			
Van Buren Co., TN	177.00	1.23	267
White Co., TN	168.00	0.99	168
Kras region, Slovenia	434.00	2.26	167
N of Planinsko polje, Slovenia	25.00	14.00	n/a

Table 8.3 Basic information about the Canary Islands, their lava tubes, and troglobiotic fauna. From Oromí and Martín (1992) and Oromí and Izquierdo (in press).

Island	Area (km²)	Maximum elevation (m)	Age (my)	Lava tubes	Number of troglobionts
El Hierro	277	1501	0.8	abundant	20
La Palma	728	2426	1.6	abundant	33
La Gomera	378	1487	12	absent	9
Tenerife	2058	3717	12	abundant	70
Gran Canaria	1534	1950	15.5	scarce	21
Fuerteventura	1731	807	22	scarce	3
Lanzarote	796	670	16	abundant	0

especially La Gomera, previously had lava tubes, but they have all eroded. We return to this question when we consider the reason for the presence of troglobionts in the MSS on La Gomera (Gilgado et al. 2011). The Azores, the other area where lava tube fauna is relatively well studied, are intermediate in age between the Canary and Hawaiian Islands. Islands in the Azores range in age from 0.7 to 8.1 million years (Oromí 2004, Borges and Hortal 2009).

Because of the highly porous nature of lava, lava tubes almost never have any permanent water, and no stygobiotic fauna (Oromí and Izquierdo in press). Occasionally lava tubes are flooded with seawater, like Corona lava tube on Lanzarote. Lava tubes can be modified by the erosion caused by water (Kempe et al. 2003), but the active stream in the cave they studied is long gone. Apparently water is retained in some potholes.

8.2.2 Physico-chemistry of lava tubes

Temperature fluctuations in lava tubes are generally less than those of the MSS and much less than those on the surface. An example of temperature variation at an MSS site and a nearby small lava tube is shown in Fig. 4.10 and Table 4.3. At a depth of 70 cm, hourly temperature of an MSS clinker site ranged between 4.0 °C and 15.0 °C, and the small cave ranged between 9.7 °C and 13.6 °C. In the much larger Cueva del Viento, variation was even less, with a maximum temperature range of slightly more than 2 °C, based on hourly temperatures over a year (Table 8.4, Fig. 8.3). Even less variable was Pahoa Cave (Hawaii, USA), which varied less than1 °C throughout the year (Table 1.1, Fig. 1.8). These values for Pahoa Cave are comparable to those reported by Cigna (2002) for the hypogenic Karchner Caverns, Arizona, USA. Hypogenic caves such as Karchner Caverns lack an active stream. Caves with active streams tend to be more variable, with annual variation sometimes up to 15 °C, which Kranjc and Opara (2002) observed in Škocjanske jame in Slovenia.

Few physico-chemical measurements have been reported from lava tubes. Stone et al. (2012) suggested that lava tubes are saturated with respect to relative humidity, due to restricted air flow, with

Table 8.4 Basic statistics for temperature (°C) in Cueva del Viento, Canary Islands, and the entrance.

Description	n[1]	Mean	Standard deviation	Coefficient of variation	10% Quantile	90% Quantile	Minimum	Maximum	Range
Entrance	10,029	13.28	2.36	17.77	10.29	16.67	7.51	18.59	11.08
Cueva del Viento	10,029	13.92	0.70	5.03	12.99	14.86	12.99	15.02	2.03

[1] 8760 hours in a year.

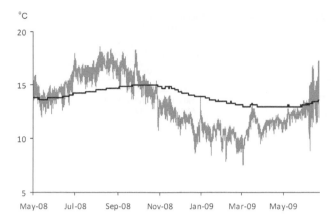

Figure 8.3 Annual temperature variation in Cueva del Viento, Canary Islands, 400 m inside the entrance (black line) and in rocks in the shade outside the entrance (grey line). Temperature measurements were from May 2008 to June 2009. See Table 8.4 for details.

the possibility of reduced oxygen and elevated CO_2, but they provided no data, with the exception of Bayliss Cave, a lava tube in Australia. Howarth and Stone (1990) found an area in the cave with elevated CO_2 (up to ten times ambient levels). They hypothesized that troglobionts are adapted to this extreme environment, in part to avoid competition with other species, and that this 'bad-air' environment is rather common in lava flows wherever there is restricted air flow. The bad air is likely the result of production of CO_2 by decaying organic matter (James 1977). Arechavaleta et al. (1996) provided data on CO_2 concentrations in Cueva del Viento and Cueva Felipe Reventón. Most values were close to ambient levels (0.01 per cent) but they found several cul de sacs in Cueva Felipe Reventón where CO_2 concentrations were 13 times higher than normal.

8.3 Biological characteristics of lava tubes

8.3.1 Organic carbon and nutrients in lava tubes

There are several possible sources of organic matter in lava tubes—tree roots penetrating into the cave, organic matter falling vertically, sometimes carried by water, through the thin ceiling of lava tubes, and chemoautotrophic bacteria using reduced forms of iron and manganese as an energy source. Of these, tree roots have received the most attention both because they can be very dense (Fig. 8.4), and because they harbour a community of troglobiotic species

Figure 8.4 Roots of *Metrosideros polymorpha* coming through the ceiling of Lanikai Cave (Hawaii, USA). Photo by H. Hoch, with permission.

that depend on root herbivores such as *Oliarus* (Hemiptera: Cixiidae).

Tree roots penetrate the substrate to obtain water and nutrients, and such roots are not limited to lava tubes but are also present in MSS habitats as well. In Hawaii, a predominant tree, especially on new lava flows, is the `ōhi'a' (*Metrosideros polymorpha*), and theirs are the predominant tree roots penetrating lava tubes. *M. polymorpha* is one of the first colonists on new lava flows. Presumably, the advantage to the `ōhi'a' is to obtain water, perhaps from the water table. As succession on lava flows proceeds, *M. polymorpha* becomes less abundant. There are fewer obvious roots in Canary Island lava flows, not necessarily because they are older, but rather because the climate is less humid than Hawaii and without a water table directly beneath the lava. The Azores Islands have a wetter climate, and roots in lava tubes are much more common than on the Canary Islands (P. Oromí pers. comm.). Hutchins et al. (2013) looked at the water and energy balance of tree roots in limestone caves in Texas. They were especially interested in the extent to which trees could exploit the water stored in epikarst, a reversal of the usual downward movement of water. They found that the vegetation was able to exploit epikarst water to a very limited extent. A similar study with *M. polymorpha* would be very interesting. Whatever the advantage, if any, to *M. polymorpha*, *Oliarus* is able to exploit the resource in the lava tubes, and is in turn a keystone resource for other troglobionts in these caves (Fig. 8.5). Planthoppers such as *Oliarus* are not only a keystone resource in Hawaiian lava tubes, they also serve that role in lava tubes in Australia (Stone et al. 2012), and in some Croatian limestone caves (Bilandžija et al. 2012). However, Oromí and Izquierdo (in press) doubt that it is the only or even major source of food in Canary Islands lava tubes. They point out that there are many caves without roots where the main input of organic matter is from percolating water carried in from the surface. Given the very thin roofs on most Canary Islands lava tubes (less than 5 m [P. Oromí pers. comm.]), this input may be considerable.

The formation of microbial mats (organic slime), which takes place in Hawaiian lava tubes and may also be chemoautotrophic (Howarth 1973, Northup et al. 2004), is largely absent from lava tubes in the Canary Islands because of the arid climate there. It is remarkable that the region with the fewest obvious sources of organic carbon is the one with by far the richest fauna, although not the most abundant

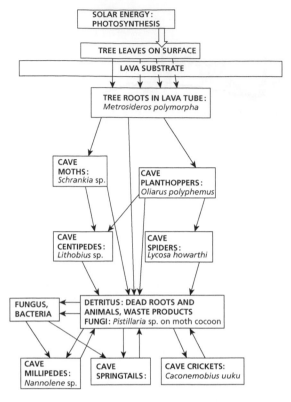

Figure 8.5 Food web of tree root communities on the island of Hawaii. Arrows indicate the direction of energy flow. Modified from Stone et al. (2012), used with permission from Elsevier.

fauna. The presence of nitrifying and ammonia-oxidizing bacteria in microbial mats in lava tubes suggests an important role for microbial mats in nitrogen cycling (Hathaway et al. 2013). Curiously, there are few if any invertebrates living on these mats (Northup et al. 2004).

Hathaway et al. (2013) provided data on carbon present in lava tubes in the Azores, Portugal (Table 8.5). They reported total carbon in cave soil, measured by percentage weight loss on high-temperature combustion, and total organic carbon (TOC) in percolating water entering the lava tubes. Both are highly variable. Percolating water is the functional equivalent of epikarst water except that it is not retained in the cave. Values tend to be somewhat higher than those reported for epikarst (see Table 3.6), and the Azorean values averaged 2.36 mg L^{-1}.

Table 8.5 Soil organic carbon (% C) and total organic carbon (mg C L⁻¹) in percolating water (TOC) in lava tubes in the Azorean Islands, Portugal. Modified from Hathaway et al. (2013).

Cave	TOC	% C in soil
Algar do Carvão	0.86	3.01
Gruta da Branca Opala	1.24	8.49
Gruta da Madre de Deus	5.40	2.06
Gruta das Agulhas	1.10	1.40
Gruta do Natal	2.32	2.00
Gruta da Terra Mole	1.19	10.15
Gruta da Malha	1.45	3.86
Gruta dos Buracos	8.10	4.56
Gruta dos Principiantes	0.78	4.36
Gruta dos Balcões	1.17	7.87
Mean	2.36	4.78

8.3.2 Origin of the lava tube fauna

With Howarth's (1972) announcement of a unique troglomorphic lava tube fauna came several consequences. As Howarth pointed out, the paradigm of animals being isolated in caves because of the climatic vicissitudes of the Pleistocene could not explain the Hawaiian fauna because the Pleistocene had no impact on Hawaii. He identified several components of Hawaiian lava tubes that he believed were key to understanding how the fauna evolved.

The first is the appearance of a new habitat or resource. There are in fact several new habitats created by lava flows. The lava flow itself is a new habitat, and it is colonized by wind-blown (aeolian) organisms, especially mites and Collembola (Ashmole et al. 1996) and eventually by shrubs and trees from elsewhere on the island. On wet islands like Hawaii, trees such as *Metrosideros polymorpha* colonize within 25 years (Howarth and Hoch 2012), but in the much drier Canary Islands, barren lava flows persist much longer (Fig. 8.6). The other new habitats are lava tubes as well as what Howarth (1980, 1987a) calls mesocaverns, and others have called the MSS. Important resources in the subterranean system of MSS and lava tubes are the roots of *Metrosideros polymorpha*.

A second component of lava flows is the pattern of relative humidity (Howarth 1980). Compared to karst caves, especially in temperate regions where entrances are small, Howarth suggests that newly formed lava tubes have lower relative humidity, but that as time proceeds, humidity increases as the soils develop on the surface. His bioclimatic model (Howarth 1980) is based on the importance of high relative humidity in subterranean habitats, and the barrier it poses to colonization, at least by arthropods. The barrier is because of the difficulty terrestrial arthropods have in surviving high humidity, which as Howarth points out, gives the habitat some features of an aquatic habitat. Unfortunately, he does not provide any extensive numerical data.

Figure 8.6 Centuries-old lava flow in Teide National Park, Tenerife (Canary Islands). Photo by T. Pipan.

A third component is the absence of climate change during the Pleistocene, which caused Howarth (1987a) to put forward the Adaptive Shift Hypothesis (ASH), to replace the Climate Relict Hypothesis (CRH) of Pleistocene changes in climate being the forcing agent of cave colonization. According to the ASH, speciation is parapatric, with the continued persistence of the surface-dwelling component of the population and strong selection along the surface–subterranean gradient. According to the CRH, environmental change forced organisms into subterranean habitats and caused the extinction of the surface-dwelling population (Culver and Pipan 2009).

These different components come together in the following scenario (Howarth and Hoch 2012). The pioneer species living on the surface of lava flows (lavicoles), as a result of behavioural changes, colonize subsurface habitats, initially ones with less than saturated humidity, in order to exploit the new resource of tree roots. The population, now distributed in both the surface and subsurface, is subject to a strong selective gradient, not only due to changes in light intensity, but also to changes in relative humidity (Howarth's bioclimatic model). Parapatric speciation is initiated by these different selective regimes. The best empirical example of this scenario is the dermapteran genus *Anataelia* (Martín and Oromí 1988, Ashmole et al. 1992). *A. lavicola* is a lavicole that is abundant on lava flows and occasionally in lava tubes, particularly in the drier parts. On the surface, it hides in cracks during the day and only appears at night. Its sister species, *A. troglobia*, is present in deeper parts of the lava tubes which are also wetter. *A. lavicola* is thus a pioneer species on both new lava and in lava tubes. While it is a compelling example of Howarth's ASH, it is the only case of 150 troglobionts on the Canary Islands where such a clear association between surface and derived lava tube species exists.

While Howarth rejects the CRH as inapplicable to the Hawaiian lava tubes, others (Leys et al. 2003, Oromí 2004) have broadened the concept to include any environmental changes, not just those associated with the Pleistocene, that both force animals into subterranean habitats and cause extinction of surface populations. These environmental changes could include periods of drying and could occur at any time during the history of the lineage.

The important difference between the CRH and the ASH is the persistence of surface populations in the case of the ASH, but not in the case of CRH (Fig. 8.7). Howarth and Hoch (2012) listed examples of species pairs with parapatric distributions (Table 8.6). While these cases are not consistent with the CRH, some of their patterns are puzzling and the ASH does not seem to provide a complete explanation. Among these is the situation with the moth *Schrankia howarthi*. Medeiros et al. (2009) pointed out that on both Hawaii and Maui there are genetically similar populations that have a wide morphological range, including surface and cave morphs. It is also the only troglobiotic species known from two

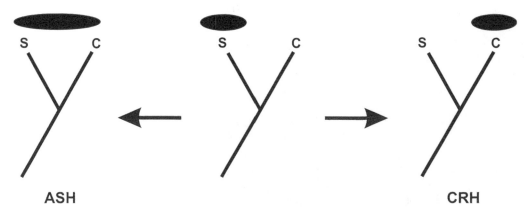

Figure 8.7 Hypothetical pattern of colonization and phylogeny for a species colonizing a subterranean habitat under the adaptive shift (ASH) and climate relict hypotheses (CRH). S denotes a surface habitat and C denotes a subterranean habitat. Ellipses indicate habitats occupied.

Table 8.6 Parapatric cave and surface populations occurring on Hawaii Island. Species marked s. lat. are represented by several populations or species. From Howarth and Hoch (2012).

Lava tube species	Surface relative	Ancestral habitat
Littorophiloscia sp.	*L. hawaiiensis*	Marine littoral
Oliarus makaiki	*O. koanoa*	Mesic forest
Oliarus polyphemus s. lat.	*Oliarus* species	Rain forest
Oliarus lorettae	*Oliarus* species	Dry shrubland
Schrankia howarthi s.lat.	*Schrankia howarthi* s.lat. *Schrankia altivolans*	Rain forest Dry to wet forests
Nesidiolestes ana s.lat.	*N. selium*	Rain forest
Caconemobius varius s. lat.	*C. fori* s. lat *C. sandwichensis*	Barren lava flows Marine littoral
Anisolabis howarthi	*A. maritima A. hawaiiensis*	Marine littoral Barren lava flows?
Lycosa howarthi	*L.* cf. *hawaiiensis*	Barren lava flows

islands, and the pattern observed seems as much involved with the evolution of flightlessness as it does with cave adaptation. If *Oliarus polyphemus* is actually many species, then the lava tube species may result from a single invasion followed by subterranean dispersal or by multiple invasions of a surface-dwelling ancestor (Hoch and Howarth 1999). Distinguishing between these two hypotheses (vicariance vs subterranean dispersal) is extremely difficult without the presence of all surface ancestors (Culver et al. 2009).

In the case of the amphipod *Palmorchestia* on La Palma (Canary Islands), it would appear to be a good candidate for the ASH hypothesis since one surface-dwelling species (*P. epigaea*) and one lava tube-dwelling species (*P. hypogaea*) are present. However, detailed genetic analysis, using mitochondrial genes, indicated that each species was actually a complex of species and that they did not form reciprocally monophyletic DNA clades (Villacorta et al. 2008). Geographical proximity was more important than habitat (surface vs lava tube) in determining relationships, and it appears that *Palmorchestia* invaded lava tubes multiple times on La Palma, and that nearby surface and subterranean populations were not usually the closest relatives. Villacorta et al. (2008) concluded that CRH is a more likely explanation.

Likewise, the absence of a surface ancestor does not necessarily indicate that the ASH is wrong. Rivera et al. (2002) argued that the absence of any

surface ancestors for the four species of the terrestrial isopod *Hawaiioscia* does not necessarily mean that the ASH hypothesis is wrong, just that the extinction of surface ancestors may have been unconnected with subterranean colonization and speciation. While this may well be the case, it hardly constitutes a test of the two hypotheses. Oromí (2004) pointed out that the subterranean fauna of Macaronesia (including the Canary Islands) includes a number of species with surface-dwelling species in the same genus and a number of species with no obvious surface-dwelling ancestors, and that the ASH and CRH are likely both to hold for different lineages.

Whatever the relative importance of ASH and CRH, Howarth performed an important service by injecting a strong dose of selectionist thinking into the discussion of colonization of subterranean habitats, in a way similar to the injection of selectionist ideas by Barr, Christiansen, and Poulson in the 1960s into understanding the morphology of cave animals.

Howarth and Hoch's (2012) scenario of colonization of lava tubes implies that the MSS may be an intermediate step, in part because it is more likely to have saturated relative humidity in the early phases of a lava flow, and in part because the MSS may be a physical gateway to lava tubes. A definitive answer to the question of the evolutionary relationship between cave and MSS species must await information on both phylogenetic positions and morphological differences of cave and shallow subterranean species that belong to the same clade. The most thorough study is that of Arnedo et al. (2007). Nine troglobiotic species of the spider genus *Dysdera* are known from Tenerife, seven of which each have their own epigean sister species, and the other two (*D. esquiveli* and *D. hernandezi*) are a pair of troglobiotic sister species (Arnedo et al. 2007). *Dysdera madai* is exclusively found in the MSS, is the least troglomorphic, and is the only one coexisting with its closest relative, the epigean *D. iguanensis* which also occurs in the same MSS station. The other eight troglobionts are restricted to lava tubes, with the exception of a unique record of the lava tube-dwelling *D. esquiveli* in the MSS, and the lava tube species rank from intermediate to highly marked troglomorphic. It might be assumed that cave *Dysdera* species are more troglomorphic than typical MSS inhabitants because *D. madai* is not strongly troglomorphic. However, in the consensus parsimony trees obtained by Arnedo et al. (2007), *D. madai* is not the most basal among subterranean species. Moreover, the phylogenetically distant, mostly cave-dwelling, and more troglomorphic *D. esquiveli* was found in the MSS together with *D. madai* in an area many kilometres from any cave. Therefore, we cannot conclude that any species occurring in the MSS or in caves is characterized by a degree of troglomorphism or by a particular phylogenetic position (Pipan et al. 2011a).

8.3.3 Adaptations to lava tubes

As is the case with the inhabitants of other SSHs, many obligate lava tube dwellers have no or reduced eyes and pigment (Howarth 1972, Hoch 2002). Most of the detailed work on adaptation has been done on planthoppers, from Australia, Hawaii, and the Canary Islands. Wessel et al. (2007) pointed out that there are convergent spine configurations of the tibia and tarsi, that apparently enhance walking ability on wet surfaces, an adaptation similar to the one described by Christiansen (1961) for cave Collembola. In the troglomorphic cixiid species *Solonaima baylissa* from Australia, on the clypeus as well as the second antennal segment there are tiny hairs which may have an additional sensory function or serve as protection against moistening in the saturated atmosphere of their habitat (Hoch 2002).

Lava tube-dwelling planthoppers are in lineages where the ancestral habitat is likely roots (Remane and Hoch 1988), and Hoch (2002) argued that pore size is an unimportant constraint in the evolution of the lava tube fauna. This is not surprising, since habitat space is larger in lava tubes than it is around roots in the soil. Thus, lava tube species are not smaller in size than their ancestors.

A remarkable aspect of the biology of lava tube planthoppers is that males and females communicate with vibratory signals transmitted along the roots (Hoch 2000). These vibrations are the most important and perhaps only means by which potential mates can recognize each other. Compared to surface-dwelling species, the signals are simplified, perhaps because there is usually only one species per lava tube, and hence no need for a complex signal. In courtship in the lava tubes, the female emitted the first calls. If a male answered, the female would remain stationary, emitting calls at regular intervals until the male located her. In all observed cases, courtship lasted about an hour before copulation occurred (Hoch 2000).

Wessel et al. (2013) reported some interesting patterns in morphology occurring in the Hawaiian planthopper *Oliarus polyphemus*. Different populations showed differing morphology, based on wing size and shape as well as tibia length and mesonotum width. These features did not show convergence with time, but rather a reduction in within-population variation with time (Fig. 8.8). Wessel et al. (2013) suggested an important role

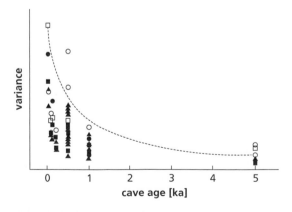

Figure 8.8 Scatterplot of variance in seven morphometric characters of *Oliarus* planthoppers against cave age. Different characters are denoted by a different symbol. From Wessel et al. (2013).

for stochastic processes rather than convergent selection in this pattern, based in part on small sizes of colonizing populations (see chapter 13).

8.3.4 Age of lava tube species

Since lava flows, especially on the Hawaiian Islands, are relatively young compared to the ages of limestone caves (and epikarst) as well as other SSHs, it is of interest that extensive changes in morphology can take place within these time constraints. For example, the island of Hawaii is approximately 430,000 years old, and most lava flows where tubes are present are much younger than that. Wessel et al. (2013) made the case that populations of *Oliarus polyphemus* on different lava flows (of which there are many on Hawaii) are descendants of separate founding populations, and it logically follows that they can be no older than the lava flow itself, the age of which is known (e.g. Medville 2009). Using a molecular phylogeny, together with the geological history of the island, Wessel et al. (2013) concluded that speciation rates are likely greater than 500 per million years, i.e. less than 10,000 years for speciation, rates higher than the highest previously reported for African cichlid fish in Lake Victoria. If this rapid speciation rate is correct, then Wessel et al. (2013) pointed out that mutation rates for the mitochondrial COI gene must be much greater than previously reported. They suggested that this apparent mutation is greatly increased by the stochastic nature of founder events in new lava flows.

Not all speciation rates are necessarily this rapid. Villacorta et al. (2008) estimated that speciation events in the amphipod genus *Palmorchestia* on La Palma in the Canary Islands were between 0.34 and 0.94 million years ago. La Palma is a much older island than Hawaii; it is approximately 1.6 million years old (Oromí and Izquierdo in press). Finally, there is no necessary connection between the rate of genetic change, typically measured using sequence differences of the COI gene, and morphological change. Wessel et al. (2007) compared the level of troglomorphy for a series of *Oliarus* species from different islands in Hawaii, and concluded that there was no connection between the two because age and indices of troglomorphy were uncorrelated.

It is possible then, for species like *Palmorchestia*, that morphological change occurs rapidly and genetic change occurs more slowly.

8.3.5 Faunal composition and species richness in lava tubes

The overall taxonomic distributions of troglobiotic genera and species for Hawaii and the Canary Islands are shown in Fig. 8.9. The number of species known is constantly increasing for both faunas, but the number of genera is more stable, and may also be a more accurate representation of the number of separate lineages that invaded lava tubes, because the number of species is inflated as different islands have different species all in the same lineage.

Other areas are much less well studied; some, such as the lava flows in northwestern USA, have very few species (Peck 1973). Approximately 20 species of troglobionts have been reported from the Azorean Islands (Oromí 2004, Borges and Hortal 2009, Borges et al. 2012), and from the Galapagos (Peck 1990).

For the Canary Islands, two groups not often found in temperate zone subterranean habitats are prominent—Homoptera and Blattaria (Oromí and Izquierdo in press). It is the planthoppers in the Homoptera that are the primary consumers in the food web based on roots. Homoptera are also important components of the Hawaiian fauna (Fig. 8.9). Based on Wessel et al.'s (2013) analysis of reproductive isolation among population in *Oliarus polyphemus*, there are many more species of *Oliarus* than indicated in Fig. 8.9. For example, they found seven groups of populations on the Island of Hawaii that were reproductively isolated from each other. Hawaii and the Canary Islands share the presence of several species and genera of pseudoscorpions, spiders, and planthoppers, although the genera and species are different in the two sets of islands. The discordances include a large number of Coleoptera species and genera in the Canary Islands but very few in the Hawaiian Islands, the presence of only one troglobiotic Collembola and no troglobiotic Orthoptera in Canary Island lava tubes, and the absence of Blattaria, Zygentoma, and Opiliones from the Hawaiian Islands, as well as other differences.

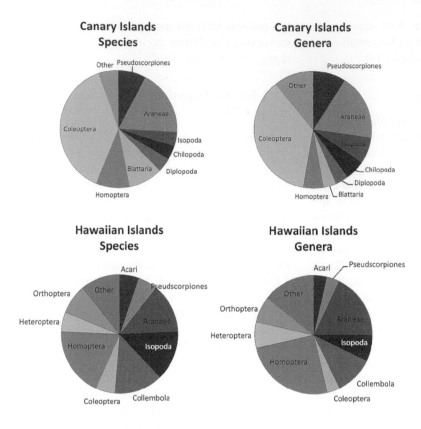

Figure 8.9 Pie charts of species and generic richness of troglobionts from the Canary Islands and Hawaiian Islands. There were 128 species and 55 genera from the Canary Islands, and groups listed as 'other' were Opiliones, Amphipoda, Collembola, Zygentoma, Dermaptera, and Heteroptera. There were 37 species and 27 genera from the Hawaiian Islands, and groups listed as 'other' were Amphipoda, Diplopoda, Dermaptera, and Lepidoptera. Data for the Canary Islands is from Oromí and Izquierdo (in press), and data for Hawaii is from Howarth (1987b) and Hobbs and Culver (unpublished).

A major difference between the two faunas is in overall species richness. While a number of undescribed species await description in both areas, it is most unlikely that species richness in the Hawaiian Islands will ever come close to rivalling the Canary Islands, even if all species were described in both areas. A number of factors may well explain this difference, including the particularities of the potential colonizing species that live on the island surface. Given the greater isolation of Hawaii, there may be a reduced number of potential colonists. The Canary Islands have a more arid climate, and this may have forced more species into lava tubes (Rando et al. 1993), at least according to the climate relict hypothesis (CRH). Additionally, and perhaps most importantly, the two island chains differ in their age. The Canary Islands are much older, ranging in age from 0.8 to 22 million years. By contrast, the Hawaiian Islands range in age from 0.4 to 5 million years (Palmer and Palmer 2009). Furthermore, the most biologically diverse lava tubes in Hawaii are on the island of Hawaii, which is only 400,000

years old, while the most diverse of the Canary Islands is Tenerife, 12 million years old.

In the case of the Canary Islands, there are relatively extensive data on the numbers of species on the different islands (Table 8.3). Some of the low numbers of species can be explained by the absence or rarity of lava tubes (MSS is still present), as is the case for La Gomera, Gran Canaria, and Fuerteventura. The remaining four islands have no discernible pattern with respect to either age, area, or a combination of the two. Tenerife, the island with the most troglobionts, is the largest island and most ecologically diverse. The statistical outlier is Lanzarote, which has no troglobionts even though it is intermediate in size and age among the Canary Islands. This may be because its caves are too dry to support troglobionts (P. Oromí pers. comm.). Island age may also be a factor in the case of the Azorean Islands, where the number of troglobiotic single-island endemics was marginally negatively correlated with island age (Borges and Hortal 2009). As islands age, lava tubes disappear and unless new flows create new lava tubes, the available habitat declines.

More insight can be gained by looking in detail at the faunal composition of individual caves (see also Box 8.1). Cueva de Felipe Reventón is one of the six most diverse caves for troglobionts, including limestone caves, anywhere in the world (Culver and Pipan 2013). Its species and generic composition (Fig. 8.10) is very similar in taxonomic proportions to the overall fauna of the Canary Islands. Beetles and spiders are especially diverse at both the species and generic levels. Especially interest-

---

Box 8.1 The perils and promise of single-site subterranean biodiversity studies

The global pattern of subterranean biodiversity has proven to be extraordinarily difficult to describe. The generally accepted pattern for caves is that species richness is highest at mid-temperate latitudes, and that the global hotspot(s) of subterranean biodiversity is the Dinaric karst of Italy, Slovenia, Croatia, Bosnia & Hercegovina, Serbia, and Montenegro, followed closely by the fauna of the French Pyrenees. However, this is by no means universally accepted. Some biologists such as Howarth (1972) think the tropics are under-represented and under-studied, and some Australian biologists hold that diversity is highest in Australia (Guzik et al. 2010), especially in calcrete aquifers (see chapter 5). Most descriptions of global subterranean biodiversity are largely based on species lists for continents (see Gibert and Culver 2009), but these species lists are incomplete, and based on wildly different collecting intensities. Except for soils (see chapter 7), no one has even suggested a global pattern of species richness for SSHs.

The primary difficulty with estimating species richness in subterranean habitats is that, with the notable exception of soils, single-site diversity (α-diversity) is a small proportion of regional diversity. For example, Malard et al. (2009) found that α-diversity in European aquatic subterranean sites (including caves, interstitial, and epikarst) was about 5 per cent of total European subterranean diversity, and that even within a small region (e.g. Belgian Wallonia) α-diversity was typically only about 20 per cent of the total. For most habitats, probably including the soil, α-diversity is much higher, and can be used as a reliable surrogate for overall diversity. This was how global patterns of soil nematode diversity were determined (Fig. 7.12, Boag and Yeates 1998).

Patterns of regional subterranean diversity can be determined with relative accuracy if a large number of samples of species richness are available. For example, Zagmajster et al. (2008) were able to produce a species richness map at the scale of a 20×20 km grid of cave beetle species richness, including 276 species and 1709 localities. This is an enormous data set, and this kind of detailed information is available for only a handful of cave regions and no SSHs.

An option is to use α-diversity for a few sites. α-diversity will work as a surrogate for regional diversity for the subter-

ranean fauna (or any biota for that matter), if the curves for the increase in species with area do not cross one another. Gibert and Deharveng (2002) showed that, for terrestrial species in different regions of the French Jura, local and regional species richness are strongly correlated ($r^2 = 0.79$). However, this is probably not the universal case. Dole-Olivier et al. (2009) showed that species accumulation curves (not the same as species area curves but in some ways analogous) did cross (see Box 2.2).

Based on the hypothesis that α-diversity can serve as a surrogate, Culver and Sket (2000) assembled a preliminary list of caves and wells with more than 20 stygobionts plus troglobionts. The pattern of these 20 sites more or less followed conventional wisdom—hotspots were concentrated at mid-latitudes, especially in the Dinaric karst. As is often true for lists, it generated a great deal of interest in the speleobiological community, and resulted in many species lists being assembled, and existing lists being updated. It may seem strange that these lists were difficult to assemble, but this was because most records were kept in taxonomic categories, rather than sites. The widespread digitization of distribution data resulted in a much-modified list, presented by Culver and Pipan (2013). They separated stygobionts and troglobionts. For stygobionts, there were 9 sites with 25 or more species. Five were from the Dinaric karst, three were deep aquifers, and one was a chemoautotrophic cave. For troglobionts there were six sites with 25 or more species—two from the Dinaric karst, one chemoautotrophic cave, two Canary Island lava tubes[3], and Mammoth Cave in Kentucky, the largest cave in the world. Note that species richness in lava tubes is as high or higher than most limestone caves.

More generally, the question arises as to whether the pattern of diversity of SSHs follows that of caves. There is some reason to think that this might be the case if the limiting factors are the same. This analysis remains to be done.

[3] An additional lava tube (Cueva del Sobrado) appears on Culver and Pipan's (2013) list, but this cave is connected to Cueva del Viento (P. Oromí pers. comm.)

**Cueva de Felipe Reventón
Species**

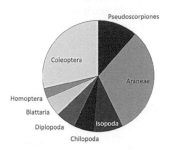

**Cueva de Felipe Reventón
Genera**

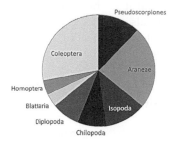

**Cueva de Felipe Reventón
troglophiles and trogloxenes**

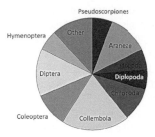

Figure 8.10 Pie diagrams of troglobiotic species, troglobiotic genera, and troglophilic plus trogloxenic species from Cueva del Felipe Reventón in Tenerife, Canary Islands. Data from Arechavaleta et al. (1999) and unpublished updates by Oromí.

ing is that there are six species of *Dysdera* spiders in the cave, adding significantly to overall species richness. What makes Cueva de Felipe Reventón a hotspot is not the large number of major taxonomic groups represented, but rather the number of species (and genera) within the Coleoptera and Araneae. This contrast can be seen when the fauna of Cueva de Felipe Reventón is compared with that of Kazumura Cave in Hawaii, the world's longest lava tube, but with a not particularly rich fauna (Fig. 8.11). Kazumura Cave has nearly the same number of arthropod orders (7) as Cueva de Felipe Reventón (8), but with only eight compared to 35 troglobiotic species.

Finally, the distribution of troglophiles and trogloxenes from Cueva de Felipe Reventón (Fig. 8.10) provides some insight into potential new colonists if they are exploiting a new resource (ASH) or an environmental catastrophe causes the surface population to go extinct (CRH). Most of the troglophiles and trogloxenes are in taxa that are well represented in subterranean habitats (e.g. Pseudoscorpiones, Araneae, Collembola) but some taxa have few or no troglobionts (e.g. Diptera and Hymenoptera). These

**Kazumura Cave
Species**

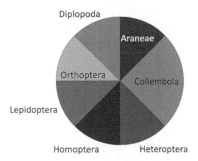

Figure 8.11 Pie diagram of troglobiotic species from Kazumura Cave, Hawaii. Data from Chapman (1985) and various literature sources.

troglophilic species are not 'troglobionts in training' but, at least based on their connection to subterranean habitats elsewhere, are important trogloxenes and troglophiles. Overall, the number of troglophilic species reported by Arechavaleta et al. (1999) is half that of the number of troglobionts (17 compared to 35). The number of trogloxenic species was 36, but this may include some accidental species as well.

8.4 Summary

Lava tubes are shallow caves, often with roots penetrating the ceiling, and with a lifetime much shorter than that of limestone caves. The original discovery of troglobionts in lava tubes caused Howarth to challenge the existing paradigm of a temperate zone cave fauna isolated by climate change during the Pleistocene. An even more diverse lava tube fauna in the Canary Islands was subsequently discovered by Oromí.

Lava tubes are concentrated at plate boundaries of the Earth's crust, and formed by surface cooling of lava or by sequential lava flows. The longest lava tubes are tens of kilometres in length with complex patterns, but most lava tubes have a simple structure. Density of lava tubes can be very high, higher than cave densities reported for karst areas.

There is rarely water in lava tubes, although dripping water may pass through in areas of high precipitation. Even though they are very close to the surface, environmental fluctuations, especially temperature, are often very small.

Organic carbon is brought into the cave by percolating water (where it exists), organic matter entering through cracks, and tree roots (especially *Metrosideros polymorpha* in Hawaii). Organic mats form in the wetter lava tubes and may play an important role in nitrogen cycling.

Species may colonize lava tubes either as the result of some climate change causing extinction of the surface populations (Climate Relict Hypoth-esis) or by active invasion and parapatric speciation (Adaptive Shift Hypothesis). The ability to survive the saturated atmosphere of lava tubes is likely an important selective factor. Strong evidence from phylogeographical studies exists for both hypotheses. It may be that colonization of lava MSS is an important intermediate step in the evolution of the lava tube fauna.

Of the various lava tube-dwelling species, planthoppers in the genus *Oliarus* have been most thoroughly studied. In addition to loss of eyes and pigment, planthoppers show convergent morphological changes in the tibia and tarsus associated with the ability to move on wet surfaces. Wessel et al. (2013) suggested that adaptation can be quite rapid, taking perhaps 10,000 years, and may be examples of the most rapid speciation reported for any animal group. They also claimed that the pattern of morphological differentiation among populations was the result of stochastic processes, rather than natural selection.

The two most well-studied and biologically diverse lava tube regions are the Hawaiian Islands and the Canary Islands. Both have groups not commonly found elsewhere, especially earwigs, sandhoppers, and thread-legged bugs. Overall, the Canary Island fauna is several times richer than that of Hawaii. Several factors may account for this, including the Canary Islands displaying a greater pool of potential colonists, greater age, and greater environmental severity, resulting in more cases of colonization via extinction of the surface ancestor (Climate Relict Hypothesis).

CHAPTER 9

The role of light in shallow subterranean habitats

9.1 Introduction

Almost by definition, subterranean, in a biological context, is taken to mean an aphotic environment. It is this absence of light that is the unifying feature of all subterranean environments. Thus, it is important to justify the inclusion of the most shallow of subterranean habitats, the hypotelminorheic, by demonstrating that it is aphotic, and we presented data from seeps on Nanos Mountain in Slovenia that showed the habitat was without light, at least to the detection limits of the data-logger. There are other habitats, ones that we have not discussed except briefly, that are best described as twilight habitats. Some of these were alluded to in chapter 7 (Fig. 7.1), including mosses, humus layer of the soil, ant nests, vertebrate burrows, and perhaps most importantly, leaf litter. Some of these may be aphotic, but in general they are characterized by low light intensity, or with patches of light and darkness—a variegated habitat. Some authors, especially Heads (2010), have even argued that these twilight habitats, especially the leaf litter (the stramnicolous habitat) are the site of the evolution of troglomorphy, including loss of eyes and pigment.

In this chapter, we consider the distribution of light with depth in a variety of situations, and summarize the distribution and behaviour of animals in response to light.

9.2 Gradients with respect to light

9.2.1 Physical gradients of light with depth

For two of the SSHs (calcrete aquifers and epikarst), there is no vertical gradient of light, and furthermore, there are no immediately adjoining photic habitats. Both calcrete aquifers and epikarst are buried in the soil, so that the habitat boundary does not correspond to a light–dark boundary. Lava tubes may fall into this category as well. In a lava tube, well away from an entrance, the boundary of the habitat (the top of the lava tube) is not at a light–dark boundary. The light–dark boundary is at the surface of the lava flow several metres above the tube. Lava tube entrances are a light gradient, but most invertebrates living in lava tubes are a considerable distance from an entrance, and are unlikely to encounter it. For organisms living in these habitats, light is not something they are likely to encounter during their lifetime.

For the other four SSHs, there is a gradient of light, with varying degrees of sharpness, that corresponds to the boundary of the SSH.

For two of the SSHs (hypotelminorheic and soil), there is a sharp gradient, and this occurs at the boundary of the habitat. For example, on Nanos Mountain (Slovenia), stygobionts were found at a depth of only 5 cm beneath the surface in a hypotelminorheic habitat, and there was no light

Shallow Subterranean Habitats. David C. Culver and Tanja Pipan.
© David C. Culver and Tanja Pipan 2014. Published 2014 by Oxford University Press.

detected for a period of nearly six months. Light gradients in the soil are even sharper, because of the relatively small dimension of air spaces in the soil. Ciani et al. (2005) reported that 99 per cent of light at a wavelength of 700nm disappeared (penetration depth) at between 120 and 300 μm. This sharp soil gradient may be modified if there is leaf litter. In this case, light likely penetrates a greater distance into leaf litter than it does in the soil. We have not explicitly included leaf litter as an SSH, but it is a boundary with a relatively shallow gradient. However, for soil inhabitants and for hypotelminorheic inhabitants, photic habitats are in close physical proximity, and are likely to be encountered during the animals' lifetime. In seepage springs, only the thickness of decaying leaf separates individuals (including stygobionts) from light.

For intermediate-sized terrestrial habitats, including the MSS, the gradient may be sharp or gradual, depending on the size of the rocks and debris that constitute the habitat, the amount of sediment clogging of the spaces between the rocks, and whether or not there is a moss covering on the surface of the MSS. At a moss-covered MSS site at Mašun (Slovenia), a data-logger failed to detect any light at a depth of 10 cm over a period of eight months (see chapter 1), and this site is undoubtedly at the sharp end of the gradient spectrum. As was the case with soil, leaf litter may create a gentler gradient of light, in this case between the MSS and the surface.

We have no data on light penetration of the hyporheic, but it seems likely that light penetrates only a few cm into stream and river beds. Rocks in a stream (the stream benthos) may be analogous to leaf litter in the sense it is variegated habitat with respect to light, and a twilight habitat.

9.2.2 Faunal gradients with respect to light

More than any other subterranean taxon, the subterranean amphipod genus *Niphargus* has been reported from surface waters (Ginet and David 1963, Fišer et al. 2006). Ginet and David (1963) discussed the presence of *N. rhenorhodanensis* in forest drainage canals in Dombes forest in France. Care must be taken about whether this is a surface-dwelling population or perhaps a hypotelminorheic population, because they were likely unaware of the hypotelminorheic, first described by Meštrov in 1962. A later description (Mathieu et al. 1997) of the Dombes forest habitat as 'composed of fine clay sediment from glacial deposits in association with leaf litter' suggests the possibility of a hypotelminorheic habitat immediately above the clay layer, with drainage canals acting to concentrate the fauna. This is similar to tile drains found in the USA (see chapter 2).

Fišer et al. (2006) discussed the occurrence in surface habitats of four narrowly endemic epigean species of *Niphargus* occurring in the Istrian Peninsula (Croatia) and nearby portions of the northern Adriatic Coast. These four species are not especially closely related phylogenetically, and thus represent four separate reinvasions of surface habitats. Their ecological distribution is summarized in Table 9.1. The most striking feature of the data in Table 9.1 is that surface habitats are not the most common habitat for any of the species except *N. vinodolensis*, which is only known from five localities. The

Table 9.1 Summary of ecological data for four narrowly endemic *Niphargus* species from the Istrian Peninsula and adjoining north Adriatic Coast. Data from Fišer et al. (2006).

Species	Geological basement (%)			Habitat type (%)						Total
	Limestone	Flysch	Unknown	Cave	Tunnel	Well	Spring	Surface stream	Unknown	
N. krameri	0.24	0.53	0.24	0.29	0.03	0.04	0.44	0.15	0.04	68
N. spinulifemur	0.06	0.78	0.15	0.08	0.02	0.00	0.52	0.25	0.14	65
N. timavi	0.50	0.50	0.00	0.19	0.00	0.00	0.63	0.19	0.00	16
N. vinodolensis	0.40	0.40	0.20	0.00	0.00	0.00	0.20	0.40	0.20	5

distribution areas of the species range from 60 km^2 (*N. vinodolensis*) to 1800 km^2 (*N. krameri*). Thus, they are surface-dwelling species but not primarily surface-dwelling species. The region itself is a complex landscape of karst and flysch (alternating layers of shale and sandstone) and some species occur on both rock types and some on one, but the limiting factor seems to be conductivity and hardness (Fišer et al. 2006). According to C. Fišer (pers. comm.), flysch may also be important as a limiting factor because it offers a number of cracks and voids where individuals can hide; in a sense, a type of SSH or at least twilight habitat. All of the species except *N. timavi* occur only in areas of intermediate to high conductivity and hardness, although this species is also found in caves.

Perhaps the most enigmatic example of a surface-dwelling population is a population of *Niphargus timavi* (Fišer et al. 2007), already discussed in chapter 2. *N. timavi* occurs in an upper stream (between two springs) without competitors or predators, but also shares a downstream section (below the second spring) with the amphipod *Gammarus fossarum*. *Gammarus* species are typically strong competitors with subterranean species, even being predators in some circumstances (Culver et al. 1991). But in this habitat, *N. timavi* was actually a more efficient predator than *G. fossarum* in laboratory experiments (Luštrik et al. 2011), and juvenile *N. timavi* were especially vulnerable to cannibalism and predation by both species. In the lower stream, *G. fossarum* dominated but both species were present for a distance of nearly 1 km.

It would seem to be self-evident that any SSH species, or more generally any subterranean species, that lacks eyes or pigment would be at a severe disadvantage in any photic environment. Direct exposure to light (and UV radiation) may have a direct negative effect on organisms such as planarians whose internal organs would be directly exposed to light, but this effect is unlikely to be important for most arthropods, whose exoskeleton shields the internal organs. Probably the main reason for the disadvantage of troglomorphic species in photic environments is the difficulty it would have in avoiding predators and competitors, the situation Culver et al. (1991) found with respect to the interaction between *Gammarus minus* and strongly troglomor-

phic amphipods and isopods, such as the isopod *Caecidotea holsingeri* (see also Culver 2012). Luštrik et al. (2011) found both the troglomorphic and non-troglomorphic species to be predaceous, but the troglomorphic *Niphargus timavi* still faced more predation, especially on juveniles, when *Gammarus fossarum* was present.

Sket (1986), in his study of the anchialine cave, Šipun near Cavtat (Croatia), provided some especially compelling examples of the role of competition and predation from surface-dwelling species in limiting the distribution of stygobionts. Šipun is a small cave (120 m) with a direct connection to and entrance on the shore of the Adriatic Sea (Ozimec 2012). Sket (1986) observed the stygobiotic amphipod *Niphargus hebereri* in intensely illuminated brackish water, but without surface-dwelling species. He went on to state that when surface-dwelling species are present, stygobionts like *N. hebereri* disappear. Sket (1986) argued that the failure of a subterranean population or species to survive in surface waters when surface-dwelling predators and competitors are present is a defining characteristic of troglobionts and stygobionts.

Given the proximity to light of species in most SSHs (with the exception of epikarst and calcrete aquifers), it would be advantageous for SSH species to avoid light in order to avoid predators, competitors, and UV radiation. Although light–dark responses of stygobionts and troglobionts is, in a sense, an old-fashioned topic (e.g. Vandel 1965), it is interesting with respect to SSH species. If avoidance of lighted surface habitats is adaptive for subterranean species, then a behavioural avoidance of light is advantageous for SSH species, but not necessarily for deep subterranean species, ones that will never encounter light. We know of no data that bear on this point.

In general, cases where troglomorphic species are found in surface habitats are unusual, and seem connected to the absence of predators and competitors in the surface habitat. The occurrence of non-troglomorphic species, ones that form reproducing populations (stygophiles and troglophiles) in subterranean habitats, one might suppose, would also be uncommon, because at least species without extra-optic sensory elaboration would be at a disadvantage with respect to foraging, mate finding, etc.

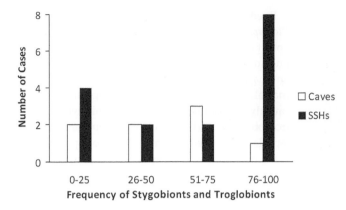

Figure 9.1 Bar graph of the relative frequency of different proportions of stygobionts and troglobionts in SSHs and caves, based on extensively sampled sites. Data from Pipan and Culver (2012a).

We have repeatedly seen that this is not the case. In the case of the hypotelminorheic fauna of the lower Potomac River basin, 55 per cent of the species regularly found in seepage springs are stygobionts (Table 2.6). In Slovenian epikarst sites, between 73 and 100 per cent of the species are stygobionts, and in Organ Cave (West Virginia, USA) the percentage drops to 40 (Table 3.10). In an MSS site in the Canary Islands, troglobionts only made up 14 per cent of the species (Table 4.8). In the Roseg River plain in Switzerland, the frequency of stygobiotic species compared to all permanent resident species was at most 40 per cent, and probably less because stygophiles were only listed according to family (Table 6.5). In the Danube Flood Plain National Park in Austria, a hotspot of stygobiotic diversity, 41 per cent of the resident species were stygobionts (Danielopol and Pospisil 2001, Culver and Pipan 2009). In soil sites in the Pyrenees, 40 per cent of the species were edaphobionts (Table 7.2). In la Cueva de Felipe Reventón (Canary Islands), a global hotspot of terrestrial subterranean biodiversity, 67 per cent of the resident species were troglobionts (chapter 8). Only in the case of calcrete aquifers (chapter 5) is the entire resident community almost invariably stygobionts.

One explanation for this pattern is that because of close surface connections, there are high levels of dispersal into SSHs, which explains the presence of non-specialized species in SSHs. This also is in accord with the idea of SSHs as ecotones (see section 1.4.1). However, this hypothesis cannot explain the widespread occurrence of stygophiles and troglophiles in deep subterranean habitats, which are not ecotones and for which dispersal is low (Pipan and Culver 2012a, Fig. 9.1). The similarity of patterns from caves and SSHs led them to conclude that light was a predominant selective factor in all these subterranean environments, more so than food resource scarcity or absence of environmental cycles (see chapter 12).

9.2.3 Biological clocks, light, and visual systems in SSHs

The biology of the subterranean beetle *Ptomaphagus hirtus* raises interesting questions about the relationship between troglobiotic and stygobiotic species and light, and whether light avoidance is the whole story with respect to troglobionts and stygobionts (Friedrich et al. 2011, Friedrich 2013). *P. hirtus* is a carrion beetle known exclusively from caves in and around Mammoth Cave National Park. It is part of the community that is dependent on the organic matter brought in by the active movement of animals in and out of cave entrances (see Culver and Pipan 2009), especially guano of the cave cricket, *Hadenoecus subterraneus*, that leaves the cave at night to feed on the surface (Peck 1975). Typically, *P. hirtus* is found near cave entrances, but in the aphotic zone. It is small in size (2.0–2.8 mm) and shows relative little modification for subterranean life in terms of life cycle (Peck 1975) or appendage lengthening (Peck 1986). However, it is certainly troglobiotic and moderately troglomorphic, especially with loss of the hindwings and the near complete reduction of the compound eye to a small lens patch.

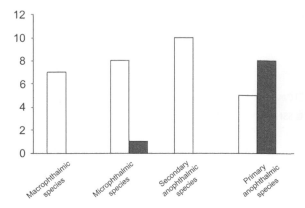

Figure 9.2 Retention of biological clock control of activity patterns across cave-dwelling species. Secondary anophthalmic species refers to adult-specific and population-specific anophthalmic species. White bars: species with circadian rhythms. Shaded bars: arrhythmic. Data from Friedrich (2013).

Using a variety of techniques, Friedrich et al. (2011) found that *P. hirtus* retained (1) a reduced but functional visual system, (2) a negative phototactic response to light–dark choice tests, and (3) expression of the complete circadian clock network. These were surprising results for a troglobiotic species in the old Mammoth Cave system (at least 3.5 million years [Brucker 2012]). The retention of a functional visual system and a circadian clock network suggests adaptation, rather than simply regressive evolution, is involved. Friedrich et al. (2011) suggested that it may be advantageous for *P. hirtus* to avoid light and hence surface predators, but also to be attracted to diffuse light, because their main food supply is concentrated near the cave entrance. In addition, the retention of expression of the clock genes suggests an adaptive advantage of entrainment of locomotory activity (Oda et al. 2000, Lamprecht and Weber 1992). Curiously, *P. hirtus* has lost the seasonal aspect of reproduction (Peck 1975) and the role of the circadian clock is unclear. What is clear is that *P. hirtus* encounters light, and that it is certainly not a deep cave species, but a species of the dark zone near entrances. Its small size makes it tempting to hypothesize that it is also a species of the MSS. The difficulty is that that no MSS sites are known in the area, but no one has really looked. The Mammoth Cave area is a series of plateaus with potential for erosional MSS along incised river valleys.

Friedrich (2013) suggested that the case of *Ptomaphagus hirtus*—a microphthalmic species with an intact circadian clock system—was a common one among troglobionts and stygobionts. Based on an analysis of 40 cavernicolous species, he found that almost all species with macrophthalmic eyes (mostly troglophiles), microphthalmic eyes (mostly troglobionts), and secondarily anophthalmic eyes (species with eyed juvenile stages or species with some populations with eyes) had a circadian clock system (Fig. 9.2). Primary anophthalmic species were mixed with respect to circadian clock control of activity, with slightly more than half being arrhythmic. The differences in the amount of eye degeneration may reflect differences in age of isolation in subterranean habitats, but it may also reflect a difference in habitats. It is possible that many of the microphthalmic species are SSH species and many of the anophthalmic species are deep subterranean species. In any case, Friedrich (2013) has revived interest in circadian clocks in subterranean species.

9.2.4 Distribution of eyelessness among habitats

Peck (1990) provided a unique insight into the role of light in structuring communities with his complete enumeration of the eyeless species found on the Galapagos Islands, and categorizing the habitats in which eyeless species were found. He appropriated the word Cryptozoa to include all those species which 'live hidden lives in spaces at or beneath the earth's surface'. This is a useful categorization and implicitly includes as obligate subterranean inhabitants all the eyeless arthropods. Habitats include deep subterranean habitats (although they are largely missing on the Galapagos), SSHs, and some twilight habitats with eyeless species. One could quibble that some eyeless species are eyeless because they descended from eyeless species rather than are adapted to their environment, but these are certainly exceptions rather than the rule. Peck (1990) identified a total of 51 eyeless terrestrial arthropods and divided them into four categories (Table 9.2):

1. Widespread eyeless species, probably introduced by humans in soil; none is endemic to the Galapagos.

Table 9.2 Summary of terrestrial eyeless species on Galapagos Islands. A species may be found in more than one habitat. Data from Peck (1990).

Category	Number of species	Habitats			
		Soil	Lava tubes	Littoral	Litter
1. Introduced eyeless species	16	16	3	1	0
2. Eyeless species in widespread eyeless taxa	10	10	4	2	0
3. Eyeless species with eyed Galapagos relatives	10	6	5	0	(a) 3
(b) 4. Endemic and relict eyeless species	(c) 15	(d) 4	(e) 9	(f) 4	(g) 0

2. Eyeless species in widespread eyeless taxa; ancestors dispersed to Galapagos in eyeless condition.
3. Eyeless and reduced eyed fauna with eyed epigean relatives in the Galapagos (often the same island) and possible examples of parapatric speciation.
4. Eyeless fauna with no close Galapagos relatives; eyelessness evolved *in situ*; endemics and relicts.

Nearly all of the first two categories are soil species, but the last two include a number of species found in lava tubes and to a lesser extent, leaf litter. This evolutionary connection between the soil fauna and the lava tube fauna points out the advantage of considering all subterranean species, not just those that fit some arbitrary habitat definition. Parapatric species (and adaptive shift) are important in category 3, and some of category 4, but category 4 also fits the climate relict model with allopatric speciation.

9.3 Summary

Absence of light is the defining feature of subterranean environments. Different SSHs have different spatial relationships with light. For epikarst and calcrete aquifers, there is no light, even at the boundary of the habitat; the same holds for lava tubes except near an entrance. For the other four, there is a light–dark gradient at the boundary of the habitat. For the MSS and the hyporheic, the gradient may be less abrupt. There are also twilight habitats, especially leaf litter.

Most eyeless and depigmented animals are rarely found in photic habitats. Some reports in the literature of surface occurrences are actually occurrences from SSHs, that were unrecognized as such. In those cases where they do occur in surface habitats, competitors and predators are rare or missing. Surface-dwelling populations of the amphipod *Niphargus*, a subterranean genus sometimes found in surface habitats, retain close connections to SSHs, where they also occur.

The converse, the presence of eyed species in SSHs, is a common occurrence, as is the occurrence of eyed species in deep subterranean habitats. The retention of some visual apparatus may even be advantageous when organisms are in a situation where they need to avoid light as well as to retain a circadian clock. This is apparently the case with the Mammoth Cave carrion beetle, *Ptomaphagus hirtus*, which retains the ability to detect light and has a circadian clock.

Peck (1990) grouped together all eyeless and depigmented species into the cryptozoa, which occur in aphotic and in some cases dimly lit habitats. In a complete enumeration of such species on the Galapagos Islands, Peck showed that their origin is from soil, often brought by human transport, and that they subsequently invade other SSHs, especially lava tubes.

CHAPTER 10

Environmental fluctuations and stresses in shallow subterranean habitats

10.1 Introduction

The literature on the role of the physical environment on the evolution, ecology, and physiology of subterranean organisms is in many ways self-contradictory. Firstly consider the hypotheses put forward to explain the colonization of subterranean habitats, and historically this has meant the colonization of caves in particular. According to the Climate Relict Hypothesis (CRH), subterranean populations become isolated because surface environmental conditions no longer can support populations of the species. The subterranean environment is a refuge and in this most fundamental of meanings, a more benign environment than the surface. Originally, the CRH was developed to explain the occurrence of species in temperate zone caves that became isolated following Pleistocene glacial advances as a result of rising temperatures (Barr and Holsinger 1985). Since the temperature of caves and other subterranean habitats approaches the mean annual temperature, and with less variability (see section 10.2.1), subterranean habitats are a refuge from temperature extremes. The CRH can also be extended to cover increases in aridity, as Leys et al. (2003) did for the dytiscid beetles that colonized calcrete aquifers in western Australia. Likewise, the Adaptive Shift Hypothesis (ASH) includes a component of the subterranean environment that is more favourable than the surface environment. The classic example of this is the availability of subterranean roots of *Metrosideros polymorpha* for planthoppers in the genus *Oliarus*

(Howarth 1987a, Wessel et al. 2013), a food source not available on the surface. In the case of lava tubes and perhaps calcrete aquifers, SSHs that occur in an environmentally extreme surface landscape (see Figs. 1.4 and 1.13), physical environmental conditions in the SSHs are considerably more benign than those of the surface, and as a result, species may actively colonize them.

In sharp contrast to the view that the subterranean environment is a refuge is the observation that it is environmental parameters that make for an extreme environment, one to which it is difficult for species to successfully adapt. The obvious extreme parameter is absence of light (discussed in chapter 9), but there are others. In his discussion of the colonization of lava tubes, Howarth (1980) argued that the relative humidity of lava tubes represents a formidable barrier to colonization, because terrestrial arthropods must have a way of maintaining water balance, or else they run the risk of becoming waterlogged. Hadley et al. (1981) provided empirical data for *Lycosa* spiders that demonstrated that the troglobiotic species they studied was modified for life at high humidities. This was the result of cuticular thinning and decreased lipids in the cuticle, both of which increase rates of water transfer. In a somewhat similar vein, Christiansen (1961, 1965) discussed the behavioural and morphological changes required for Collembola to successfully move across wet surfaces and water pools, habitats that exist because of the high humidity of caves.

Not only are there individual factors, such as absence of light and high relative humidity, that are

Shallow Subterranean Habitats. David C. Culver and Tanja Pipan.
© David C. Culver and Tanja Pipan 2014. Published 2014 by Oxford University Press.

barriers to successful adaptation to subterranean life, but also the pattern of fluctuation, and specifically the reduction of cyclicity, may make successful adaptation to subterranean life difficult. Early speleobiologists, e.g. Racoviţă 1907, emphasized the environmental constancy of subterranean habitats, and this led later authors, including Poulson and White (1969), to emphasize the barrier such constancy (and absence of cyclicity) posed to colonization of subterranean habitats.

Circadian rhythms (daily endogenous rhythms) play a central role in the organization of physiology and behaviour of most organisms. Lamprecht and Weber (1992) pointed out that this advantage includes external (ecological) and internal factors. External factors include the concentrating of interactions, such as mating. Internal factors, in particular the separation of temporally incompatible metabolic reactions such as synthesis and utilization of particular compounds (Lamprecht and Weber 1992), may pose a formidable barrier to successful colonization of subterranean habitats (Jegla and Poulson 1968). The loss of environmental cues that reset the circadian clock may be a barrier to colonization. Thus, there has been considerable interest in determining whether or not stygobionts and troglobionts have retained a circadian clock (Oda et al. 2000, Friedrich 2013). While some species (such as the beetle *Ptomaphagus hirtus* studied by Friedrich et al. [2011]) have an entrainable circadian clock, other species, such as the cave fish *Amblyopsis rosae* and *Typhlichthys subterraneus*, displayed no entrainable circadian clock (Poulson and Jegla 1969). In other species, such as the crayfish *Orconectes pellucidus* (Jegla and Poulson 1968), some, but not all, individuals had a circadian clock. While there are other mechanisms to organize metabolic activities, such as rest homeostasis (Deboué and Borowsky 2012) and 'stochastic regularities' (Lamprecht and Weber 1992), that serve to organize rest–activity patterns, circadian rhythms are widespread and important for many subterranean species (Lamprecht and Weber 1992, Friedrich et al. 2011, Friedrich 2013). The presence of daily cycles in physical parameters which can reset the circadian clock (zeitgebers) is of considerable interest in this context.

Not all speleobiologists emphasized environmental constancy. In particular, Hawes (1939), based on observations of caves and poljes in the Dinaric karst, argued that annual flooding was vital to timing of reproduction, availability of food, and dispersal and colonization into caves. In essence, Hawes (1939) argued that the annual cyclicity of subterranean environments was important, and by implication, a circadian clock was not important in the context of annual cycles. Hawes' direct observations of flooding were also unusual because caves are rarely visited by explorers or biologists during times of flood. Typically, access is difficult and dangerous, and collecting is nearly impossible due to high water. The absence of direct observation of flooding and other seasonal changes led to their being underemphasized.

10.2 Environmental variation in SSHs

10.2.1 Overall patterns in SSHs

Table 10.1 summarizes temporal variation in SSHs for hourly data, generally for one year or more. Most of the data are for temperature, but one long record for conductivity and discharge is included. Overall variability is represented by the coefficient of variation, which is independent of the mean. When a comparator habitat was available (either the surface, a shallower SSH site, or a cave), the coefficient of variation was expressed as a percentage of the comparator habitat. Extreme variability is measured by the range, in particular the percentage of the range relative to the comparator value. Examples are listed for all SSH types except for calcrete aquifers and the soil, for which no data were available.

For all SSHs, except for one record for terrestrial epikarst (Jama v Kovačiji in Slovenia), coefficients of variation were less than 100 per cent, usually considerably less. In a few cases (lava tubes), coefficients of variation were less than 10 per cent. Relative to surface habitats and shallower subterranean habitats, coefficients of variation were smaller, with the exception of two temperature records for hypotelminorheic habitats on Nanos Mountain, Slovenia. In this case, the surface habitat comprised entirely the runoff of a series of seepage springs, and water had been on the surface less than an hour.

Table 10.1 Summary of environmental variability of SSHs, based on hourly records, mostly of temperature. Cyclicity was determined by spectral analysis and the occurrence of a peak at 24 hours (see Pipan et al. 2011a). For aquatic epikarst sites, depths are cave ceiling thicknesses, an overestimate of the depth of the epikarst because it also includes the upper vadose zone.

Site	Location	Variable	Depth	Mean	CV	% of surface	Cyclicity	% of surface range	n	Source
Hypotelminorheic										
Nanos	Slovenia	temperature	< 10 cm	10.65	37.12	106.21	Yes	80.98	8626	unpublished
Nanos	Slovenia	temperature	< 10 cm	9.45	56.56	161.83	Yes	96.02	8626	unpublished
Nanos	Slovenia	Temperature	< 10 cm	11.4	30.30		Yes		3773	Table 2.1
Nanos	Slovenia	conductivity (μS cm^{-1})	< 10 cm	280.6	19.60		Yes		3773	Table 2.1
Prince William Park	Virginia, USA	temperature	< 10 cm	12.8	38.15	76.68	Yes	72.92	7274	Fig. 2.4
Epikarst										
Pivka jama	Slovenia	Temperature	50 m	9.21	1.17	3.59[1]	No	2.83[1]	19,394	Table 3.3
Postojnska jama	Slovenia	Temperature	100 m	9.6	1.69	1.89	No	1.91	8408	Table 3.2
Postojnska jama	Slovenia	discharge (ml min^{-1})	100 m	833.7	89.42		No		8242	Table 3.2
Jama v Kovačiji	Slovenia	temperature (terrestrial)	20 cm	1.23	220.3	100.13[2]	Yes	123.09[2]	9889	Table 4.4
MSS										
Jama v Kovačiji	Slovenia	Temperature	20 cm	5.38	64.27	29.21[2]	Yes	123.30[2]	9889	Table 4.4
Mašun	Slovenia	Temperature	10 cm	7.46	71.29		Yes		9889	Table 4.5
Mašun	Slovenia	Temperature	20 cm	7.38	64.54	90.53[3]	Yes	87.77[3]	9889	Table 4.5
Mašun	Slovenia	Temperature	30 cm	5.38	64.27	90.15[3]	No	72.75[3]	9889	Table 4.5
Mašun	Slovenia	Temperature	50 cm	6.91	55.61	78.01[3]	No	67.55[3]	9889	Table 4.5
Teno	Canary Islands	Temperature	70 cm	12.5	17.00		No		10,055	Table 4.2
La Guancha	Canary Islands	Temperature	70 cm	10.38	31.17		No		10,029	Table 4.3
Hyporheic										
Méant	France	Temperature	10 cm	13.1	23.7		Yes		10,637	Table 6.2
Méant	France	Temperature	50 cm	13.0	21.4	90.30[4]	Yes	73.49[4]	10,637	Table 6.2
Morcille	France	Temperature	10 cm	10.9	32.6		Yes		10,617	Table 6.2
Morcille	France	Temperature	50 cm	10.9	25.8	79.14[5]	Yes	70.29[5]	10,617	Table 6.2
Lava tubes										
Cueva del Mulo	Canary Islands	Temperature	< 5 m	11.49	9.80		Yes		10,029	Table 4.3
Cueva del Viento	Canary Islands	Temperature	< 5 m	13.92	5.06	28.46	No	18.32	10,029	Table 8.4
Pahoa Cave	Hawaii, USA	Temperature	< 10 m	19.91	1.38	9.41	No	4.20	8570	Table 1.1
Pahoa Cave	Hawaii, USA	Temperature	< 10 m	20.01	1.50	10.23	No	4.29	8570	Table 1.1

[1] Comparison is to underground Pivka River.
[2] Comparison is Jama v Kovačiji, a cave with perennial ice.
[3] Comparison is to Mašun at − 10 cm.
[4] Comparison is to Méant at − 10 cm.
[5] Comparison is to Morcille at − 10 cm.

A particularly consistent and striking pattern is that the reduction in the range of values is always greater than the reduction in coefficients of variation for SSHs, compared to surface or shallower subterranean sites. The only exception was temperature in an epikarst drip in Postojnska jama, and in this case both range and coefficient of variation were very small relative to surface variation. SSHs are good places, then, to avoid temperature extremes. It is not known if it is short spikes and dips in temperature or other environmental parameters that either push organisms into subterranean habitats or cause surface extinctions, but it is a pattern worth following up.

Not surprisingly, given their superficial location, many SSH sites showed a daily cycle of temperature. The shallowest habitat types, the hypotelminorheic and hyporheic, invariably had daily cycles. Shallow MSS sites (20 cm or less in depth) showed cycles and deeper ones did not. Most of the epikarst and lava tube sites did not show daily variation. The exceptions were a terrestrial epikarst site at Jama v Kovačiji (Slovenia), which was a very shallow site that was the equivalent to an MSS site, and an old lava tube (Cueva del Mulo, Canary Islands) without an aphotic zone.

Annual cycles are not apparent from the summary in Table 10.1, but examination of the graphs of the data (see Figs. 1.8, 1.9, 2.4, 2.5, 4.10, 4.11, 4.12, 6.7, and 8.3) shows a clear annual cycle with the exception of Pahoa Cave, Hawaii (Fig. 1.8), but there is only a very weak seasonal cycle on the surface around Pahoa Cave, and on the surface daily fluctuations are much greater in amplitude than seasonal differences.

In the following sections, we explore the details of environmental values and fluctuations for the different SSH types, except for calcrete aquifers and soil, for which we have no data.

10.2.2 Environmental variation in hypotelminorheic habitats

Even though this is the most superficial of SSHs, there is still a significant reduction in temperature variation, most clearly seen in the comparison between a seep and surface stream 5 m away, in Prince William Forest Park, Virginia, USA (Fig. 2.4). Although limited data are available, it would seem that conductivity is also quite stable (Table 10.1). What is not so stable is the long-term persistence of individual seepage springs. Given their superficial nature and lack of connection with bedrock, it is difficult to imagine an individual seepage spring lasting centuries let alone millennia. We have observed the disappearance or repositioning of several seepage springs over the decade or so during which we have been observing them. Even persistent hypotelminorheic and seepage spring habitats experience fluctuations in water level, as was documented in Fig. 2.5 for a seepage spring on Nanos Mountain, Slovenia. In the lower Potomac River basin (USA), most seeps have little or no water flow during the growing season when evapotranspiration is high. During this time, sampling is nearly impossible. On the other hand, the hypotelminorheic upstream of the seepage spring may be more stable. The existence of a hypotelminorheic fauna, and a rather rich one at that, in areas like the lower Potomac River watershed (Culver et al. 2012a) which are far from any other SSHs or caves, strongly suggests the ability of the inhabitants to survive episodic drying. The clay layer underlying hypotelminorheic habitats likely serves a dual role of aquiclude and habitat that retain enough water for species to survive drought, resulting in a local perched aquifer.

10.2.3 Environmental variation in epikarst habitats

Temperature data from both Pivka jama and Postojnska jama (Table 10.1) demonstrate that water in epikarst shows very little temporal variation. Coefficients of variation are less than two per cent. The length of time epikarst water has been underground varies, but it is measured in weeks and months, or even years (Kogovšek 2010), so such stability is not unexpected. What does vary is the discharge rate from epikarst drips, and by implication, the velocity of water moving through the epikarst. Annual variation in precipitation results in changes in the size of the area drained by an individual drip (Table 3.1, Kogovšek 2010) so population sizes are likely to change as well. Pipan and Culver (2007a) hypothesized, based on travel distances and times, that velocities are sufficient to dislodge copepods, in a manner similar to particle entrainment and mobilization in stream flow, although active drift may be involved as well (Moldovan et al. 2011). What seems impossible to explain by any mechanism other than miniature-scale catastrophic flooding is the appearance of terrestrial species in drip waters (see Table 3.11). Water chemistry, including the amount of dissolved organic carbon, also varies spatially and temporally, but this variation seems to be small relative to the variation in water velocity and the fractions of the epikarst aquifer that are water-filled and air-filled.

Other factors vary temporally as well, including most cations and anions (Pipan 2005a). Musgrove and Banner (2004b) showed that for some drips, the stable isotope ratio $^{87}Sr/^{86}Sr$ was nearly constant over a four-year interval while for other drips it varied considerably. The ratio reflects, among other things, residence time of the water underground, and indicates different residence times for different drips. Ban et al. (2008) demonstrated that dissolved organic carbon concentrations in Shihua Cave, China, vary between 0.1 mg L^{-1} and 3.0 mg L^{-1} and that peak concentrations occurred after precipitation events. While these precipitation events were annual and were not floods in surface habitats (although they may have flooded the epikarst), they pointed to the importance of episodic 'floods' in the input of organic carbon and nutrients, in a way analogous to Hawes' (1939) suggestion of the importance of floods. In temperate zone caves like Postojnska jama in Slovenia studied by Kogovšek (2010), there is typically an annual cycle of precipitation, but one that varies from year to year.

In contrast with the hypotelminorheic, where individual habitats probably persist on the scale of decades or centuries, epikarst is much more persistent. Eventually epikarst will erode (Šušteršič 1999) but its persistence is orders of magnitude longer than seepage springs.

10.2.4 Environmental variation in intermediate-sized terrestrial SSHs

MSS sites vary both temporally and spatially. The deeper the position in the MSS (see Mašun in Table 10.1), the less temperature variation occurs, and the weaker the daily cycle becomes, until it disappears at around 30 cm. Temperatures were generally less variable with depth, with the coefficient of variation reduced approximately 20 per cent and the range 30 per cent at a depth of 50 cm, compared to a depth of 10 cm (Table 10.1). The damping effect of depth is also very striking for uncovered MSS sites, such as the talus slope at on Lovoš Hill in the České Středohoří mountains, Czech Republic (Fig. 4.8). Surface temperatures varied much more than air temperatures, and ranged between 28 °C and nearly 50 °C. Temperatures at depths between 45 cm and 120 cm only varied between 16 °C and 19 °C.

On an annual basis, temperatures in the MSS of the two Slovenian sites dropped below 0 °C at least for the shallower sites (Tables 4.4 and 4.5). This is in accord with Crouau-Roy's (1987) finding that the beetle *Speonomus hydrophilus* disappeared from upper layers of the MSS at the Tour Laffont site in the French Pyrenees in the winter months (Fig. 4.21A). It is likely that seasonal vertical movements of the MSS fauna are typical for temperate zone MSS sites.

It is difficult to put any age limits on MSS habitats, but Gers (1992) hypothesized that, at least in the high elevation sites in the French Pyrenees that he studied, the spaces of the MSS became clogged with soil after a few centuries. On the other hand, some of the clinker sites on the Canary Islands (Fig. 4.9) have persisted since the original lava flow, certainly at least tens of thousands of years old, and probably more.

10.2.5 Environmental variation in interstitial habitats along rivers and streams

Temperature variation at both French interstitial sites—Méant and Morcille—was generally low (Tables 6.2 and 10.1), with coefficients of variation less than 33 per cent. As was the case with MSS, variability declined with depth, but a daily cycle was still detectable at sites at a depth of 50 cm (Table 10.1).

What is clearly important is the annual cycle, especially the episodes of flooding. For animals in a stream and in the shallow sediments, a flood is a major environmental challenge, and the deeper sediments of the hyporheic and phreatic zones are a refuge from the flood. Current velocities during floods are often enough to mobilize sediments the size of small rocks (see Fig. 3.11 for the Hjulstrom curve which determines velocities required), and hence refuges are especially important. As discussed in chapter 6, the hyporheic refuge hypothesis (HRH) proposes advantages for hyporheic colonization, but in practice it has been difficult to demonstrate, in part because of the high degree of spatial heterogeneity of the hyporheic (e.g. upwellings and downwellings, Dole-Olivier 2011). In any case, floods have a major impact on distribution of animals, even in the hyporheic (Fig. 6.14). As this example shows, floods drive stygobionts such as *Niphargus* and *Niphargopsis* deeper into the sediments, while typical benthic species like *Gammarus* invade the upper layers of the hyporheic zone. A particularly striking example of the effect of flooding on the hyporheic zone was the change in size of upwellings of the Roseg River, Switzerland, a glacial-fed river. The upwelling zone, which brings stygobionts closer to the surface, grew in extent as melt and flooding increased in the summer, and then contracted in the fall (Fig. 6.6).

In general, the hyporheic will persist as long as the stream or river persists. Some of the richest hyporheic sites are in cutoff meanders of rivers and these of course eventually fill in. However, the interstitial habitat in general is both ancient and persistent. Many stygobionts in sandy interstitial aquifers near the Mediterranean are ancient lineages, dating back to the Tethys and before, suggesting habitat longevity as well (Coineau and Boutin 1992).

10.2.6 Environmental variation in lava tubes

Although they are very shallow, lava tubes are among the most stable of SSHs. Probably only calcrete aquifers, about which we have no data, are as stable. In the three caves monitored in Hawaii and Tenerife (Canary Islands), coefficients of variation were less than 10 per cent over the year (Table 10.1). Relative to surface temperatures, coefficient of variation and especially the range were a small fraction of the surface range. Given this stability, it is not surprising that only in small Cueva del Mulo was any daily cycle detected. Both Tenerife lava tubes had an annual cycle, while Pahoa Cave in Hawaii did not. The surface annual cycle was also stronger at the Tenerife sites.

Long-term stability of the habitat is moderate compared to other SSHs. Palmer (2007) states that most lava tubes are several thousand years old, but this is largely based on data for Hawaii, and caves in the Canary Islands may be older. Nonetheless, in geological terms lava tubes are highly transient. The food resources in lava tubes, especially *Metrosideros polymorpha*, which are a critical component of the food web (Stone et al. 2012), are an early successional tree on lava flows (Wessel et al. 2013), and may disappear before the lava tube is eroded away or covered by new flow. Just because lava tubes are generally young, this does not mean that the habitat has not been present much longer, and as new flows cover old lava tubes, new lava tubes are created. Thus, one cannot say that the fauna of a 500-year-old lava tube, such as Kazumura Cave in Hawaii is only as old as the cave in terms of the length of time the lineage has been underground. It could well have been colonized by species from older lava tubes.

10.3 Are SSHs extreme environments?

An extreme environment can really only be defined by the organisms that are potential inhabitants, a point raised by Levins and Lewontin (1985). It is a common observation that most SSHs harbour a number of species that are transient, and, at least in many cases, were not forced into SSHs. For example, at the Teno MSS site on Tenerife, Canary Islands, the majority of species were generalists, neither troglobionts nor edaphobionts (Table 4.9). Most of the individuals of these generalist species were in the MSS because in some sense it was a favourable environment, relative to other choices available. The only exceptions to the widespread presence of non-specialists in SSHs are calcrete aquifers, presumably because the habitat is largely inaccessible from the surface.

Whether these non-specialized species represent potential permanent residents of the SSH is another matter entirely. It is likely that many of these species only utilize SSHs for part of their life cycle, or even part of the day, and their transition to stygobionts or troglobionts would seem unlikely. Such temporary residents also avoid what are the two most extreme aspects of SSHs. The first of these is the absence of light (chapter 9), and the second of these is the scarcity of organic carbon and nutrients (chapter 11). Other environmental conditions, especially relating to cyclicity and severity of conditions, seem unlikely to present a barrier to colonization and survival.

10.4 Summary

The subterranean environment is typically viewed in two contradictory ways. One is that the subterranean environment is a refuge from unfavourable surface conditions, or an opportunity to exploit new resources. The other is that it is an extreme environment, with many environmental barriers to successful colonization. These barriers include not only the absence of light but the absence of cycles that allow synchronization of physiological processes. The emphasis of most early speleobiologists was on the constancy of the environment, rather than any variability, cyclic or otherwise. A notable exception was Hawes' emphasis on the importance of annual flooding of caves.

As measured primarily by detailed temperature records, SSHs showed low overall variability, considerably less than that of nearby surface sites. Temperature range was reduced by a greater amount than was the coefficient of variation, and SSHs effectively buffered against temperature extremes. The shallower SSH sites showed a daily temperature fluctuation. All sites with the exception of Pahoa Cave, Hawaii, showed an annual cycle. There was only a weak annual cycle on the surface at Pahoa Cave.

One important temporally varying feature of the hypotelminorheic is drying and re-wetting, and impermeable clay layers are most likely to serve as refuges during episodes of drying. As with temperature in the hypotelminorheic, epikarst temperatures are very constant, but water chemistry and current velocity show relatively large variation. The presence of terrestrial animals in drip collections also indicates that small epikarst cavities and solution pockets periodically flood and dry out. Organic carbon seems to enter the epikarst largely after precipitation events.

Temperature fluctuation in MSS sites was much less than that of the surface, even at depths of 10 cm. However, MSS temperatures during winter months dropped below 0 °C at several Slovenian MSS sites. This likely resulted in seasonal vertical movement of beetles and other arthropods, as was reported by Crouau-Roy for sites in the Pyrenees.

For SSHs along and under rivers and streams (e.g. hyporheic), temperature varied little, but there was still a daily cycle at depths of at least 50 cm. A major environmental factor in the hyporheic is flooding, and it has been proposed that the hyporheic is a refuge from surface flooding (hyporheic refuge hypothesis), but evidence for this is ambiguous (Dole-Olivier 2011). In any case, flooding results in major distributional changes, especially the downward movement of stygobionts.

Lava tubes are among the least variable of SSHs, although an individual lava tube has a relatively short life time, lasting on average only a few thousand years.

SSHs harbour a number of species that are either transient, or also occur in surface habitats. SSHs for these species are likely refuges from temporarily unfavourable surface conditions, but these species may or may not be incipient colonists.

CHAPTER 11

Organic carbon and nutrients in shallow subterranean habitats

11.1 Introduction

Along with the absence of light (chapter 9), reduction in cyclicity, and extreme values of some physico-chemical parameters (chapter 10), reduction in food resources is widely viewed as a major barrier to adaptation to subterranean life. One of the most well-known examples of adaptation comes from the amblyopsid cavefish, for which Poulson (1963) showed that the most morphologically modified (troglomorphic) species had the lowest metabolic rates. Hüppop (2000) summarized information, not on availability of organic carbon and nutrients, but on metabolic rate, starvation resistance, and other physiological comparisons between subterranean and surface-dwelling species. Hüppop (2000) listed a number of cases where physiological differences can best be explained by adaptation to low food. This pattern was by no means universal, with counter examples of no differences in metabolic rates in subterranean and surface-dwelling species, including lava tube species (Ahearn and Howarth 1982). Even the classic example from the amblyopsid fish is not so straightforward because recent molecular phylogenetic analysis has shown that neither morphological change nor metabolic reduction is correlated with subterranean age (Niemiller et al. 2012).

Development of techniques for the automated analysis of dissolved and total organic carbon (DOC) using the persulphate digestion method (American Public Health Association 1999) has led to quantitative, comparative assessments of available organic carbon, at least that component in water. We use those studies that have been done in SSHs to begin to make some sense of the pattern of organic carbon availability.

One shared characteristic of all SSHs is that they rely on allochthonous organic carbon and nutrients since by definition, there is no photosynthetic production in SSHs. There may be cases of chemoautotrophic production in SSHs, but none have been documented, and we focus on the organic matter brought from photic habitats. Movements of material or energy across habitats are called spatial subsidies (Polis et al. 1997). For example, spatial subsidies are important in detritus-based ecosystems, and can serve to organize and constrain the detritivore community.

Schneider et al. (2011) pointed out that subterranean habitats are useful model systems to study spatial subsidies because they have no autochthonous production. Their model system was a series of twelve shallow pits (5–10 m deep) in Greenbrier County, West Virginia, USA, with short lateral passages, less than 25 m in length, at the bottom. Technically, these are SSHs because they are shallow, but they do not fit into any of the categories we have discussed (but see chapter 16). Firstly, they removed all visible organic matter, including scrapings of the floor of the pit, from all 12 sites. This totalled 1.5 tonnes (metric tons), which is evidence that, in their natural state, rather large quantities of organic matter were present. They then added either leaf litter packs or frozen, dead rats to the shallow pits (in six pits rats or litter packs were added at two separate sites), and censused the invertebrates at one-month intervals for 23 months. This was a large-scale study, and a total of 20,000 individuals

Shallow Subterranean Habitats. David C. Culver and Tanja Pipan.
© David C. Culver and Tanja Pipan 2014. Published 2014 by Oxford University Press.

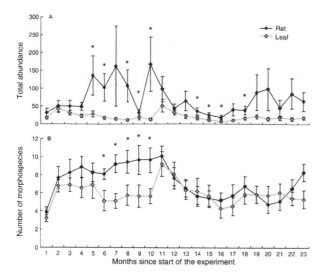

Figure 11.1 Overall trends in abundance (A) and number of morphospecies (B) of cave invertebrates collected during a resource-manipulation experiment in Greenbrier County, West Virginia, USA. The numbers are averaged by treatment over the experiment. Data are means ± standard error. Asterisks denote significant differences (p<0.05) between the two treatments based on Bonferroni corrected t-tests. From Schneider et al. (2011).

belonging to 102 morphospecies were enumerated. Given that the rats had a higher nutritional value than the leaf packs, it is not surprising that more individuals were found near the rat carcasses than near the leaf packs (Fig. 11.1), but it does demonstrate the response of the detritivore community to different spatial subsidies of food. Diplopoda and Collembola dominated the leaf packs, while Diptera and Collembola dominated the rat carcasses. Non-troglobionts dominated both treatments, with troglobionts only accounting for 11 per cent of the total number of species (Pipan and Culver 2012a). Although the species composition was different, the total number of species for the most part did not differ between the two treatments.

Species composition was similar within treatments, as indicated by the clustering of Jaccard indices (which measure species overlap), both for the species present on the last day of the experiment (Fig. 11.2A) and the species occurrences summed over the entire experiment (Fig. 11.2B). For the species present on the last day of the experiment, there was one rat community that clustered with leaf communities, and a rat and leaf community that clustered together, but separate from other treatments. The situation for overlaps of species over the entire experiment is cleaner, with two major clusters, one with all of the rat treatments, and another with seven of nine leaf treatments. Two leaf treatments are basal, representing the occurrence of species throughout the experiment (Schneider et al.

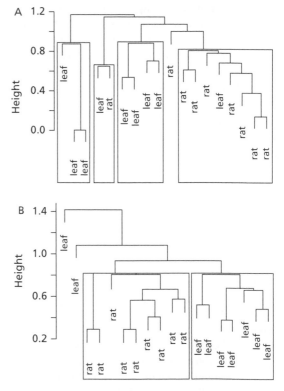

Figure 11.2 Dendrograms depicting the hierarchical cluster of presence/absence of the resource-manipulation experiment in shallow pits in Greenbrier County, West Virginia, USA, using Jaccard indices [J = c/(a + b)] where a is the number found in site 1, b the number found in site 2, and c the number found in both. A shows the data for the last day of the experiment and B shows the data for all species ever recorded during the experiment. Grey-outlined boxes indicate significant hierarchical clusters. From Schneider et al. (2011).

2011). Except for these two basal leaf treatments, all the treatments showed changes during the nearly two-year experiment, and succession was clearly occurring. These similarity patterns also indicate how spatial subsidies can organize communities.

11.2 Sources of organic carbon and nutrients in SSHs

Culver and Pipan (2009) identified six potential sources of energy entering subterranean systems: percolating water, flowing water, gravity and wind, active animal movement, roots, and chemoautotrophy. All but chemoautotrophy are important in one or more SSHs (Table 11.1) and chemoautotrophy may be important as well—it just hasn't been studied. The energy source for the hypotelminorheic habitat is primarily from leaf fall at the exit of the hypotelminorheic (chapter 2). Additionally, some DOC and nutrients may be brought into the habitat from leachate from leaves by water sinking into the small, shallow drainage basins. Because many hypotelminorheic habitats are in forested areas, this source is relatively abundant, depending primarily on net primary productivity.

The major input of organic carbon and nutrients into epikarst is from water percolating through the soil, with the attendant changes in quantity and quality of DOC (see chapter 3). The DOC percolating into the epikarst is dominated by compounds such as proteins that are readily used by microbes (Simon et al. 2010). Within epikarst, there is lateral as well as vertical movement of nutrients, but we are mostly ignorant about the details, except for estimates of residence time (e.g. Kogovšek and Urbanc 2007), estimates of current velocities (Pipan and Culver 2007a), and the observation that there is considerable movement of animals within the epikarst (Simon 2013).

Intermediate-sized terrestrial SSHs, which occur in a large variety of morphologies with differing levels of connection with the soil (Fig. 4.3) have multiple potential sources of spatial subsidies. These include percolating water, gravity and wind (especially aeolian fallout), active animal movement (e.g. earthworms), and roots. All of these sources are not present for any one site and their importance depends on the geomorphological context (see chapter 4). Percolating water is especially important in areas with more precipitation, and sites that are covered with soil or moss to slow the movement of water. Gravity and wind have been demonstrated to be important in barren volcanic sites on the Canary Islands with few other sources of nutrients (Ashmole and Ashmole 2000). Active animal movement is important in sites with close contact with soil and where many spaces between rocks are at least partially soil-filled. Roots may be important in a variety of contexts, but the best-known one is lava tubes and associated MSS, where the roots of *Metrosideros polymorpha* are the base of a subterranean food web (Stone et al. 2012).

Our ignorance of energy sources for calcrete aquifer communities is nearly complete. Since rainfall, vegetation, and water are scarce in western Australia where calcrete aquifers are known to

Table 11.1 Major sources of energy for different SSHs. Asterisks denote the relative importance of the contribution.

Energy source	Hypotelminorheic	Aquatic epikarst	MSS	Calcrete aquifers	Hyporheic	Soil	Lava Tubes
Percolating water		**	**	?		**	*
Flowing water	*				**		
Gravity and wind	**		*				*
Active animal movement		*	**			**	
Roots			*			*	**
Chemoautotrophy				?			

occur, it is tempting to speculate that chemoauto-trophy is important, but it is only speculation.

Organic carbon and nutrients enter the hyporheic primarily through downwelling of water in surface streams and rivers (see chapter 6). Flood pulses may be especially important, and may even rearrange the habitat. The hyporheic is also highly suscepti-ble to environmental pollution, and anthropogenic enrichment of a stream with organic carbon rapidly enters the hyporheic (Buss et al. 2009).

The downward movement of water in the soil carries with it organic carbon and nutrients. As we show in section 11.3 below, the decline in concen-tration of DOC can be very rapid, so rapid in fact that Gers (1998) asserted that the movement of animals, including some of the beetles he studied, was a more important spatial subsidy. Roots, where they penetrate the soil layers under consideration, can obviously deliver organic carbon and nutrients at depth.

The best-known sources of food in lava tubes are roots of *Metrosideros polymorpha* (see Fig. 8.4), but overall are probably no more important than per-colating water and gravity in transferring organic matter, because not all lava tubes have roots, in part because *M. polymorpha* is an early successional tree.

11.3 Amount and pattern of spatial subsidies of organic matter in SSHs

Overall, there is a reduction in the amount of or-ganic matter in SSHs compared to surface habitats because all organic matter must be transported into the SSHs, as there is no primary productivity. There is however little quantitative evidence to support this assertion in part because organic carbon typi-cally varies temporally and spatially, but especially varies with depth. The best quantitative information available is for aquatic habitats. We can use the DOC concentration for two karst rivers as surface com-parisons (Pivka River in Slovenia [4.36 mg C L^{-1}] and the entrance stream to Organ Cave in West Vir-ginia, USA [7.67 mg C L^{-1}]). Average values of epi-karst drips are around 1.0 mg C L^{-1} (Table 3.6, Fig. 3.5); values for seepage springs are around 2.75 mg C L^{-1} (the average of all seeps in Table 2.4); and values for the hyporheic are around 2.5 mg C L^{-1} (Fig. 6.9). Thus, hyporheic and hypotelminorheic have about

half the DOC concentration of surface streams, and epikarst has about 15 per cent of the DOC concentra-tion of surface streams. These are all estimates of the amount of organic carbon (the standing crop) and not the flux, for which we have no comparative data (but see Simon 2013).

Data for terrestrial SSHs are scarcer and more difficult to compare directly with surface values. If we take the Welsh soil data for the A horizon as a standard (Table 7.1), organic carbon concentration is 101 mg C L^{-1}. Gers (1998) reports values for the MSS in the French Pyrenees of less than 1 mg C L^{-1} at a depth of 2 m, and Hathaway et al. (2013) report means in Azorean Islands lava tubes of 47.8 mg C L^{-1} in cave soil. While a comparison of Welsh soil with Azorean Islands lava tube soil is not an especially appropriate one, it does indicate a roughly 50 per cent reduction in the standing crop of organic carbon relative to the surface.

A common pattern of organic carbon in SSHs is a decline with depth. One of the most striking exam-ples is Gers' (1998) study of soil and MSS in the Bal-longue forest in the French Pyrenees at an elevation of 1050 m. Soil water at a depth of 20 cm in the A horizon had 82 mg C L^{-1}, and only 2 mg C L^{-1} in the BC horizon at a depth of 1.3 m. At the depth of the MSS (2 m), there was no detectable organic carbon in the water, less than 1 mg C L^{-1}. This sharp gradi-ent in carbon content of soil water led Gers (1998) to propose that the activities of invertebrates were important mechanisms for the transfer of organic carbon to deeper soil layers and to the MSS. Job-bágy and Jackson (2000) reported a sharp decline in organic carbon through the top 3 m of Welsh soil (Fig. 7.4).

Hyporheic habitats can have organic carbon val-ues as high as surface values, as Marmonier et al. (2000) demonstrated for Vanoise brook, a Rhône River tributary in France (Fig. 6.9). Samples taken at a depth of 40 cm had slightly more DOC (approxi-mately 1.8 g C L^{-1}) than surface samples. Because of the pattern of upwellings and downwellings (Fig. 6.3), longitudinal variation is greater than shallow vertical variation (see Fig. 6.10). At greater depths, DOC concentration declines due to the absence of any autotrophic production.

Both hypotelminorheic and epikarst habitats have a pattern of organic carbon concentration that shows occasional spikes in values. Ban et al. (2008) showed that DOC concentrations increased approximately threefold in Shihua Cave, China, immediately following precipitation events (Fig. 3.5). The quantitative impact of these events is even greater because more water with a higher concentration of DOC enters epikarst during these episodes of precipitation. Pipan and Culver (2013) reported a similar pattern of temporal spikes in DOC for hypotelminorheic and other SSH habitats on Nanos Mountain, Slovenia (Fig. 11.3). This area had a series of seepage springs, small springs, and small streams resulting from the confluence of several seepage springs and small springs. There is a remarkable and consistent pattern of relatively low values of DOC (typically between 1 and 4 mg C L^{-1}) with occasional spikes of DOC up to nearly 11 mg C L^{-1} for the hyporheic of a small stream and nearly 10 mg C L^{-1} for a seepage spring. The causes of these spikes are not at all clear, although the extreme values of seepage springs and the hyporheic occurred at the same time (in May 2011).

The examples above illustrate the patterns in total organic carbon availability. It is important to note that organic carbon is actually a large, complex pool of organic molecules that vary considerably in their biological availability, or 'quality'. This variation in the composition or quality of the organic matter pool can influence microbial production

(e.g. Cooney and Simon 2009), but little is known about how organic carbon composition varies in space and time in subsurface systems.

11.4 Are SSHs carbon or nutrient limited?

Because of the general absence of autotrophic production and lowered levels of organic carbon standing crops and fluxes in subterranean habitats, there is the strong possibility that carbon, rather than nitrogen or phosphorus, is limiting. The first serious attempt to separate these possibilities in a subterranean system was done in streams in Organ Cave, West Virginia, USA. Simon and Benfield (2001) tested for nitrogen and phosphorus limitation of microbial films on wood by using nutrient diffusion experiments. They found no evidence that either nutrient-limited microbial activity. They also found that whole-stream demand for nitrogen, as NH_4^+, was rather low in cave streams compared to surface streams (Simon and Benfield 2001), suggesting low demand for nitrogen. In contrast, they found that the rate of carbon turnover was quite high in cave streams suggesting organic carbon was limiting. Experiments with microbial films from those cave streams showed that microbes responded strongly to experimental DOC enrichment (Cooney and Simon 2009), indicating clear carbon limitation.

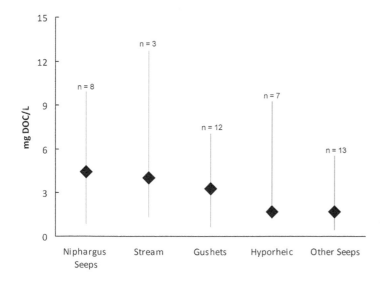

Figure 11.3 Medians, maxima, and minima of dissolved organic carbon (DOC, in mg C L^{-1}) for five habitats on Nanos Mountain, Slovenia, listed in order of decreasing median DOC: seepage springs with the stygobiotic amphipod genus *Niphargus*, surface stream, gushets (small springs), hyporheic, and seepage springs without *Niphargus*. From Pipan and Culver (2013).

Stoichiometric imbalances (mismatches between elemental ratios of consumers and their food) can be a useful tool in examining stresses caused by nitrogen or phosphorus shortages (Moe et al. 2005). In general, a resource with a low N or P content compared to the nutrient content of its consumer is a poor quality resource; organisms with low N or P ratios are stressed, and in some cases may have evolved a low ratio as an adaptation. Schneider et al. (2010) looked at various predictions about stoichiometric ratios in terrestrial cave animals from shallow pits in the Buckeye Creek drainage basin in West Virginia. Since these are shallower habitats with strong surface connections, carbon limitation is less likely than it was in the case of Organ Cave, which is deeper subterranean habitat.

Schneider et al. (2010) examined several hypotheses related to phosphorus limitation. If phosphorus is limiting, then molar carbon:phosphorus (C:P) ratios should be higher for subterranean detritus relative to surface detritus, and for subterranean species relative to surface-dwelling species. They indeed found that organic material in the pits had a C:P ratio of twice that reported for the fermentation layer of the soil. In addition the per cent phosphorus content of a troglophilic millipede (*Pseudotremia fulgida*) was slightly higher than the per cent phosphorus content of a troglobiotic millipede (*P. hobbsi*). Since both millipedes are living in the same habitat, Schneider et al. (2010) attributed the difference to adaptation to phosphorus-poor food by the troglobiont. The C:P mismatch between cave detritus and troglobiotic millipedes was nearly 50–75 per cent more than the mismatch between that of stream detritus and shredders or between terrestrial plants and herbivores reported in the literature. This is in spite of the higher C:P ratio for troglobiotic compared to troglophilic millipedes.

Schneider (2009) suggested that, via constraints on the availability of N-rich amino acids, long-term N-limitation may drive aspects of protein evolution, with a reduction in N-content of free amino acids. Compared to the troglophilic species *Pseudotremia hobbsi*, she found that the troglobiotic *P. fulgida* had decreased amounts of N-rich free amino acids and decreased amounts of essential and nonessential amino acids. In contrast, no such patterns were found when stygophilic and stygobiotic amphipods were compared, suggesting that N-limitation is not important in aquatic subterranean environments.

Neither Simon and Benfield (2002) nor Schneider et al. (2010) worked on typical SSHs, although Schneider et al.'s study was conducted on shallow pits within the zone of SSHs. An interesting research question is whether most SSHs are N and P limited, as suggested by Schneider's study, or carbon limited, as suggested by Simon and Benfield's study.

11.5 Summary

Reduced food is often considered one of the major features of the cave habitat and an important selective constraint. We summarize the available information for SSHs, especially with respect to dissolved organic carbon (DOC), the most frequently measured parameter. In common with caves, nearly all SSHs depend entirely on allochthonous carbon and nutrients. That is, SSHs require spatial subsidies. From experiments on detrital communities in shallow pits (technically SSHs), it was clear that the allochthonous input of carbon, and the nature of that organic matter, organized the community in terms of species composition and abundance.

Different SSHs have different energy sources (percolating water, flowing water, etc.), but in general the amount of organic carbon is intermediate between that found on the surface and in deeper subterranean habitats. In general, the amount of organic carbon declined, often rather sharply, with depth. Some SSHs, especially epikarst and the hypotelminorheic, showed highly variable temporal patterns, with much of the DOC entering the system over short periods of time—rainfall events in the case of epikarst.

The little amount of data that are available suggest that caves may be carbon-limited, but that SSHs may be nutrient limited, but much more work is needed.

Guano is often cited as the counter-example to the evolution of troglomorphy and to food limitation in subterranean habitats (see Box 11.1). We argue for a re-examination of these communities, particularly with respect to the presence of troglomorphy, and provide an example of a troglomorphic species living on guano. Just as with the soil fauna, the fauna of guano in caves should be included as part of the subterranean fauna.

Box 11.1 Guano

If caves and subterranean habitats are characterized by scarce food resources, there is one subterranean habitat that is its antithesis—guano accumulations in caves. Throughout the world, bats, and occasionally birds such as the South American swiftlet, congregate in sufficient numbers that persistent piles of guano accumulate (Hutchinson 1950). This includes lava tubes, perhaps the only SSH with the possibility of having bat colonies and guano. Because it is an energy-rich habitat and in darkness, it has been used as a comparison with food-poor subterranean habitats. There is the frequent claim (e.g. Poulson 1972) that the species of guano communities are without troglomorphic characters, and retain eyes and pigment. If this is the case, and there are no other complicating factors, then this would seem to refute the centrality of lack of light as a selective force, a claim we have put forward in this book and elsewhere (Culver and Pipan 2009, Pipan and Culver 2012a).

We first consider guano as a food resource and also as an extreme environment, not because it is subterranean, but because of the chemical conditions in and around the guano. Then, we briefly consider the evidence for the occurrence of at least some troglomorphic species on guano.

Freshly deposited guano (based on insectivorous bats) comprises approximately 10 per cent nitrogen and 7 per cent phosphorus (Hutchinson 1950). Much of the nitrogen is present as ammonia, and the air immediately around such piles is rich in ammonia. Through time, the guano becomes less alkaline, and relatively enriched in phosphorus. In tropical caves, bats are often present year round, but in temperate zone caves they are typically present seasonally and with seasonal drying of the guano as well. In South America, there are a number of caves with vampire bat guano, which is a black, viscous mass with low oxygen and high ammonia concentration (Palacios-Vargas and Gnaspini-Netto 1992). Although guano caves have sometimes been associated almost exclusively with the tropics, this is not the case, as even the now very outdated list of guano caves by Hutchinson (1950) indicates. In addition, most tropical caves are not guano caves. According to Gnaspini and Trajano (2000), no more than five per cent of Brazilian caves can be considered as guano caves (eutrophic caves).

A set of species specialized for life on bat guano in caves has been described, largely from tropical caves. These species, which can be given a parallel ecological classification to the Shiner–Racoviţă system of subterranean species, are separated into guanobionts[1], guanophiles, and guanoxenes (Gnaspini 1992). Whether yet another set of terminology is useful is debatable, but Gnaspini's real point is that these species are also subterranean species, and that species found only on bat guano should not be excluded from lists of cave fauna (Gnaspini 2012). Deharveng and Bedos (2000) also forcefully argued that bat guano should be considered as a resource in a cave, not a habitat independent of caves. Gnaspini (2012) pointed out that, if guano-eating species are excluded on this basis, then other resource specialists, like planthoppers feeding on *Metrosideros polymorpha*, should also be excluded. Additionally, guano, in nonconcentrated form, is an important food resource both in caves with small and in caves with large bat populations. Deharveng and Bedos (2000) pointed out that even in bat guano caves, there are regions of the cave that have only scattered amounts of guano with more typical troglobionts.

Finally, we take a preliminary look at the morphology of guanobionts. Deharveng and Bedos (2012) stated that the frequency of troglomorphic species in bat guano caves increases with distance from the guano piles, and that it is clear in tropical caves that troglomorphy is associated with food scarcity. However, this does not mean that frequency of troglomorphic species is effectively zero on guano piles. The most common Collembola species on vampire bat guano is *Acherontides eleonorae,* reaching densities of hundreds per cm^2. All but two of the thirteen species in the genus are associated with caves (Palacios-Vargas and Gnaspini-Netto 1992), and all are eyeless and depigmented. One could argue that these traits are result of phylogenetic inertia or exaptation, but there is a unique troglomorphic feature of *A. eleonorae* that seems adaptive and troglomorphic. *A. eleonorae* must navigate on the viscous surface of vampire bat guano, and the unguis is modified to be a narrow, smooth structure that provides good purchase on the substrate (Fig. 11.4). The ungual modifications share features with the parallel and convergent changes Christiansen (1961, 1965)

---

[1] The system is not exactly parallel because Gnaspini (2012) defined guanobionts as obligate inhabitants of guano *when in caves.* As Deharveng et al. (2011) made clear, guanobionts can be widespread outside of caves. One could use the phrase troglobiotic guanobionts to describe obligate cave and guano inhabitants.

continued

Box 11.1 *Continued*

described for another family of Collembola (the Entomobryidae), that enable them to walk on the surface of pools and on wet mud. The guano and mud share the feature of being a slippery surface, and the morphological solution to the problem has similarities.

Figure 11.4 Tibiotarsus and unguis of leg I of the collembolan *Acherontides eleonorae*, a troglomorphic guanobiont from vampire bat caves in Brazil. From Palacios-Vargas and Gnaspini-Netto (1992). Copyright © Kansas Entomological Society, used with permission.

We are not claiming that all or most guanobionts are troglomorphic, even guanobionts only found in caves, but then most of the permanent inhabitants of SSHs and caves aren't either (Pipan and Culver 2012a). It would also seem likely that the frequency of troglomorphy is lower on guano. What is possible is that guano allows for new guilds of species, non-troglomorphic ones that are able to exploit guano. Deharveng and Bedos (2000) described the 'giant arthropod community' which in the southeast Asian caves they studied included one primary consumer (a large rhaphidophorid cricket). One of these species, *Rhaphidophora oophaga*, is apparently an omnivore and also eats swiftlet eggs (Bullock 1965). Other species of *Rhaphidophora* from Thai caves have some degree of eye reduction (Deharveng and Bedos 2000). There are several predators of the crickets, including giant Arachnida and Chilopoda, but they show no morphological sign of adaptation to caves, although some species may leave the cave at night. In addition to the 'giant arthropod community', there are normal-sized species, including *A. eleonorae.*

Finally, it is worth noting that guano, more so than caves, is an extreme environment. Palacios-Vargas and Gnaspini-Netto (1992) reported that, unlike most species including other species living on vampire bat guano, *A. eleonorae* can thrive, even in the extreme conditions in the guano. When placed in sealed containers containing guano, *A. eleonorae* can survive for months, growing and reproducing in this oxygen-poor, ammonia-rich environment. The morphology and adaptation of species associated with guano is worthy of much more attention.

CHAPTER 12

Evolution of morphology in shallow subterranean habitats

12.1 Introduction—the status of troglomorphy

For 50 years, troglomorphy has been an important concept in the study of the evolutionary morphology. Standard classification schemes of subterranean organisms, such as the Schiner–Racoviță system of trogloxene–troglophile–troglobiont are habitat based, but associated with the habitat categories are morphological changes. Christiansen (1962) proposed the word 'troglomorphy' to describe those features (both regressive and progressive) associated with cave life. As D. Fong (pers. comm.) pointed out, troglomorphy is really a hypothesis that troglobionts (and stygobionts) have a distinct morphology. Christiansen (1961) also used the term cave-dependent for these troglomorphic, parallel and convergent characters, as opposed to cave-independent characters, which he suggested could help determine lineages. Cave-independent characters bear a close resemblance to synapomorphic characters useful in phylogenetic analysis (see Desutter-Grandcolas 1997). In Christiansen's original usage of the term, it was closely connected with parallelism and convergence, as well as adaptation to subterranean life. However, he included features like loss of eyes and pigment, which are not necessarily either adaptive or under the control of natural selection (Culver and Pipan 2009).

After 50 years of use, the concept of troglomorphy has been expanded to include behaviour and physiology, as well as morphology, and may include characters that occur in a small number of lineages (Christiansen 2012). Christiansen's (2012) list of troglomorphic characteristics is shown in Table 12.1. The morphological features fall into three groups.

- Reduced (regressed) characters like eyes and pigment, that may or may not be adaptive (i.e. under the control of natural selection, as opposed to neutral mutation).
- Elaborated characters that appear in small numbers of lineages, and thus with limited parallelism and convergence, such as the crop-empty live weight ratio in crickets.
- Elaborated characters that appear in large numbers of lineages, such as appendage elongation.

Although 50 years have passed, the number of characters for which the action of natural selection has been demonstrated is very small in number, but somewhat remarkably it does include eye and pigment loss (Yamamoto et al. 2009, Bilandžija et al. 2012).

Although Christiansen (2012) did not explicitly mention SSHs, he did mention that (1) not all cave organisms are troglomorphic, (2) many of the regressive features associated with the absence of light are found in other habitats, such as soil and the abyssal benthos, and (3) some habitats, such as soil and microcaverns, may serve as recruiting grounds for cave organisms.

The idea that organisms in aphotic or low light habitats, such as the leaf litter and soil, are pre-adapted or exapted to cave life is widespread (e.g. Barr 1968, Heads 2010). In this point of view, species in twilight habitats and in SSHs are expected to have a morphology intermediate between that of caves and the surface, in part because of the intermediate nature of the habitat (the case for leaf litter)

Shallow Subterranean Habitats. David C. Culver and Tanja Pipan.
© David C. Culver and Tanja Pipan 2014. Published 2014 by Oxford University Press.

Table 12.1 List of troglomorphic features, from Christiansen (2012), used with permission of Elsevier.

Specialization of sensory organs (e.g. touch, chemoreception)
Elongation of appendages
Pseudophysiogastry
Reduction of eyes, pigment, and wings
Compressed or depressed body form (Hexapoda)
Increased egg volume
Increased size (Collembola, Arachnida)
Unguis elongation (Collembola)
Foot modification (Collembola, planthoppers)
Scale reduction or loss (teleost fishes)
Loss of pigment cells and deposits
Cuticle thinning
Elongate body form (teleost fishes, Arachnida)
Depressed, shovel-like heads (teleost fishes, salamanders)
Reduction or loss of swim bladder (teleost fishes)
Decreasing hind femur length, crop-empty live weight ratio (crickets)

or because of differences in time since isolation in the subterranean realm. In any case, the overall focus of this hypothesis is on convergence, and we call it the convergence hypothesis.

An alternative view is that all of these habitats, SSHs and twilight habitats like leaf litter, have selective factors that lead to specializations to that particular habitat, not subterranean habitats in general. For example, the soil (see chapter 7) is an unlikely staging ground because the environments are so different. Soil has small pores and caves have large pores; soil has high levels of organic matter and caves have low levels of organic matter (chapter 11). This argument can be expanded to hypothesize that adaptations to SSHs are more or less unconnected to adaptations to caves, and under this hypothesis, divergence is at least as important as convergence. We call this hypothesis the divergence hypothesis.

12.2 Morphology of SSH species

12.2.1 Aquatic SSH species

The subterranean amphipod genus *Stygobromus* is widespread in SSHs, including hypotelminorheic and epikarst, and in cave habitats, including streams and phreatic pools in the eastern USA. A total of 56 species, all either described or redescribed by Holsinger (1978, 2009) occur in these habitats. In the course of description and redescription, Holsinger measured body length, antenna I length, and number of segments in antenna I flagellae for the type series (holotypes and paratypes). Body length should be correlated with pore size of the habitat, if that is an important selective factor. One or both antennal measurements should be correlated with an increase in extra-optic sensory fields, if this is an important selective factor.

As expected under the divergence hypothesis, body size varied according to pore size of the habitat (Fig. 3.21). Epikarst species were the smallest (averaging 5.3 mm), followed by the hypotelminorheic, phreatic, and cave stream habitats (averaging 9.5 mm). This is in agreement with the pore size of the SSHs (Fig. 1.14). The smaller size of amphipods from phreatic waters compared to cave streams was unexpected, and reflects our ignorance about the nature of these deep phreatic habitats. Trontelj et al. (2012) also found anomalous morphologies (in their case giant sizes) in *Niphargus* amphipods in phreatic habitats in the Dinaric karst. Although no *Stygobromus* are known from hyporheic sites in eastern North America, there are species known from western streams and rivers (Ward 1977, Wang and Holsinger 2001). Measurements are available for two of the hyporheic species—*S. coloradensis* and *S. pennaki*—and female body size for the type series was even smaller than those for the hyporheic, averaging just 3.8 mm. This is also in accord with the reduced pore size of the hyporheic (Fig. 1.14).

Culver et al. (2009) found that stygobiotic copepods in epikarst in Slovenia were smaller than either stygobiotic species in other habitats or stygophiles (Table 3.14). This implies selection for reduced body size for copepod species living in epikarst, as well as the amphipod species discussed above.

The final study of size of stygobiotic SSH species is that of Vergnon et al. (2013), who examined sizes of the dytiscid beetles living in calcrete aquifers in western Australia. Pore size for calcrete aquifers is not known so it is difficult to make a prediction about expected maximum size. What they found was that in 34 calcrete aquifers beetles were never larger than 4.5 mm. Although pore size may play some role in setting this maximum, the largest-sized beetle differed greatly among calcrete aquifers, ranging down to less than 1.5 mm. Pore size may vary among calcrete aquifers, but there was no direct evidence for this, and other factors such as food sizes for the beetles may also vary. What was clear was that interspecific competition among beetle species plays a major role in determining size, resulting in a nearly constant ratio of sizes (1.5) of beetles of adjacent size classes (Fig. 5.9). This is one of the best examples in all of the ecological literature of the limiting similarity of competing species. The resulting pattern is one of divergence of sizes within each calcrete aquifer, not convergence, as is the expectation of troglomorphy.

In general, size appears to have been strongly selected, and this resulted in divergence rather than convergence. Habitat size (pore size) and animal size were strongly correlated for both amphipod and copepod examples. The result was convergence within an SSH type, but divergence among SSHs. The presence of competitors, in the case of dytiscid beetles in calcrete aquifers, had a strong effect on body size, resulting in divergence within an SSH.

Shape factors are somewhat harder to predict. Fišer et al. (2012) and Trontelj et al. (2012) predicted

that both habitat size and current velocity affects morphology. Increased current velocity should result in reduced appendage length while increased habitat size should result in increased body size (Trontelj et al. 2012). This is in fact what they found, and different morphological groups of the genus *Niphargus* occurred within a cave (Fig. 12.1). This is also a pattern of within-habitat divergence, as found by Vergnon et al. (2013), and in agreement with the divergence hypothesis.

By extrapolation to SSHs, Trontelj et al.'s predictions can be extended—SSHs with low currents might be expected to have animals with shorter appendages. Certainly, cave stream species encounter stronger currents than species living in epikarst or the hypotelminorheic. However, in their study of morphology of the type series of *Stygobromus* species, Culver et al. (2010) were unable to detect any differences in shape among cave stream, phreatic, epikarst, or hypotelminorheic inhabitants (Tables 12.2 and 12.3). In the case of first antennal length (Table 12.2), cave stream species had the longest antennae (after correction for size), rather than the smallest, as predicted by the Trontelj et al. (2012) model, although the differences were not significant. In the case of the number of segments of the first antennae (a potential measure of increased sensory hairs etc.), cave stream species had the most segments and epikarst species had the least (Table 12.3). The differences were

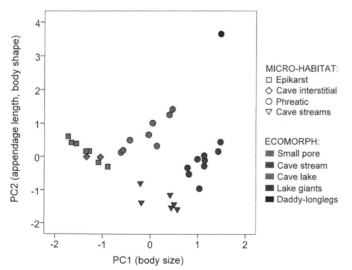

MICRO-HABITAT:
□ Epikarst
◇ Cave interstitial
○ Phreatic
▽ Cave streams

ECOMORPH:
▪ Small pore
▪ Cave stream
▪ Cave lake
▪ Lake giants
▪ Daddy-longlegs

Figure 12.1 Principal Components Analysis on morphometric traits (mean values) of 33 *Niphargus* species and populations from seven cave communities in the Dinaric karst. The first two axes (PC1 and PC2) together explain 97.5 per cent of the total variation. Cave microhabitats (symbols) and proposed ecomorphs (grey shading, or colours in Plate 14) are only partly in agreement. There is no morphological distinction between inhabitants of the epikarst and the cave interstitial, and there are three distinct morphological groups within the phreatic habitat. From Trontelj et al. (2012) with permission from John Wiley & Sons. (See Plate 14)

Table 12.2 Analysis of Covariance for *Stygobromus* from eastern North America. A. Analysis of Covariance of ln of first antennal length, with ln size as controlling variable. B. Least squares means of ln first antennal length at the mean ln female length (1.75), and observed mean ln first antennal length for the four habitat types. From Culver et al. (2010).

A.

Source	Df	SS	F	p
ln length	1	8.9557	412.76	<0.001
habitat	3	0.0948	0.39	0.240
ln length * habitat	3	0.0442	0.68	0.57

B.

Habitat	Least squares mean	SE	Observed mean
Cave streams	1.122	0.074	1.432
Hypotelminorheic	1.005	0.052	1.458
Phreatic	1.012	0.044	1.077
Epikarst	0.988	0.029	0.773

Table 12.3 Analysis of Covariance for *Stygobromus* from eastern North America. A. Analysis of Covariance of ln of first antennal flagellar number, with ln size as controlling variable. B. Least squares means of ln first antennal flagellar number at the mean ln female length (1.75), and observed mean ln first antennal flagellar number for the four habitat types. From Culver et al. (2010).

A.

Source	Df	SS	F	p
ln length	1	3.7248	85.83	<0.001
Habitat	3	0.2305	2.41	0.08
ln length * habitat	3	0.0125	0.13	0.94

B.

Habitat	Least squares mean	SE	Observed mean
Cave streams	2.835	0.061	3.046
Hypotelminorheic	2.769	0.050	2.833
Phreatic	2.657	0.088	2.797
Epikarst	2.652	0.034	2.537

not significant and there was no model that predicted the observed order of number of antennal segments. The *Stygobromus* specimens provided no support for the divergence hypothesis since no shape differences were statistically significant (Culver et al. 2010), and because they weren't

significant the convergence hypothesis is support for shape.

Brancelj (2007, 2009) described a character—very strong spinules at the base of the caudal rami—for some epikarst copepods that may represent an adaptation to strong currents. These spicules may help epikarst copepods from being displaced by water flow, which is likely sufficient to dislodge copepods from the substrate (Pipan and Culver 2007a).

12.2.2 Terrestrial SSH species

Terrestrial SSHs present a wide variety of habitat sizes, ranging from the size of caves (lava tubes), typically easy for humans to enter, to the soil, probably with the smallest pore sizes of all SSHs. The MSS lies somewhere in the middle in terms of size.

The body size of the fauna of lava tubes typically ranges to above 10 mm. Species much larger than this, e.g. Chilopoda and Diplopoda, are frequently found in lava tubes throughout the world. Soils harbour extremes in small size (Fig. 7.13) and thinness (Fig. 7.14). Generally, the soil fauna is smaller than its surface relatives. Coiffait (1958) provided paired comparisons for 74 edaphobiotic beetles from the Pyrenees with their closest surface ancestor. It is important to keep in mind that the surface ancestor, given that this work was done in 1958, was not the result of any DNA sequence or even a cladistics analysis. Even taking this into account, his finding that only one of the edaphobionts was larger than its surface relative is strong evidence for miniaturization of the soil

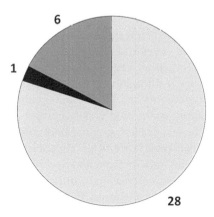

Figure 12.2 Pie diagram of the size relationship between closely related soil and surface beetle species from the Pyrenees. Light grey are species pairs where the surface species is larger; dark grey are species pairs where soil and surface species are the same size; black are species pairs where the soil species is larger. Data from Coiffait (1958).

fauna, and support for the divergence hypothesis (Fig. 12.2).

The relationship of size of animals in MSS habitats relative to caves, surface, and soil depends to a large extent on the particular site, because there is great variability among MSS sites (see Fig. 4.3). Perhaps the most typical case is the one described by Růžička et al. (2011) for the spider *Porrhomma myops* and a closely related, but undescribed species found in the soil. The soil species is smaller (as measured by the width of the cephalothorax) with relatively shorter appendages (as measured by the length of the metatarsus) (Fig. 7.15). Individuals from caves and screes did not differ in either character, suggesting that in this case, habitat size in scree was not an important selective factor. Thus, the divergence hypothesis was supported for shape but not for size.

J. Růžička (1998) did a more extensive morphometric analysis of the *Choleva agilis* species group of Leiodidae beetles from the Czech Republic. Růžička (1998) distinguished three major habitats: caves, talus, and epigean (surface) habitats. Using Canonical Variate Analysis, the three habitats are quite distinct (Fig. 4.23). On the first CVA axis, talus populations are intermediate and on the second CVA axis, talus populations are below cave and epigean samples. Loadings on the first CVA axis are primarily appendage lengths and eye size, and talus populations are intermediate with respect to both groups of variables. The second CVA axis is primarily a measure of size. Somewhat surprisingly, the talus slope populations have somewhat larger body sizes than either cave- or surface-dwelling populations. In this study, divergence in shape and size occurred.

In the case of the spider genus *Dysdera* from the Canary Islands, an analysis of their reported size data for females in the type series failed to yield any difference among the three habitats studied (lava tubes, MSS, and surface). Instead, there was broad variation within each habitat type (Fig. 12.3), suggesting that perhaps interspecific competition or simply habitat differences within the major habitat types were important. However, at least among habitats, there was no divergence, but also no convergence with respect to subterranean habitats.

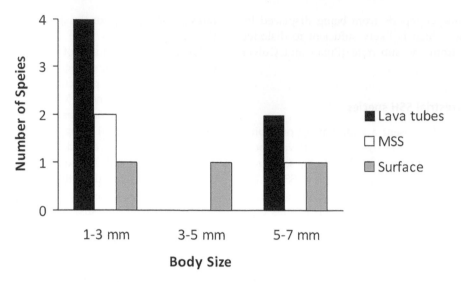

Figure 12.3 Distribution of body sizes of spiders in the genus *Dysdera* from lava tubes, MSS, and surface habitats on the Canary Islands. Data from Arnedo et al. (2007).

12.3 A new look at troglomorphy

The coining of the phrase troglomorphy by Christiansen (1962) came at the time of the rise of neo-Darwinism, and what Gould (2002) called the hardening of the synthesis. Christiansen, Barr, and Poulson, the three dominant American speleobiologists, all contributed to the paradigm of the morphology, behaviour, and life history of cave animals being moulded by convergent selective pressures, resulting in convergence of their biological characteristics. Christiansen (1961, 1965) demonstrated convergence in claw characters of cave Collembola; Poulson (1963) demonstrated life history adaptation of amblyopsid cave fish; and Barr (1968) developed models of speciation and dispersal based on cave beetles. This selectionist view was expanded to include an ever more sophisticated analysis of genetics of adaptation by Culver et al. (1995) and Carlini et al. (2009), working with the amphipod *Gammarus minus*; and by Jeffery and his colleagues (Jeffery 2009, Yoshizawa et al. 2012, Yamamoto et al. 2009, Bilandžija et al. 2012), primarily working with the cavefish *Astyanax mexicanus*. The synthesis was hardened by the frequent reference to any subterranean species with reduced eyes or pigment as troglomorphic, thereby bypassing any traits that might be elaborated or expanded, and

by the dismissal of the occurrence of any non-troglomorphic species in subterranean habitats as either phylogenetically young and therefore not yet troglomorphic, or somehow just an accidental occurrence in subterranean habitats.

In the last several years, cracks have appeared in the facade of the paradigm of convergent evolution and its subterranean manifestation, troglomorphy. Pipan and Culver (2012a) reviewed data on the frequency of stygobionts and troglobionts (nearly all of which were troglomorphic) for some well-studied caves and SSHs (Fig. 9.1). It was not common for either caves or SSHs to have only stygobionts and troglobionts. No cave had exclusively stygobionts and troglobionts, although the aquatic fauna of the biodiversity hotspot cave Vjetrenica (Bosnia & Hercegovina) had only a single stygophile (Sket 2003), and among SSHs, only four epikarst sites (all in Slovenia) were exclusively stygobiotic. It is increasingly difficult to dismiss the non-specialized species in these habitats as either accidentals or somehow phylogenetically young species, in some sense imperfectly adapted to subterranean life, especially due to a species removal/recolonization experiment done by Schneider et al. (2011) in some shallow pits in West Virginia, USA.

Schneider et al. (2011) obtained an especially thorough inventory of the terrestrial invertebrates when they monitored colonization of 12 small West Virginia caves following defaunation (see chapter 11). In the course of their study, they counted 18,396 adult invertebrates, belonging to at least 91 species. If abundance is a measure of fitness, then there was no evidence that troglobionts were more fit than troglophiles. Ten of the species (11 per cent) and 4342 of the individuals (24 per cent) were troglomorphic (they did not identify most species, but rather relied on morphology). Of the troglomorphic species, two Collembola species (one unidentified Entomobryidae and the entomobryid *Pseudosinella gisini*) were the second and fourth most abundant species overall. The troglophilic millipede *Pseudotremia hobbsi* was third most abundant and an unidentified Dipteran in the family Calliphoridae was the most abundant, and they are likely able to fly in and out of the cave. As a final example from their study, they found 113 individuals of the troglomorphic and troglobiotic millipede *Pseudotremia fulgida* and 1486 individuals of the non-troglomorphic and troglophilic *P. hobbsi.* With available evidence, it is hard to conclude that the troglomorphic *P. fulgida* is more adapted to subterranean life.

In their study of *Niphargus* communities in the Dinaric karst, Fišer et al. (2012) and Trontelj et al. (2012) found that both size and shape were controlled by a combination of convergence and divergence (Figs. 6.19 and 12.1), but with divergence being the more obvious trend within any given subterranean community. Body size in dytiscid beetles in Australian calcrete aquifers was best explained by divergence, as a result of limiting similarity of competing species (Fig. 5.9). In both of these cases, divergence is more obvious than convergence.

Other studies raise questions about the extent of convergence, especially of eye and pigment reduction. Friedrich et al. (2011) found a tiny, but functional eye in the cave beetle *Ptomaphagus hirtus*, a species previously thought to have a vestigial eye. Romero and Green (2005) claimed that, among stygobiotic cave fish, eye loss, pigment loss, and scale reduction are at different levels in different species and that there is no apparent convergence. Heads (2010) claimed that troglomorphy, to the extent that it exists, did not evolve in caves, but rather in

leaf litter, for the case of crickets. Finally, Desutter-Grandcolas (1997) and Desutter-Grandcolas et al. (2003) claimed that there is no more convergence (homoplasy) among subterranean lineages than other lineages.

Amidst the support for and attack on the concept of troglomorphy, several generalities are emerging. The first is that species without eyes or pigment and with varying degrees of other morphological modification are present in large numbers in most SSHs as well as aphotic habitats in general. Additionally, there is no documented tendency for SSH species to be less troglomorphic than species from caves and other deeper subterranean habitats. The link between losses (especially eyes and pigment) and gains (especially extra-optic sensory structures as well as appendage shortening or lengthening) seems widespread but there are relatively few studies that demonstrate the gains in a quantitative way.

The second generality to emerge is that both convergence and divergence occur in subterranean communities. Convergence is largely with respect to surface-dwelling populations while divergence is both relative to different SSHs and within a habitat, as we outline later in this section. The more broadly, in a taxonomic sense, that we look for convergence, the more it is limited to eye and pigment loss. In some subterranean habitats, animals evolve larger sizes and in other subterranean habitats animals evolve smaller sizes, and the same holds for shape characteristics as well.

The third generality is that natural selection is ubiquitous or nearly so, even in the case of eye and pigment reduction. The group of evolutionary-developmental biologists studying the Mexican cavefish *Astyanax mexicanus* has demonstrated that the blockage of pathways leading to eye development leads to overexpression of the *sonic hedgehog* gene which in turn is implicated in increased numbers of taste buds (Yamamoto et al. 2009) and superficial neuromasts (Yoshizawa et al. 2012), both of which are advantageous to fish living in aphotic environments. Even pigment, long the stronghold of a likely selectively neutral loss even at the molecular level (see Protas et al. 2006), appears likely to be at least partly under the control of natural selection (Bilandžija et al. 2012). Blockage of the melanin synthesis pathway at the L-tyrosine to L-Dopa step may allow for increased production of catecholamines (e.g. dopamine and norepinephrine), which could modify activity patterns. The availability of a complete sequence for *Astyanax mexicanus* (Gross et al. 2013, Hinaux et al. 2013) should rapidly accelerate research in this area.

If eye and pigment loss were not linked to selection, then eye and pigment loss would solely be the result of the accumulation of selectively neutral mutations that reduce eyes and pigmentation. Forcefully championed by Wilkens (1971, 1988), the rate of loss is determined by intrinsic genetic factors, such as number of loci, population size, and mutation rate (Culver 1982). While the debate about its importance continues (Wilkens 2010, Jeffery 2010), it would seem that situations where neutral mutation is dominant to selection are becoming fewer and fewer (see Protas et al. 2006). If it were dominant, then species with reduced eyes and pigment but few or no constructive gains would be common. However, it seems to be very unusual to find a subterranean species or population that is identical to a surface population except for the loss of eyes and pigment. This is certainly not the case for *Astyanax mexicanus*, which has many constructive characteristics (Jeffery 2005b). Lewis (2013) provided a potential example of a species with only losses with an isopod, *Caecidotea insula*, found in caves on an island in Lake Erie in North America, in a region covered by Pleistocene ice sheets. It differs from the surface-dwelling *Caecidotea forbesi* primarily by its lack of eyes and pigment.

In Table 12.4 we summarize the selective pressures that diverge and the selective pressures that converge in the subterranean realm. When one looks at subterranean environments as a whole, or

just SSHs as a whole, it is hard to identify any shared selective factor except for the absence of light. In those situations where there is little if any morphological difference between species in different SSHs or between species in SSHs and caves, then this suggests that absence of light is a major selective factor. The work of evolutionary-developmental biologists on *Astyanax mexicanus* also suggests a major role for light.

As outlined in the preceding chapters, SSHs differ both among themselves and with respect to caves in terms of pore size (Fig. 1.14), organic flux, cyclicity, and surface connections. In short, the bigger the pore and the less the flux, the more the selective pressures are cave-like. The smaller the pore and the greater the flux, the more the selective pressures are like upper soil layers for terrestrial habitats and stream bottoms for aquatic habitats.

The one factor we identify as promoting divergence within habitats is interspecific competition. Those places where it has been found are biological hotspots (the Dinaric karst [Sket 1999] and Australian calcrete aquifers [Guzik et al. 2010]), and whether similar interspecific effects are present in less diverse areas remains to be seen.

One factor that does not appear in Table 12.4 is time. In earlier studies of adaptation to subterranean life, especially those of Poulson (1963), the amount of eye degeneration was shown to be negatively correlated with features such as the elaboration of extra-optic sensory structures, reduced metabolic rate, and life history modifications. In the absence of any rigorous phylogeny, let alone any phylogeny based on molecular data, it seemed logical to assume that the amount of eye degeneration was a measure of time of isolation in caves. There

Table 12.4 Hypothesized pattern of selective pressures with respect to habitat. Habitat is taken to mean different shallow subterranean habitats as well as caves.

Convergent among and within habitats	Divergent among and convergent within habitats	Divergent among and within habitats
Absence of light	Pore (habitat) size	Interspecific competition
	Organic carbon and nutrient flux	
	Cyclicity and connections with surface	

are few cases where age and troglomorphy are correlated, and that of Derkarabetian et al. (2010) on harvestmen in western US subterranean sites is one of the best examples (see chapter 14). By contrast, Wessel et al. (2007) found no such correlation for *Oliarus*. Also, recent molecular phylogenies indicate that this is not the case (Niemiller et al. 2012) and neither degree of eye reduction nor troglomorphy in general is correlated with age of the lineage in caves.

Even the length of time required either to lose eyes and pigment or to acquire increased extra-optic sensory structures is not at all clear. The age of subterranean lineage may range from tens of thousands of years (Wessel et al. 2013) to tens or hundreds of millions of years (Poore and Humphreys 1998), but the acquisition of troglomorphy may happen relatively quickly, even if the subterranean lineage itself is ancient.

12.4 Summary

In the 50-year history of the concept of troglomorphy, it has come to include a list of morphological, behavioural, and ecological characteristics that are adaptive for subterranean organisms either in general or in particular, as well as features that are lost, especially eyes and pigment, whose loss may or may not be adaptive. There is a substantial body of evidence that supports the hypothesis that elaboration of extra-optic sensory structures and the pleiotropic tradeoffs associated with subterranean life have a selective advantage. SSH species often show morphological changes associated with subterranean life, including loss of eyes and pigment. Body size for SSH inhabitants is often associated with pore size, especially for SSHs with small pore sizes. For terrestrial species, body size of MSS species is often intermediate between soil and surface-dwelling species. In addition, size may also be affected by interspecific competition, and result in divergence of morphology, rather than convergence. Some shape factors that are unique to particular SSHs have also been identified.

In spite of several major studies affirming the basic validity of the concept of troglomorphy, there have also been a number of challenges. In both SSHs and caves, many species are not troglomorphic and there is not a sufficient reason to dismiss these species as either accidentals or phylogenetically young species. Cases of divergence have also been documented in the hotspot subterranean faunas of the Dinaric karst and the Australian calcrete aquifers. Finally, there are claims that the troglomorphy (convergence) of the subterranean fauna is either an illusion or an exaptation.

Several generalities do emerge:

• species with reduced eyes and pigment are present in large numbers in SSHs and other aphotic habitats;
• both convergence (troglomorphy) and divergence occur; and
• natural selection and its effects are nearly ubiquitous in the subterranean realm.

Of the selective factors of particular importance in subterranean habitats, only lack of light is ubiquitous and unambiguously convergent in its effect. Habitat (pore) size, organic carbon flux, and environmental cyclicity all vary among SSHs. Interspecific competition is a divergent selective factor. Finally, time since isolation in subterranean habitats may also be important in affecting morphology.

CHAPTER 13

Colonization and dispersal in shallow subterranean habitats

13.1 Introduction

Aside from taxonomic descriptions of subterranean species, more has been written about the biogeography of subterranean animals than any other topic in speleobiology. Of the approximately 20,000 described species of stygobionts and troglobionts, it is fair to say that the 'typical' species, whether from caves or SSHs, is known from only a handful of specimens from nearby localities (often only one), whose closest living relative is another stygobiotic or troglobiotic species. Given this limited information, the natural extension of the species description is to consider biogeographical questions rather than ecological or evolutionary questions, for which there is little or no information. Two recurring questions have been (1) why are ranges so restricted, or alternatively, given apparent limitations on dispersal, why are ranges so large? and (2) is a subterranean species' immediate ancestor a living subterranean species, a living surface species, an extinct surface species, or even an extinct subterranean species?

Conceptually, we can divide the process of colonization and dispersal in SSHs into four phases. Firstly, what causes animals to enter (colonize) SSHs? Secondly, what factors contribute to the success or failure of these colonizations? Thirdly, what is the role of extinction of surface populations in isolation and adaptation of the subterranean populations (allopatric vs parapatric speciation)? Fourthly, how much subsequent subsurface dispersal occurs?

The interest in colonization and dispersal stems from the apparent difficulty in colonizing and dispersing among deep subterranean habitats, i.e. caves. This interest persists for SSHs but with some modification. There are two potential important conceptual differences in this regard between caves and SSHs. The first is that the barrier to colonization of SSHs from surface habitats should be less challenging than the barrier to colonization of caves and other deep subterranean habitats. There are four potential barriers to subterranean colonization:

1. Absence of light (chapter 9).
2. Available organic matter and nutrients (chapter 11).
3. Environmental cyclicity (chapter 10).
4. Access and physical connections to the habitat.

While the first barrier is the same for caves and SSHs, the other three are diminished in SSHs. The last barrier, access and physical connection to the surface, is clearly much reduced for soil, MSS, interstitial, and hypotelminorheic habitats, all with broad ecotones with surface habitats. For epikarst, even though we access it through caves, it is more continuous than caves themselves, and so the barrier to epikarst colonization should also be less than that of caves. For lava tubes, access and degree of physical connection is equivalent to that of limestone caves. Calcrete aquifers are perhaps more isolated from the surface than any other SSH, given the high aridity of the surface environment.

The second difference is the spatial distribution of SSHs compared to the distribution of caves. Limestone caves occur on roughly 12.5 per cent of the earth's land surface (Jones and White 2012), and the fraction of the earth's surface occupied by epikarst should be similar (although more continuous). Lava tubes occur in a small fraction of the earth's surface,

Shallow Subterranean Habitats. David C. Culver and Tanja Pipan.
© David C. Culver and Tanja Pipan 2014. Published 2014 by Oxford University Press.

less than three per cent (Palmer 2007). Hyporheic habitats also occupy a small fraction of the earth's surface, but occur basically as long strings. The extent of hypotelminorheic habitats is unknown, but its requirement of a clay layer, among other features, suggests it is uncommon. MSS habitats are only common in areas of moderate to high elevational relief, and thus also restricted in extent. Only the soil is truly ubiquitous or nearly so.

13.2 What causes animals to enter (and colonize) SSHs?

In many subterranean habitats, there is a continuing flux of invaders and migrants. The fauna of surface streams gets swept into the hyporheic and interstitial through downwellings, particularly in times of flood (Figs. 6.13 and 6.14). The hyporheic refuge hypothesis (Dole-Olivier 2011) makes explicit the potential importance of this route of colonization. Given the generally broad areas of contact between the soil and MSS, it is not surprising that samples of MSS fauna typically contain large numbers of soil species (Tables 4.8, 4.9, and 4.10). Likewise, soil samples contain surface-dwelling species, listed as edaphoxenes in Table 7.2. In hypotelminorheic habitats, during times of heavy rainfall, sheet flow of water across the land surface can occur, allowing opportunities for mixing of surface and hypotelminorheic faunas (Tables 2.5 and 2.6). Lava tubes have both more or less direct surface connections via roots, but also via MSS habitats (small cracks and fissures, clinkers, etc.). Calcrete aquifers stand alone among SSHs in receiving few if any immigrants at present.

There is a rain of individuals both into and out of the epikarst. Epikarst receives a supply of colonists, especially copepods living in leaf litter above (Reid 2001). Moldovan et al. (2011) suggested that the appearance of the surface-dwelling copepod *Bryocamptus caucasicus* in the epikarst depressed species richness in those parts of Peștera Ciur Izbuc in Romania where *B. caucasicus* was present. The losses of animals out of epikarst via drips can be considerable, and Pipan et al. (2006a) found that an average of one copepod per day exited each drip in Organ Cave, West Virginia, USA.

In general, the reasons that animals enter, either actively or passively, and remain in subterranean

habitats are threefold (Belles 1991): (1) avoidance of environmental extremes on the surface, (2) avoidance of predators and competitors on the surface, and (3) exploitation of new resources in subterranean habitats. Even the most superficial of SSHs, the hypotelminorheic and associated seepage springs provide a significant refuge from temperature extremes (Fig. 2.4). Salamanders such as *Eurycea lucifuga* are more frequent in caves in the southeastern USA in summer (Camp and Jensen 2007), apparently to avoid high summer daytime temperatures. The avoidance of temperature extremes is the basis of the Climate Relict Hypothesis (CRH) of colonization of subterranean habitats (chapter 10). It is possible that the successful colonization of caves and SSHs by species during the Pleistocene was the result of an initial colonization of species to avoid rising temperatures resulting from retreating ice sheets. Belles' second category is usually associated with the first, in the sense that subterranean habitats are refuges from deteriorating surface conditions. For both of these, the implication is that the mean fitness of individuals in surface populations is declining due to abiotic and biotic conditions.

Belles' third category, exploitation of new resources, is conceptually quite different. In this case, it is not a decline in suitability of the surface habitat, but rather the availability of a subterranean habitat with superior resources. Thus, the mean fitness of an individual in a population of both surface- and subterranean-dwelling individuals, is increased in the subterranean habitat, rather than decreased in the surface habitat. This is what Howarth (1980, 1987a) calls the Adaptive Shift Hypothesis (ASH), in which some individuals from a population exploit a new habitat or food resource (Howarth and Hoch 2012). In his studies of the fauna of Hawaiian lava tubes, Howarth noted that food was in very short supply on the surface of lava flows, especially recent ones. In this environment, much of the organic carbon on the surface comes from wind-blown debris (Ashmole and Ashmole 2000). In contrast, subterranean habitats in lava flows, including MSS habitats which Howarth calls mesocaverns, as well as lava tubes enterable by humans, have many tree roots (Figs. 8.4 and 8.5) which are an abundant source of organic carbon. Howarth and Hoch

(2012) suggested that Adaptive Shift Hypothesis can explain the presence of many troglobionts in tropical caves.

13.3 What factors contribute to the success or failure of colonizations?

The process of colonization is a filter, and most colonizations are doomed to extinction and failure. This was the case even in the transition between epikarst and the drip pools that collected epikarst water (Pipan et al. 2010). Drip pools would seem to be a very similar habitat and are the favourite collecting place for epikarst fauna. However, the two faunas are quite different (Table 13.1). Drips are dominated by stygobiotic epikarst specialists while pools have more non-stygobionts and stygobionts not limited to epikarst. These differences suggest that the drip pools are more transient and hence have more generalist species. It is certainly a habitat distinct from the epikarst.

The fauna of any subterranean site is not a random sample of species occurring on the surface or even of the colonizing organisms. Any subterranean community is 'disharmonious' in this sense. Some of this is easy to understand. Aquatic insects, except for aquatic beetles, are generally missing from the stygobiotic fauna, presumably because the winged adults would have difficulty mating.

In general, strongly visually oriented organisms do not successfully colonize caves because they cannot initially overcome the penalty imposed by visual orientation. However, reduced visual orientation is not sufficient to explain the pattern of species since not all species with reduced visual apparatus occur in subsurface habitats.

Speleobiologists have often used the word pre-adapted to describe successful colonists of subsurface water. Pre-adaptation is used in the sense of possession by an organism of the necessary properties to permit a shift into a new habitat. A structure is pre-adapted if it can have a similar function in a new habitat. For example, long antennae in arthropods foraging in litter may aid in food location, and have the same function in caves. A similar concept and one without the connotation of destiny is exaptation, an adaptation for one function serving for another function. Apparent examples of pre-adaptation (and exaptation) are forest litter arthropods that colonized caves. Regions without forests, such as portions of the Black Hills region in South Dakota, USA (Culver et al. 2003) have a depauperate cave fauna, at least in part because there are few surface species pre-adapted to caves. Species living in the dimly lit, humid environment of leaf litter are exapted for the aphotic, humid environment of caves even though there is little if any leaf litter

Table 13.1 Frequency of stygobiotic copepods in caves in Slovenia. From Pipan et al. (2010).

Cave	Habitat	Number of copepod species	Number of stygobiotic copepods	Number of epikarst specialists
Črna jama	drip	8	8	4
Črna jama	pool	7	5	0
Dimnice	drip	8	8	5
Dimnice	pool	8	7	2
Pivka jama	drip	11	8	4
Pivka jama	pool	10	5	2
Postojnska jama	drip	5	4	2
Postojnska jama	pool	12	6	1
Škocjanske jame	drip	9	8	5
Škocjanske jame	pool	13	6	3
Županova jama	drip	13	12	4
Županova jama	pool	9	7	2

Table 13.2 Scorpion families containing species reported to be troglomorphic in the literature, and the number of these that are found in caves. Data from Vignoli and Prendini (2009). Courtesy American Museum of Natural History.

Family	Number of genera	Troglomorphic species	Cave-dwelling species
Akravidae	1	1	1
Buthidae	3	3	1
Chactidae	2	2	0
Chaerilidae	1	3	2
Diplocentridae	2	5	3
Euscorpiidae	1	2	2
Liochelidae	2	2	1
Pseudochactidae	1	1	0
Troglotayosicidae	2	3	1
Typhlochactidae	4	9	6
Urodacidae	1	1	1
Vaejovidae	3	5	4
Total	23	37	22

present in caves. The connection between the litter and the subterranean fauna can be complex. For scorpions, there are number of genera with troglomorphic non-cave species (Table 13.2) that Vignoli and Prendini (2009) assign to leaf litter although they may in some cases be deep enough to be MSS habitats.

In spite of the importance of exaptation and pre-adaptation, they remain rather elusive concepts. We can a posteriori explain why particular groups of animals are in caves and SSHs. For example, omnivorous millipedes living in moist leaf litter would seem likely successful colonists, and they are. But we cannot always successfully make a priori predictions. For example, Symphyla, blind and eyeless inhabitants of leaf litter would seem ideal candidates for subterranean life. Yet they are rarely found in caves (Juberthie-Jupeau 1994).

Overall environmental conditions can also contribute to the success or failure of colonizations. The data of Peck (1990) on the cryptozoic fauna of lava tubes and soil in the Canary Islands, Galapagos

Table 13.3 Analysis of multivariate regression of age, area, elevation, and all interaction terms on number of cryptozoic species in the Canary, Galapagos, and Hawaiian Islands. Interaction terms in B are standardized with mean age = 7.056 million years; mean area = 1293.9 km^2, and mean elevation = 1380 m. Data from Peck (1990).

A. Analysis of variance

Source	DF	Sum of squares	Mean squares	F-ratio
Model	7	2296.263	328.037	9.85***
Error	17	565.978	33.293	
Total	24	2862.24		

*** $p < 0.01$

B. Parameter estimates

Term	Estimate	S.E.	t-ratio	p-value
Intercept	−6.7311	2.5119	2.68	0.0158
Age	0.3052	0.1763	1.73	0.1015
Area	0.0030	0.0018	1.70	0.1075
Elevation	0.0054	0.0017	3.17	0.0056
Age × Area	0.0004	0.0002	1.76	0.0971
Area × Elevation	0.0000033	0.0000001	3.45	0.0030
Age × Elevation	0.0003	0.0002	1.83	0.0842
Age × Area × Elevation	0.00000045	0.00000013	3.41	0.0034

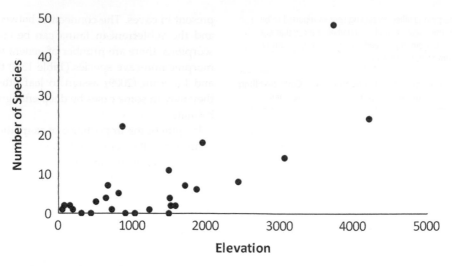

Figure 13.1 Graph of maximum elevation of islands (Canary, Galapagos, and Hawaii) and number of eyeless terrestrial arthropods in lava tubes and soil. See Table 13.3 for a statistical analysis. Data from Peck (1990).

Islands, and Hawaiian Islands are an example of this (see section 9.2.4). Peck looked at three predictor variables—island area, island age, and maximum elevation. Area and age can clearly impact species number and he suggested that elevation was important because higher elevation habitats were moister and suitable soil and lava tube environments may require more moisture. Our multivariate analysis of his data (Table 13.3, Fig. 13.1) indicated that elevation was the key factor, both alone and in combination with other variables. The effect of elevation was greater on larger islands, as is indicated by the interactions terms. Somewhat surprisingly, age and area by themselves were unimportant.

13.4 Allopatric versus parapatric speciation

Allopatric speciation is part of the Climatic Relict Hypothesis (CRH), in which surface-dwelling populations and species go extinct because of climate changes, such as those that occurred during the Pleistocene. Allopatric speciation occurs because in this case the adjoining surface population goes extinct. In these cases, the most closely related species is another troglobiont or stygobiont, typically in a nearby geographical region, which may have become isolated in caves in a separate colonization

from the surface or from subsurface colonization (see section 13.5).

The counter view is that of Howarth—the Adaptive Shift Hypothesis (ASH)—in which speciation is parapatric, with differentiation occurring between contiguous, non-overlapping populations. In these cases, the most closely related species is a parapatrically distributed surface-dwelling species. Largely developed to explain the rich, tropical lava tube fauna of Hawaii, it has been extended to other tropical subterranean faunas, especially lava tubes but limestone caves as well (Peck and Finston 1993).

A starting point to distinguish the two hypotheses is a list of troglobionts and stygobionts, indicating the most closely related species. Howarth and Hoch (2012) provided a list of examples from the Hawaiian lava tube fauna where the most closely related species of troglobionts in lava tubes are nearby surface-dwelling species (Table 8.6). Peck and Finston (1993) provided a similar list for the Galapagos Islands lava tube fauna. Peck and Finston (1993) found that in lava tubes, there were 10 troglobionts with no surface ancestor on the Galapagos Islands, and nine species with an apparent surface ancestor on the Galapagos Islands. They also found nine eyeless soil-dwelling species without a surface ancestor, suggesting that ASH is unimportant for the soil fauna.

While it is tempting to attribute all cases where there is a surface ancestor to the ASH, Villacorta et al. (2008) pointed out that this is not necessarily the case. In their study, what appeared to be two species of terrestrial *Palmorchestia* amphipods, one surface-dwelling and one lava tube-dwelling, in the Canary Islands, actually consisted of a series of cryptic taxa. Villacorta et al. (2008) argued that lava tubes served as refugia for surface populations at various times in the evolutionary history of these populations rather than sources of new resources, and Peck and Finston (1993) argued that the Pleistocene did affect tropical climates, just not as intensely as it did temperate climates.

The converse is also not necessarily the case. A surface ancestor may have gone extinct after parapatric speciation, a situation that Hoch and Howarth (1999) believe has happened in the planthopper genus *Oliarus*. While no doubt this is a real phenomenon, practically speaking it renders the ASH untestable because there are no distributional data that can falsify the hypothesis.

Accumulating evidence strongly supports both hypotheses, although obviously for different cases. One of the best examples of the CRH comes from an SSH—calcrete aquifers. Leys et al. (2003) demonstrated that the phylogenetic age of the dytiscid beetle fauna in calcrete aquifers in western Australia is strongly correlated with latitude, and hence the time of warming, generally in the Miocene (Figs. 1.12 and 5.6). Other investigators of SSHs, particularly lava tubes (e.g. Villacorta et al. 2008, Peck and Finston 1993), have argued that deteriorating environmental conditions have been an important factor isolating species in subterranean habitats in the tropics. There are other examples where the ASH holds, including some cases for the spider *Dysdera* in the Canary Islands (Arnedo et al. 2007), and the terrestrial isopod *Littorophiloscia* in Hawaiian lava tubes (Rivera et al. 2002).

13.5 Post-isolation biogeography

Once a speciation has occurred, either as a result of the Adaptive Shift Hypothesis or the Climate Relict Hypothesis, there can be subsequent dispersal into additional SSHs, and there can be fragmentation

of a species' range as a result of vicariant events[1]. Indeed, the central question for subterranean biogeography has to explain species ranges, especially disjunct distributions (Holsinger 2012, Christman and Culver 2001). The low subsurface dispersal hypothesis is that after a species colonizes a subterranean habitat, relatively little dispersal occurs and that species occupy only a single subterranean site or a few such sites that are physically connected. Closely related subterranean species are by this model the result of independent colonizations by a common surface ancestor. The dispersal hypothesis is that after colonization, species not only occupied a single site or a few connected sites, but that occasionally dispersed beyond this and that these occasional migrants were themselves isolated and speciation occurred.

A model of (1) speciation by multiple colonizations of subterranean habitats and subsequent isolation, and (2) speciation by subsurface dispersal are shown in Fig. 13.2. In the multiple colonizations model, populations of a surface ancestor colonize separate subterranean habitats, and by the CRH, surface populations go extinct (Fig. 13.2, left). Under the subterranean dispersal model there is a single initial colonization of a subterranean habitat (Fig. 13.2, right). After isolation of the subterranean population and extinction of the surface population, it disperses to new sites where populations differentiate genetically. If migration is low enough speciation occurs. The resulting phylogenies of these two processes—multiple isolation and subterranean dispersal—are shown in Fig. 13.3, and in fact they are identical (Culver et al. 2009). Thus, an appeal to the power of cladistic analysis to sort out the hypotheses is insufficient. If surface populations persist as happens in the ASH, the two scenarios can be distinguished (see Desutter-Grandcolas and Grandcolas 1996), and if ages can be put on the nodes, then the two scenarios can be distinguished. This was the approach of Leys et al. (2003) in the context of separating CRH and ASH.

In order to assess the potential for migration, it is important to know the actual range of species.

---

[1] Vicariance is the occurrence of closely related species in disjunct areas, which have been separated by the development of natural barriers (Holsinger 2012).

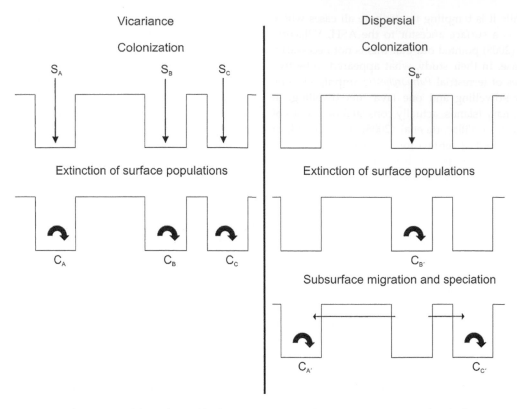

Figure 13.2 Diagram of vicariance and dispersalist models of speciation in caves. In the vicariance model three surface-dwelling populations (S_A, S_B, and S_C) of the same surface-dwelling species enter caves. S_A is more geographically distant from the other two populations. Following the extinction of the surface populations, three cave populations speciate. In the dispersal model, one surface population ($S_{B'}$) enters caves and the surface population goes extinct. Subsequently two other caves are colonized by subsurface dispersal and form separate species under the assumption that dispersal events are rare. From Culver et al. (2009), used with permission of John Wiley & Sons.

While species definitions typically involve inferences about reproductive isolation (biological species concept) or monophyly, having arisen from one ancestral population (phylogenetic species concept), in practice subterranean species are defined on the basis of morphological differences that indicate both reproductive isolation and monophyly. Cryptic species are genetically distinct but morphologically indistinguishable species. Cryptic species can occur anywhere, but they are especially common in subterranean habitats. Most studies of molecular variation in subterranean animals, beginning in the early days when the only information available was different rates of movement of soluble proteins in an electric field (allozyme analysis) to the present when DNA sequences are

becoming more and more common, have indicated the presence of cryptic species (Sbordoni et al. 2000, Trontelj et al. 2009).

Trontelj et al.'s (2009) paper provided empirical evidence that for stygobiotic cave species with linear ranges greater than 200 km, the species was in fact two or more cryptic species. Eme et al. (2013) examined five species of the isopod genus *Proasellus*, all with ranges greater than 200 km, and all of whose ranges included areas covered by Pleistocene glaciation. All of these species were primarily hyporheic in habitat, and each of these species (*Proasellus cavaticus, P. slavus, P. strouhali, P. synaselloides,* and *P. walteri*) in fact comprised between two and four cryptic species (Table 13.4), but with the exception of *P. walteri*, the pattern of range size within a

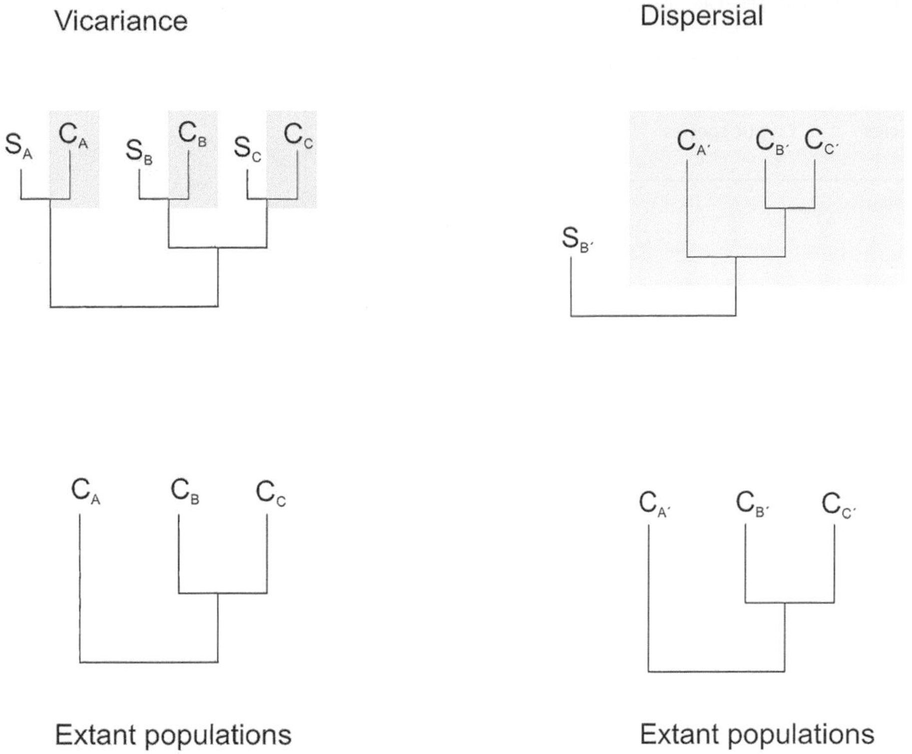

Figure 13.3 Cladograms for the two models shown in Fig. 13.2. Populations are labelled as in Fig. 13.2. Shaded parts of the cladograms are those parts of the tree where evolution is occurring in caves. The resulting cladogram of extant populations is identical for both models. From Culver et al. (2009), used with permission of John Wiley & Sons.

morphospecies was one wide ranging species and the rest limited to a very small area, often a single site. Also contrary to expectation, dispersal, with the exception of *P. cavaticus*, was pre-Pleistocene, and these species presumably survived the Pleistocene in micro-refugia within the area of glaciation (Eme et al. 2013). Dispersal apparently occurred in relatively short bursts that were asynchronous for the different species. For the three morphospecies with the largest ranges (*Proasellus cavaticus, P. slavus,* and *P. strouhali*), dispersal occurred in asynchronous bursts (Fig. 13.4), lending credence to the idea that the hyporheic and groundwater habitats below are an interstitial highway of dispersal (Ward and Palmer 1994). These species may be atypical because they are northern species with ranges that extend into glaciated areas, but Eme et al.'s study does put to rest the idea that subterranean species disperse only short distances.

Phylogeographical analyses with estimates of lengths of branches of phylogenies are not the only way to get information about dispersal and migration of the SSH fauna. Frequency of occurrence of species within a given area, the relationship between body size and frequency of occupancy, and the relationship between body size and range all provide insights into dispersal and migration. There are two datasets for which sufficient information is available to assess the role of subterranean dispersal in SSHs:

- Epikarst copepod occurrence in 35 drips in six caves in central Slovenia (Pipan 2003, 2005).
- Hypotelminorheic amphipod occurrence in approximately 400 seepage springs in the mid-Atlantic region of the USA (Culver et al. 2012a).

Among 14 stygobiotic species in the amhipod genus *Stygobromus*, five stygobiotic species in the isopod

Table 13.4 Number of cryptic species and their linear range for five morphospecies of the isopod *Proasellus* in central and northern Europe. s.s. indicates a species found at a single site. Data from Eme et al. (2013).

Described species	Cryptic species	Maximum linear extent (km)
Proasellus cavaticus	1	1312
	3	s.s.
	3	s.s.
	4	s.s.
	total	1312
Proasellus strouhali	1	704
	2	6
	total	704
Proasellus slavus	1	663
	2	36
	3	s.s.
	4	s.s.
	total	667
Proasellus synaselloides	1	s.s.
	2	s.s.
	3	64
	4	235
	total	300
Proasellus walteri	1	27
	2	192
	3	229
	total	530

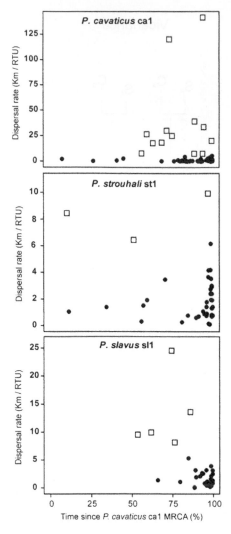

Figure 13.4 Variation in dispersal rates over the course of a species' evolution for *Proasellus* isopods in central and northern Europe. For each branch the dispersal rate is plotted against the mean age of the branch. Relative time units (RTU) correspond to the time duration of a branch divided by the age of the *P. cavaticus* ca1 most recent common ancestor (MRCA). White squares are outliers defined as branches with dispersal rates higher than 1.5 times the interquartile range (boxplot convention). From Eme et al. (2013), used with permission of John Wiley & Sons.

genus *Caecidotea,* and one stygobiotic species in the gastropod genus *Fontigens,* all but one occurred in less than 50 sites, less than 12.5 per cent of sampled seepage springs (Fig. 13.5). One species was nearly ubiquitous (*Stygobromus tenuis potomacus*), occurring in nearly 250 seepage springs (Fig. 2.9). The pattern for copepods found in epikarst drips was less extreme (Fig. 13.6, see also Fig. 3.23), but nonetheless was one of many rare species and a few common species. Of 24 stygobiotic copepod species found in drips, 18 were found in four or fewer drips, and only four species were found in half or more of the drips (*Parastenocaris* n.sp., *Speocyclops*

infernus, Parastenocaris nolli alpina, and *Elaphoidella cvetkae*).

In the case of the epikarst copepods, frequency of occurrence in drips also predicted frequency of occurrence in caves (Fig. 3.22, Culver et al. 2009). That

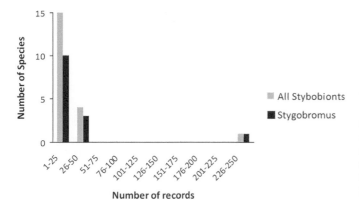

Figure 13.5 Histogram of distribution of number of seepage springs in the mid-Atlantic region of USA occupied by species in the amphipod genus *Stygobromus* (black bars) and all stygobionts (grey bars). The total number of seepage springs sampled is approximately 400. Data from Culver et al. (2012a).

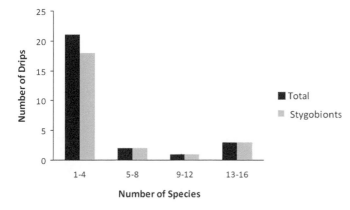

Figure 13.6 Histogram of distribution of number of drips occupied by stygobiotic copepods (grey bars) and all copepods (black bars) in caves in central Slovenia. The total number of drips sampled is 35. Data from Pipan (2003).

is, small-scale (drips) occupancy frequencies were good predictors of larger-scale (caves) occupancy frequencies. It seems likely that differences among species in occupancy frequencies are the result of differences in dispersal ability or differences in ecological preferences, e.g. specialists vs generalists. Body size is an obvious candidate to be a determinant of dispersal rate, although the direction of the prediction is not clear. Larger copepods should not be at the mercy of currents as much as smaller copepods (see Fig. 3.11) but larger copepods should be more restricted by the size of pores and cracks in the epikarst. In fact, there is no statistically significant pattern of body size and number of drips occupied (Fig. 13.7). However, epikarst specialists (stygobionts limited to epikarst) are significantly smaller than other stygobionts and stygophiles (Table 3.14), indicating a likely evolutionary trend towards the evolution of smaller body size in epikarst.

Body sizes of amphipods in mid-Atlantic seepage springs show a strikingly different pattern (Fig. 13.8). Body size is a very good predictor of number of seepage springs occupied, with larger *Stygobromus* amphipods occupying more seepage springs. Log body size explains more than 70 per cent of the variation in the log of the number of seeps occupied. This suggests that habitat size in the hypotelminorheic, unlike epikarst, is not limiting body size. That larger species are found in more seepage springs implies that they can disperse between seepage springs, probably through the litter at times of sheet flow of water, better than smaller-sized species, perhaps because of the ability to move against currents and their greater speed (and hence ability to reach another suitable habitat). Body size is not only a good predictor of frequency of occupancy, but it is also a good predictor of species range (Fig. 13.9), although R^2 is not as high as it was

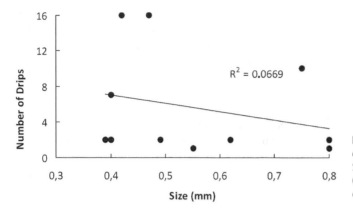

Figure 13.7 Relationship between body size and number of drips occupied by stygobiotic copepods in central Slovenian caves. Data from Pipan (2003) and Culver et al. (2009). R^2 is the fraction of variance in drip occupancy explained by body size.

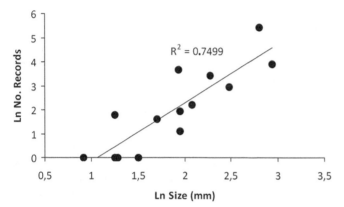

Figure 13.8 Relationship between body size and number of seepage springs occupied by stygobiotic *Stygobromus* amphipods in the mid-Atlantic region, USA. Logs were taken to normalize the data, and similar results are obtained from non-logged data. Data from Culver et al. (2010, 2012a).

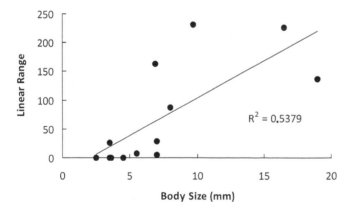

Figure 13.9 Relationship between body size and maximum linear range for mid-Atlantic *Stygobromus* species found in seepage springs. *Stygobromus tenuis tenuis* was excluded because it has very large range (approximately 500 km in linear extent), suggesting it is actually several cryptic species.

for the number of sites occupied. What we do not know is if there is a trend towards larger or smaller body size in the hypotelminorheic. All the *Stygobromus* species are stygobionts and no phylogeny is available that would allow determination of which groups are basal. The presence of a few widespread large species and a larger number of smaller species is consistent with the possibility that competition among *Stygobromus* is important, as Fišer et al. (2012) found for *Niphargus* living in high diversity caves in the Dinaric karst and Vergnon et al. (2013) found for dytiscid beetles in Australian calcrete aquifers.

13.6 Summary

Two questions have dominated discussion of subterranean biogeography: why are ranges so restricted (or alternatively, so large) and is the closest ancestor another subterranean species or a surface-dwelling species? The process of colonization and dispersal has four parts: (1) causes of colonization; (2) success of colonizations; (3) the role of extinction of surface populations in speciation; and (4) the extent of subsurface dispersal. Discussion of these questions and processes in SSHs is modified compared to deeper subterranean habitats because barriers to colonization are less and because SSHs often have a different spatial distribution relative to caves.

Animals may enter SSHs because of deteriorating surface conditions (e.g. temperature) or favourable subterranean conditions (e.g. tree roots in lava tubes). Most colonists into SSHs fail to establish reproducing populations. Species exapted to environments with some similarities to SSHs (e.g. leaf litter) are more likely to succeed, but prediction of which groups will be successful in SSH is only partially possible.

According to the CRH, speciation should be allopatric as a result of the extinction of surface populations, and according to ASH, speciation should be parapatric with a strong selection gradient. Among lava tube faunas in the tropics, there are good examples of both allopatric and parapatric speciation. However, troglobionts and stygobionts without a living surface ancestor may have had one at the time of speciation, but the surface ancestor subsequently went extinct. Likewise, parapatric species may not be the closest ancestor of each other.

On the basis of phylogeny alone, it is impossible to distinguish between multiple isolations of species in subterranean habitats from a single isolation followed by subsurface dispersal without information on ages of lineages. Alternatively, evidence for dispersal and migration can be examined. Although many ranges of stygobionts and troglobionts from both SSHs and deep subterranean environments are small, and follow Trontelj et al.'s (2009) dictum that for ranges greater than 200 km in linear extent cryptic species are present, there are exceptions in the hyporheic (Eme et al. 2013), supporting Ward and Palmer's (1994) concept of the interstitial dispersal highway along rivers and streams.

Two datasets, one on epikarst copepods from Slovenia (Pipan 2003, 2005) and one on hypotelminorheic amphipods from eastern USA (Culver et al. 2012), were extensive enough to examine with respect to the likely extent of subsurface dispersal among SSHs. For both datasets, a small number of species were widely distributed, but most had very restricted ranges. In the case of copepods, species in larger numbers of drips were also in larger number of caves, indicating the distribution was the result of subsurface dispersal. Size was not a predictor of range or occupancy, but specialized species (stygobionts limited to epikarst) were smaller in size. *Stygobromus* amphipods in seepage springs had a very different pattern. Body size was a good predictor of both occupancy and range, with larger species having larger ranges and occupying more seepage springs. This suggests that habitat size in the hypotelminorheic does not limit body size, as it apparently does in epikarst. Size differences among *Stygobromus* species may be the result of interspecific competition, but this remains to be demonstrated.

Phylogeny in shallow subterranean habitats

14.1 Introduction

Because of the long history of study of the biology of caves and because of the recurring idea that the cave fauna is in some way exapted or pre-adapted, phylogenies of cave-dwelling lineages might be expected to have SSH species as an intermediate step in adaptation to caves. This is not the only possible perspective, and in fact, there are several situations where the only subterranean habitats in a region are shallow subterranean habitats. Three prominent examples are calcrete aquifers in the Yilgarn in Western Australia where there is no surface water or other subterranean habitat (Leys et al. 2003), seepage springs on the mid-Atlantic USA Coastal Plain where deep groundwater is scarce and disconnected with surface features (Culver et al. 2012a), and MSS habitats in non-karst areas (Juberthie et al. 1980a, 1980b). An additional challenge in analysing the phylogeny of SSH species is that sometimes their habitats are not clearly identified as being SSHs, especially in the sense of being aphotic. Habitats described as epigean (surface) habitats under large boulders (Barr 1967), leaf litter (Heads 2010), or deep stony debris (Derkarabetian et al. 2010), all with eyeless species, are almost certainly SSHs as we have defined them.

In this chapter, we analyse those few cases of phylogenies that have included SSH species. The starting point is a phylogeny (Dytiscidae beetles) from the Yilgarn calcrete aquifers that only includes SSH species, without surface-dwelling or deeper subterranean species. We then consider somewhat more complex cases that involve surface-dwelling species in the lineage (Opiliones from western USA), or cave-dwelling as well as SSH species (*Proasellus*

isopods from Europe). The last three phylogenies allows us to more explicitly focus on the question of whether SSHs are gateways to the deeper subterranean realm (*Dysdera* spiders from the Canary Islands, spider crickets from the Dominican Republic amber, and New World tropical scorpions).

14.2 Diving beetles in calcrete aquifers

As discussed in chapter 5, several groups of aquatic invertebrates (Dytiscidae beetles, Oniscoidea isopods, and Chiltoniidae amphipods) invaded multiple calcrete aquifers in the Yilgarn region of Western Australia (Fig. 1.12). The best-documented case is that of the Dytiscidae (Leys et al. 2003, Vergnon et al. 2013), and is also the best-documented example anywhere of Climate Relict Hypothesis (CRH, see Fig. 5.6). What Leys et al. (2003) demonstrated is that many different calcrete aquifers were independently invaded, probably during the late Miocene and early Pliocene, between 3.5 and 9 million years ago. Except for eight sympatric species pairs each within a different calcrete aquifer, there was no geographical structure to the phylogeny (Fig. 14.1). Phylogenetic relationships among species coexisting in the same aquifer are highly variable, with sister species in some cases and phylogenetically distant species in others (Leijs et al. 2012), but the species always sort themselves in distinct size classes (Vergnon et al. 2013). There are no surface freshwater habitats in the region at present, and Leys et al. (2003) pointed out that it is unlikely there have been any for several million years.

Leys et al. (2003) suggested that at the time of colonization of calcrete aquifers, the ancestral populations

Shallow Subterranean Habitats. David C. Culver and Tanja Pipan.
© David C. Culver and Tanja Pipan 2014. Published 2014 by Oxford University Press.

may have been living in permanent rivers and streams that began drying out. Dytiscid beetles, by this scenario, could only survive by colonizing and adapting to the hyporheic zone. This is the Hyporheic Refuge Hypothesis (Dole-Olivier 2011). As long-term drying continued, the hyporheic zone became inhabitable both because of drying and sediment clogging, Leys et al. (2003) proposed a second subterranean colonization, this time a colonization of the

calcrete aquifers from the hyporheic. This two-stage model seems likely to be case for other calcrete aquifer inhabitants, including amphipods (Cooper et al. 2007) and isopods (Cooper et al. 2008).

It is clear from the phylogeny in Fig. 14.1 that each stygobiont in a calcrete aquifer is not necessarily the result of a separate invasion, since speciation within the aquifer can and does occur. While invasions occurred at different times, as is clear from Fig. 14.1,

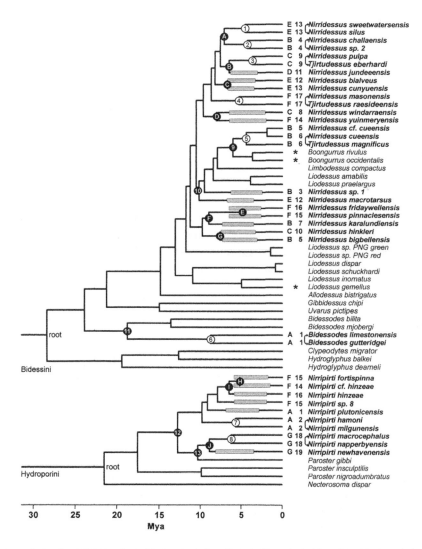

Figure 14.1 Phylogenetic tree for dytiscid beetles in Western Australia calibrated with geological time. Stygobionts are in bold, and sympatric sister species are shown in double-headed arrows. Letters at the tips of the branches show drainage systems and numbers show calcrete aquifers. Aquifer numbers are shown in Fig. 1.12. Divergence time of the modes with black dots shows maximum estimates of transition times to the subterranean environment and the open circles show minimum transition times, based on sympatric sister species. Bars on the branches represent 95 per cent confidence intervals of predicted isolation times. From Leys et al. (2003), used with permission of Blackwell Publishing.

invasion times depended on the time of onset of late Miocene and early Pliocene drying (Fig. 5.6). As a consequence of within-aquifer speciation (we hesitate to call it sympatric speciation since the calcrete aquifers can extend tens of kilometres), total regional diversity can be very high since each of the 52 isolated calcrete aquifers (Vergnon et al. 2013) may accumulate many species through evolutionary time. This has led Guzik et al. (2010) to claim that the Australia fauna is uniquely rich in species, even though the large majority of species are undescribed, and aspects of their methodology have been challenged (Culver et al. 2012c). In addition, the process of interspecific competition sets strong constraints on numbers of coexisting species, at least of amphipods, in calcrete aquifers (Vergnon et al. 2013, see chapter 12). Nevertheless, calcrete aquifers are highly diverse, especially among SSHs.

14.3 Cave and SSH opilionids in western USA

In western USA, Laniatores opilionids in the subfamily Sclerobuninae are typically found in moist, dark, surface habitats (e.g. under rocks and logs) but are also sometimes found in caves (Derkarabetian et al. 2010). They collected individuals from three genera in localities centred around the Four Corners area (Utah, Colorado, Arizona, and New Mexico), and they paid special attention to populations in caves. In addition to a DNA sequence analysis of three regions of the mitochondrial and nuclear genome, they did a morphometric study of the animals. From the point of view of the cave populations, it was clear that each of the five cave populations was an independent invasion of caves. Morphometric analyses showed that cave populations tended to have longer legs (PCA1) and larger palps and chelicerae (PCA2), and that the degree of this troglomorphy, as measured by PCA1 and PCA2, increased with time (Table 14.1 and Fig. 14.2). Each troglomorphic population is the result of a separate and independent colonization of subterranean habitats by surface-dwelling ancestors. What is particularly interesting in the context of SSHs is that the population from deep stony debris near Taos, New Mexico, is as troglomorphic as would be expected for a cave population that had been isolated from

Table 14.1 Relationship between estimated divergence time and PCA1 score (a measure of troglomorphy) for five separate subterranean lineages of harvestmen from western USA. From Derkarabetian et al. (2010).

Troglomorphic sample	Average PCA1	Estimated divergence time (mya)
Skeleton Cave, Colorado	−3.402	0.74
Fault Cave, Colorado	−1.254	1
Taos MSS, New Mexico	2.165	5.9
Cave of the Winds, Colorado	5.030	10.1
Cyptobunus caves, Montana, Nevada, and Utah	4.702	17

surface populations for the same period of time. In fact, Derkarabetian et al. (2010) did not distinguish either the quality or quantity of troglomorphy of the stony debris (MSS) population from cave populations. Its position in the phylogenetic tree (Fig. 14.2) is isolated from other troglomorphic populations and not basal. On the basis of their study, one could conclude that the selective pressures on the MSS and cave populations were identical or nearly so. This is turn suggests that absence of light is the major selective factor (see chapter 9).

14.4 *Dysdera* spiders in Canary Islands MSS and lava tubes

Arnedo et al. (2007) analysed the phylogenetic relationships of ten subterranean species in the spider genus *Dysdera* to a large number of surface-dwelling *Dysdera* all occurring on the Canary Islands. The phylogeny of Canarian *Dysdera* (Fig. 14.3) bears a certain superficial resemblance to the western USA opilionids studied by Derkarabetian et al. (2010). In both cases most subterranean species were found in caves, and there were independent invasions of the subterranean habitat, resulting in independently derived troglobiotic species. Arnedo et al. (2007) found that *Dysdera* independently colonized subterranean habitats at least eight times, and there was only one clear case of subsequent subterranean dispersal and speciation. Among the subterranean species were two that were found in the MSS. *Dysdera madai* was exclusively found in MSS habitats,

Figure 14.2 Combined data 50% majority rule consensus Bayesian tree for Sclerobuninae opilionids from western USA caves and other sites. Major clades are named and thick branches are supported by a posterior probability of 95 or higher. Troglomorphic populations are indicated by special icons. The troglomorphic population of *Sclerobunus robustus glorietus* from an MSS habitat is enclosed in a rectangle. From Derkarabetian et al. (2010).

and its sister species is a surface-dwelling species, *D. iguanensis*, which also occurred in MSS habitats. The other troglomorphic species found in the MSS was *D. esquiveli*, whose sister species is a troglomorphic species found in lava tubes—*D. hernandezi*. The phylogenetic position of *D. madai* is more basal than that of *D. esquiveli*. Several species apparently unmodified for subterranean life are also found in the MSS, including *D. iguanensis* and *D. gibbifera*.

Several other non-troglomorphic species of *Dysdera* were found in lava tubes as well.

In addition to the phylogeny generated from mitochondrial gene sequences of cytochrome oxidase subunit I (*cox1*) and 16S rRNA (*rrnL*), Arnedo et al. (2007) took extensive morphological measures in order to understand morphological variation related to subterranean adaptation in a phylogenetic context. Troglobiotic (and troglomorphic) species

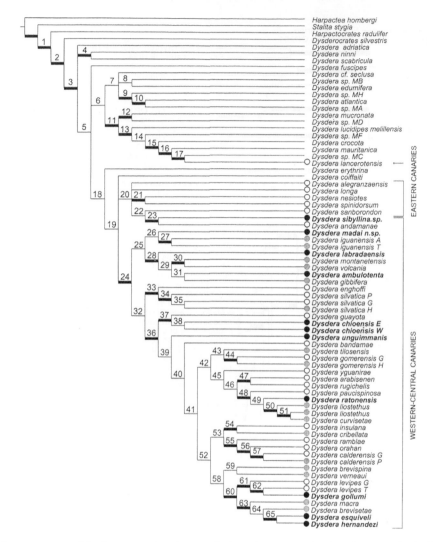

Figure 14.3   Strict consensus tree of the five most parsimonious trees for the spider genus *Dysdera* from the Canary Islands. Thick branches denote clades recovered under all alignment parameter cost combinations. Black dots identify troglobionts, white dots refer to endemic Canarian species exclusively reported from epigean localities, and grey dots show epigean species also collected in subterranean habitats. From Arnedo et al. (2007). Courtesy CSIRO Publishing (http://www.publish.csiro.au/nid/ISO7015.htm).

showed a wide spread of values for the first two principal components (Fig. 14.4). The first principal component showed positive loadings for appendage length and negative loadings for prosoma width, and the second had high loadings for cheliceral differences. Generally, the more troglomorphic species should be on the right side of the graph. The SSH species are close to the centre of the cluster of troglomorphic species, but certainly not the least morphologically modified. If the MSS is a

staging habitat for the colonization of lava tubes, as is suggested by Howarth's bioclimatic model (see Howarth and Hoch 2012), then either the MSS habitat has a selective environment very similar to that of a lava tube (a result Dekarabetian et al. [2010] found for opilionids in MSS and limestone caves [section 14.3]), or the MSS was colonized from lava tubes without subsequent modification. This second possibility is not as unlikely as it might first seem, because as lava tubes erode, the MSS

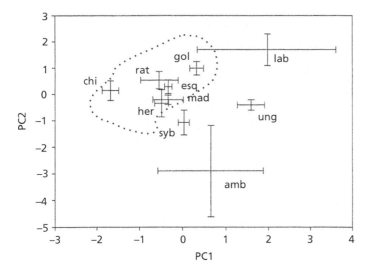

Figure 14.4 Principal Components Analysis of morphological variation in *Dysdera* species from the Canary Islands. The distribution of troglobiont morphology, based on mean principal component analysis scores and 95 per cent confidence limits for each character and each species. The cloud of points enclosed by the dotted line includes the values for troglobiotic species. Species coding: amb—*D. ambulotenta*, chi—*D. chioensis*, esq—*D. esquiveli*, gol—*D. gollumi*, her—*D. hernandezi*, lab—*D. labradaensis*, mad—*D. madai*, rat—*D. ratonensis*, syb—*D. sibylline*, and ung—*D. unguimmanis*. Modified from Arnedo et al. (2007). Courtesy CSIRO Publishing (http://www.publish.csiro.au/nid/ISO7015.htm).

persists, and in order to survive, lava tube species must colonize the MSS, just as they perhaps did earlier in their evolutionary history. For example, La Gomera has no lava tubes, but extensive MSS, and a small number of troglobionts (although no *Dysdera*). These species may either be remnants of a past lava tube fauna or more recent colonists of the MSS. Arnedo et al. (2007) suggested that the variability among troglomorphic species may be due in part to different subterranean niches and because they are not adaptations, but rather exaptations, as is also argued by Desutter-Grandcolas (1997). In any case, the degree of troglomorphy is not correlated with the time of cave colonization, contrary to the results of Derkarabetian et al. (2010).

14.5 *Proasellus* isopods along hyporheic corridors

The phylogenetic analysis of four *Proasellus* species with broad distributions that occupy regions glaciated during the Würm served as an example of the potential for subterranean species to disperse (chapter 13, Eme et al. 2013), especially along the hyporheic corridors of rivers. It is the best analysis of a case of subterranean dispersal and one of the best examples of Ward and Palmer's (1994) hyporheic highway. Not only is the hyporheic a long continuous subterranean habitat, but it may be a geologically old habitat as well. Coineau and

Boutin (1992) suggested that shallow interstitial habitats are among the most geologically stable of aquatic subterranean habitats, and hence can serve as a source of species to colonize other habitats. This idea is similar to that of SSHs as staging habitats for colonization of deeper subterranean waters, but they emphasize the importance of this type of aquatic SSH as a source of colonists in both ecological and evolutionary time.

The phylogenetic tree for *Proasellus cavaticus*, *P. slavus*, *P strouhali*, *P. synaselloides*, and *P. walteri* is shown in Fig. 14.5. Each species is monophyletic and separated by relatively deep splits (Eme et al. 2013). With the exception of *P. walteri*, the discovery of cryptic species did not significantly reduce the range of the most widespread species (Table 13.4). The distribution of the cryptic species of *Proasellus slavus* is shown in Fig. 14.6. Three additional cryptic species within the *P. slavus* morphospecies were found, each with very restricted range. Among the five morphospecies analysed (Table 13.4), a total of ten new cryptic species (seven known from a single site) were uncovered in their DNA analysis. These species were not the result of isolation by distance; else the big ranges would have been split. Rather they occurred at the periphery of the ranges, and at least to a limited extent, these peripheral isolates occurred in deeper habitats than the hyporheic, especially karst groundwater accessible only by wells and springs. Of the ten new cryptic species

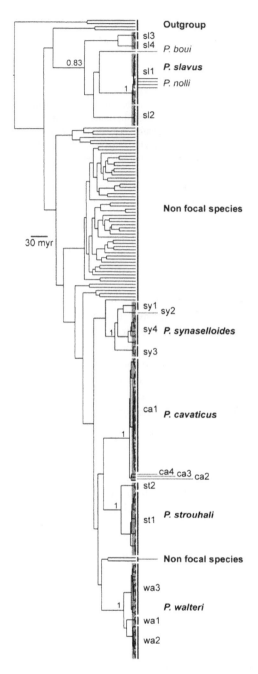

Figure 14.5 Maximum clade credibility tree of 68 Aselloidea morphospecies used to test for the monophyly of the five morphospecies (grey patterns): *Proasellus cavaticus*, *P. strouhali*, *P. slavus*, *P. synaselloides*, and *P. walteri* (BEAST analysis under an uncorrelated lognormal relaxed molecular clock model). Each terminal branch for non-focal and outgroup species corresponds to a morphospecies. Abbreviations ca, st, sl, sy, and wa indicate cryptic species. Numbers along selected branches are posterior probabilities. From Eme et al. (2013).

discovered, four were exclusively found in karst waters (D. Eme pers. comm.). In this example, the hyporheic is acting as a staging area of colonization of deeper subterranean sites.

14.6 Scorpions in caves, leaf litter, and soil

Scorpions in litter have an even more complex and puzzling relationship with species in caves. This is epitomized by the data in Table 14.2, which shows that the number of troglomorphic species (in many cases with reduced or absent eyes and pigment) exceeds the number of species found in caves (Vignoli and Prendini 2009). Most of the other species are reported from leaf litter (e.g. Botero-Trujillo and Francke 2009), and most species are quite rare. For example, the nine described species in the family Typhlochactidae are known from a total of 29 specimens (Vignoli and Prendini 2009). Given the lack of familiarity of many biologists with the MSS (chapter 4), litter habitats may actually be MSS habitats or possibly soil habitats. For example, Vignoli and Predini (2009) listed the habitat of *Typhlochactas mitchelli* as under stones buried in litter, which sounds very much like a description of an MSS habitat. This may also account for the rarity of specimens because collections have not been made in the primary habitats which may be deeper, below the leaf litter. Interestingly, two of the three typhlochactid species from non-cave habitats (*T. mitchelli* and *T. sylvestris*), are among the smallest scorpions known, with total body lengths of less than 10 mm, suggesting a small pore-size SSH. One of the typhlochactid species, *Alacran tartarus*, is found in some of the deepest caves in the world in the state of Oaxaca, Mexico, at depths of 720–916 m below the surface (Francke 1982).

The phylogeny of the family Typhlochactidae, based on 142 morphological characters (Vignoli and Prendini 2009), indicated that the cave species gave rise to three species found in litter and soil habitats (Fig. 14.7). A posteriori optimization on the most parsimonious tree resulted in the cave (hypogean in their terminology) habitat being basal for the Typhlochactidae (Fig. 14.7). In fact, one could argue that cave-dwelling typhlochactids are exapted for litter (or MSS if that is the preferred habitat). As

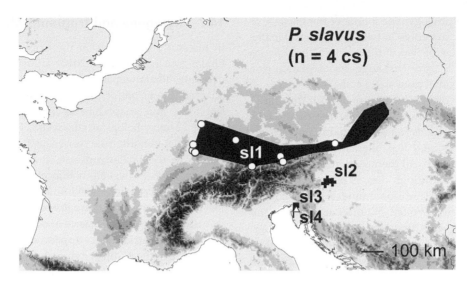

Figure 14.6 Distribution of cryptic species of *Proasellus slavus*. The geographical range of the most widely distributed species is shown by white dots and enclosed in a black envelope of all known localities. The three cryptic species of *P. slavus* are known from two karst wells along the Drava River (Bosnia & Herzegovina) and hyporheic sites along the Drava. From Eme et al. (2013).

Table 14.2 Scorpion families containing species reported to be troglomorphic in the literature, and the number of these that are found in caves. Data from Vignoli and Prendini (2009).

Family	Number of genera	Troglomorphic species	Cave-dwelling species
Akravidae	1	1	1
Buthidae	3	3	1
Chactidae	2	2	0
Chaerilidae	1	3	2
Diplocentridae	2	5	3
Euscorpiidae	1	2	2
Liochelidae	2	2	1
Pseudochactidae	1	1	0
Troglotayosicidae	2	3	1
Typhlochactidae	4	9	6
Urodacidae	1	1	1
Vaejovidae	3	5	4
Total	23	37	22

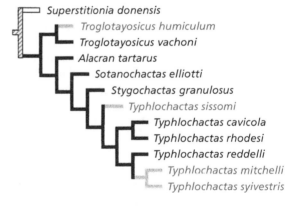

Figure 14.7 Habitat (ecomorphotype) optimized on the single most parsimonious tree obtained by cladistic analysis of 142 morphological characters scored for nine taxa in the New World scorpion family Typhlochactidae and three outgroup taxa. Branches leading to epigean taxon are indicated in white, branches leading to endogean (humicolous) taxa indicated in grey, and branches leading to hypogean (troglobitic) taxa indicated in black. Ambiguous optimization is shaded. From Prendini et al. (2009), used with permission of John Wiley & Sons.

Prendini et al. (2009) pointed out, this pattern also implies that deep subterranean life (troglobionts) are not necessarily evolutionary dead ends.

There is ambiguity about the habitat in which the three non-cave Typhlochactidae are found. Given the attenuation of light in litter, let alone soil, it seems likely that the habitat is aphotic and an SSH, but this is just a conjecture. In addition, the presence of these eyeless species in the habitat suggest it is aphotic, although it is possible that eyelessness is a result of the phylogenetic inertia of the lineage, or that eyeless can evolve in twilight habitats. Finally, the minute size of two of the species (*Typhlochactas mitchelli* and *T. sylvestris*) suggests adaptation to a habitat with small pore size, i.e. the soil.

A number of habitats are ones with dim light, including the benthos of streams, nests of mound-building ants (e.g. *Formica rufa* group), upper parts of talus slopes, and leaf litter. It would be incorrect in our view to call them subterranean habitats, because light, although dim, is present at least in close proximity. Traditional models of the colonization of caves invoked the litter as the ancestral habitat for species that became isolated in caves. For example, Peck (1980) reported that troglobiotic species only occurred in those caves in the Grand Canyon (Arizona, USA) where forests occurred on the surface during the Pleistocene. Speleobiologists have often used the phrase 'pre-adapted' to describe the litter fauna in relation to subterranean environments. The term itself is unfortunate because it implies a direction to evolution (teleology), and Romero (2009) strongly criticized its continued use.

No matter what the exact nature of the scorpion habitat is, their analysis, especially that shown in Fig. 14.7, indicates a colonization of non-cave habitats (most likely SSHs but possibly 'twilight' habitats such as deep litter) from cave populations. This is exactly the reverse of the scenario proposed for Pleistocene colonization of caves (CRH) during warming (e.g. Barr and Holsinger 1985). In fact, Mitchell and Peck (1977) proposed that the soil/litter/MSS typhlochactid scorpions were ancestral to the cave species, a proposal not borne out by subsequent cladistic analysis. Whichever direction of evolution, either from cave to non-cave or non-cave to cave, there was a change in body size, with bigger animals found in caves. If the source of cave colonists during the Pleistocene was soil-dwelling species, then considerable morphological change would be expected, including increase in body size and increase in appendage elongation. The evolution of eyelessness and pigment loss in the soil could be considered an exaptation (pre-adaptation of some authors), but other morphological features of the soil organisms are not.

14.7 Phylogeny and troglomorphy

In his study of crickets in the tropical subfamily Phalangopsinae, Heads (2010) argued that cave-dwelling Phalangopsinae are exapted, not pre-adapted, by virtue of the evolutionary history in leaf litter, and adaptation to living in leaf litter. Using 22 morphological characters, Heads (2010) produced a majority rule consensus tree for the ten genera in the Phalangopsinae, including a fossil genus (*Araneagryllus*) from Miocene amber in the Dominican Republic (Fig. 14.8). He then looked at the distribution of four troglomorphic characters (eye area, presence or absence of ocelli, presence or absence of the stridulatory apparatus, and presence or absence of the auditory tympana) and found that only eye area was congruent with the distribution of cave-dwelling species.

Two comments need to be made about the list of troglomorphic characters. The first is that, aside from fellow orthopterist Desutter-Grandcolas (1997), loss of stridula and tympana is not usually considered a troglomorphic character, and certainly not a convergent feature of subterranean life. The second is that all four troglomorphic characters used by Heads (2010) are reduced characters. Desutter-Grandcolas

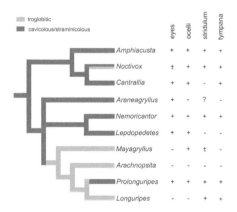

Figure 14.8 Phylogenetic inference assessment of Amphiacustina. Life histories are mapped onto the phylogeny in light grey (troglobionts) and dark grey (leaf litter epigean), and the distributions of the supposedly troglomorphic characters are shown on the right. Key: eyes large (+) or reduced (−); ocelli present (+) or absent/reduced (−); stridulatory apparatus (stridulum) present and functional (+) or absent/non-functional (−); auditory tympana present (+) or absent (−); in all cases, the ± symbol indicates polymorphism. From Heads (2010).

(1997) did consider an elaborated character (relative hindleg size) and finds it is also not congruent with subterranean species. Thus, the amount of exaptation relative to adaptation is still not completely resolved, although both Heads (2010) and Desutter-Grandcolas (1997) rejected the concept of troglomorphy. Regardless of the amount of exaptation it is clear that the cave species are derived from the litter/MSS species (Fig. 14.8), as expected by the standard paradigm of the Climate Relict Hypothesis. However, bear in mind that these are tropical species so that strictly speaking, relictualization as a result of the Pleistocene is highly unlikely.

One way to re-evaluate Heads' findings is to re-evaluate the nature and definitions of the habitats involved. If the litter-dwelling (straminicolous) species are in an aphotic habitat, then it could be argued that the spider crickets are adapted to subterranean habitats, whether they are SSHs or caves, and indeed we have argued that absence of light is a major selective factor in the evolution of subterranean organisms. In a sense, it is not exaptation because there are not really two distinct habitats. Rather there is a continuum of aphotic habitats.

However, such semantic debates do not speak to Heads' argument, which is made more forcefully by Desutter-Grandcolas (1997), that troglomorphy is not a real phenomenon. The conceptual problem for cladists such as Desutter-Grandcolas is that troglomorphy is in fact a set of homoplasies, and homoplasy is minimized in cladistic analysis (Desutter-Grandcolas et al. 2003). Phylogenetic systematics depends on the existence of shared derived characters (synapomorphies) and homoplasies are really false synapomorphies. However, Heads (2010) and Desutter-Grandcolas (1997) may be right about the difficulties associated with the concept of troglomorphy, and we have pointed out some of these as well (chapter 12, Pipan and Culver 2012a). However, analysis of two or three putative troglomorphic characters does not constitute a refutation of the concept, in our opinion. A more complete approach would be a broader morphometric study where large numbers of characters are analysed, and sets of characters are identified as being associated with the subterranean habitat rather than a node or nodes in the phylogeny. This was the approach of Christiansen (1961) in his now

classic studies of convergence in claw morphology of Collembola, and the approach of Prevorčnik et al. (2004) who analysed more than 50 characters in subterranean and surface-dwelling *Asellus aquaticus* isopods in this regard.

14.8 Summary

Some SSHs may serve as gateways to the deeper subterranean realm, but others are in geographical locations where no other subterranean habitats are present. In some cases, SSHs are not recognized as subterranean habitats and are listed as epigean, litter, or soil habitats, creating additional problems of analysis. An example of a phylogeny for SSH species in a region without other subterranean habitats is that for the Yilgarn calcrete aquifers of Western Australia. There are 52 such isolated calcrete aquifers. Leys et al. (2003) proposed that dytiscid beetles first colonized hyporheic habitats of surface streams as the continent dried in the late Pliocene and early Miocene. As drying continued, beetles colonized the calcrete aquifers. The potential number of species is larger since each species occurs in only one calcrete aquifer and speciation within the aquifer can and does occur.

A second phylogeny involving SSHs is that of opilionids in caves and MSS in western USA. The degree of troglomorphy is strongly dependent on the age of the lineage in caves, and an SSH species is as troglomorphic as a cave-dwelling species of that evolutionary age was predicted to be, i.e. there seemed to be no difference between the course of evolutions in SSHs and in caves.

Spiders in the genus *Dysdera*, including troglomorphic and non-troglomorphic species, are present in MSS habitats as well as lava tubes in the Canary Islands. There is no evidence that the MSS species are intermediate in morphology compared to lava tube and surface-dwelling species. There was no evidence for overall differences between lava tube and MSS species.

The isopod genus *Proasellus*, common in hyporheic habitats in central and northern Europe, is an example of how SSH species may be staging areas for colonization of deeper subterranean habitats. Contrary to the situation for most stygobionts, four species of *Proasellus* studied by Eme et al.

(2013) had ranges greater than 200 km, even after accounting for cryptic species. A number of cryptic species were found, but most had very small ranges, typically a single locality. Many of these cryptic species were found in karst habitats, suggesting that the hyporheic acts like an interstitial highway of dispersal, as proposed by Ward and Palmer (1994).

Tropical scorpions in the family Typhlochactidae occur in caves and in shallow subsurface habitats, which may or may not be aphotic. Whatever the precise habitat, it is clear that the cave species are basal, the near-surface species (in SSHs?) are derived from them, and that the ancestral habitat for the family was caves. Animals from non-cave habitats were very small, suggesting these species occupied a subterranean habitat of small pore size, perhaps the soil. The difference in morphology between troglobionts in soil and caves is large, and both body size and shape differ profoundly. In any case, troglobionts are not necessarily an evolutionary dead end.

Because of the apparent reversibility of evolution and the apparent lack of convergence aside from eye and pigment loss, some cladists (Heads and Desutter-Grandcolas) have suggested that there is no such thing as troglomorphy. However, their analysis was based on very small numbers of characters, and larger-scale morphometric studies show the presence of considerable convergence.

CHAPTER 15

Conservation and protection of shallow subterranean habitats

15.1 Introduction

By their very definition, SSHs are vulnerable habitats. They are hidden from view, yet shallow. Many if not most local environmental problems involve the alteration of this shallow layer of the subsurface. Development, including urban and industrial development, by necessity involves the destruction or alteration of at least the uppermost layer of the earth's surface. Toxic spills first contaminate upper layers of both terrestrial and aquatic habitats, the very habitats that are the subject of this book. Many of these threats are especially insidious because their surface manifestations may quickly disappear, and there may not be awareness that there are even SSHs present.

Conservation biologists look to two paradigms for justification for protection of species and habitat. One of these is the concept of ecosystem services (Kareiva et al. 2011)—goods and services to human populations provided by ecosystems that would otherwise have to be accomplished in some other way. Among SSHs, the most obvious example is the soil. The multi-billion dollar fertilizer industry, which provides fertilizers because impoverished soil does not provide enough nutrients itself, is an example of the cost of replacement of an ecosystem service. Other SSHs provide less obvious ecosystem services. The critical role of the hyporheic in the health of rivers and streams is a multi-faceted example of ecosystem services. Buss et al. (2009) listed the following goods and services provided by the hyporheic:

- Controlling the flux and location of water exchange between stream and subsurface.
- Providing a habitat for benthic and interstitial organisms.
- Providing a spawning ground and refuge for certain species of fish.
- Providing a rooting zone for aquatic plants.
- Providing an important zone for the cycling of carbon, energy, and nutrients.
- Providing a natural attenuation zone for certain pollutants by biodegradation, sorption, and mixing.
- Moderating river water temperature.
- Providing a sink/source of sediment within a river channel.

Other aquatic SSHs, such as epikarst, may also provide ecosystem services, especially for the natural attenuation of pollutants (Herman et al. 2001). In general, ecosystem services provided by terrestrial SSHs (except the soil) are less important. Lava tubes also provide an ecosystem service indirectly since many lava tubes harbour bat populations. Bats provide a variety of ecosystem services including insect control, seed dispersal, and pollination (Kunz et al. 2011). However, not all SSHs provide ecosystem goods and services, and, for example, it is hard to list an ecosystem service provided by seepage springs.

The second paradigm for species protection is a biocentric view, that species are worthy protecting in their own right (Danielopol 1998, Danielopol and Pospisil 2005). The morphology of subterranean

Shallow Subterranean Habitats. David C. Culver and Tanja Pipan.
© David C. Culver and Tanja Pipan 2014. Published 2014 by Oxford University Press.

animals is often unique and unusual, sometimes even bizarre, making them an important part of biodiversity.

Based on extensive surveys of regional cave fauna, with a thousand or more records (e.g. Culver et al. 2006a, Zagmajster et al. 2008), most stygobionts and troglobionts in caves are highly restricted geographically and often are numerically rare, making them vulnerable to even relatively minor disturbances. Relative to caves, SSHs are data deficient. For most SSH types, only a handful of sites have been well studied. For example, only three hypotelminorheic sites are well studied with more or less complete species lists—Medvednica Mountain in Croatia, Nanos Mountain in Slovenia, and the lower Potomac River drainage in the USA (see chapter 2). A special problem is that many SSH sites are not only insufficiently sampled, but their very existence is not known at all. Nevertheless, we examine the concept of rarity (Rabinowitz et al. 1986) in section 15.2.1, and how it applies to the SSH fauna, using examples from the cave fauna when no data are available for SSHs.

There are other biological factors in addition to rarity that may put the SSH fauna at increased risk of extinction, including low reproductive rates, high susceptibility to environmental change, and inability to withstand disturbance. Taken together, these factors may reduce SSH species' ability to respond to environmental change and stress, a topic covered in section 15.2.2. In light of these biological attributes, we consider some generally important threats to SSH communities in section 15.3. These include universal threats such as global warming and groundwater pollution (especially nonaqueous phase liquids [NAPLs]), which should be recognized as a universal threat to subterranean habitats, especially shallow ones (Loop 2012).

One unifying aspect of the conservation and protection of SSHs is that of connectivity. Threats are connected, as is the case when a spill moves laterally underground (Loop 2012), but so are populations. A common theme in speleobiology is that of endemism and restriction of dispersal, a prime example being the fauna of calcrete aquifers (chapter 5). However, populations are often connected at the local level at the scale of tens of kilometres or even larger (Christman et al. 2005) and protection strategies must take this into account. This requires

a landscape approach to protection. This will be the subject of section 15.4.

In the final section (15.5), we consider each SSH types, discuss threats, protection status, and provide examples of successful protection efforts. This includes a discussion of site management. The time is long past when it is appropriate to 'protect' an SSH by only protecting a few metres around the site.

15.2 Biological risk factors

15.2.1 Rarity

Rabinowitz (1981) stated that rarity means different things to different biologists. From a botanical perspective she suggested that rarity has three meanings—a species can be numerically rare throughout its range (numerical rarity), it can occur in a rare habitat or habitats (habitat rarity), and it can be geographically rare with a restricted range (geographical rarity). Species may be rare along one, two, or all three of her axes, leading to her 'seven forms of rarity'. Her classification works perfectly well with regard to the subterranean fauna.

There is little doubt that the majority of the subterranean fauna is geographically rare. In an analysis of stygobionts from caves and SSHs in Belgium, France, Italy, Portugal, Slovenia, and Spain, Michel et al. (2009) found that 464 of 1059 stygobionts (44 per cent) occurring in caves and interstitial habitats were limited to a single $0.2° \times 0.2°$ cell, approximately 400 km^2 in area. In the case of the USA troglobiotic cave fauna east of the Mississippi River, 211 of 467 species (45 per cent) are known from a single cave (Christman et al. 2005)! One of the hallmarks of subterranean organisms is their endemism, although this is largely based on data for the cave fauna. All of these levels of endemism are much higher than that recorded for any surface habitat. Endemism seems higher in troglobionts than stygobionts, and troglobionts in general have smaller ranges (Lamoreux 2004), but to date no pattern has been reported with respect to trophic position or body size (Culver et al. 2009).

The pattern with respect to habitat rarity is mixed. The hypotelminorheic has not been studied in any relatively large geographical area except for the middle Atlantic region of eastern USA. Figure 15.1 shows all of the hypotelminorheic localities where

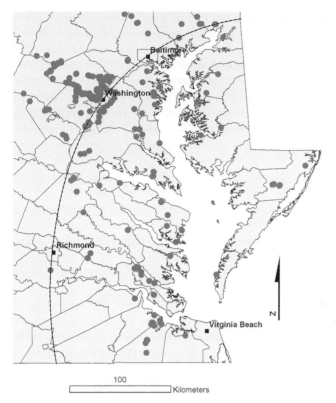

Figure 15.1 Hypotelminorheic sites with stygobionts in the mid-Atlantic region of the USA. Sites are shown as grey dots. Most hypotelminorheic sites in this region have at least one stygobiont. The approximate location of the Fall Line (the boundary between the Coastal Plain and the Piedmont with metamorphic rock) is shown as a dotted line. From Culver et al. (2012a).

one or more stygobionts have been collected, over an approximately 10,000 km² area. While not a complete set of localities, and probably biased towards localities near Washington, DC, it does represent several decades of work, and shows the patchy distribution of the habitat. This patchiness is evident at smaller scales as well (Fig. 1.11). The patchiness is likely dictated by the distribution of clay layers in the shallow (< 2 m) subsurface.

As a habitat, epikarst is probably more common than caves, epikarst being more or less continuous in karst areas, which comprise about 15 per cent of the earth's surface (see section 3.2.3) except in the tropics and glaciated areas.

Interstitial habitats, especially fluvial aquifers are common, or at least as common as surface streams and rivers. They are not really patchy in distribution (although some karst areas have few surface streams), but are one-dimensional rather than two-dimensional in their geographical coverage.

Calcrete aquifers are probably not limited to western Australia, but they have only been stud-

ied in Western Australia and Northern Territory. In Western Australia, there are over 50 such calcrete aquifers (Vergnon et al. 2013, see Fig. 1.12), ranging in size from less than 3 km² to greater than 200 km² (Leijs et al. 2012), but they are a small fraction of the entire Australian landscape (see Fig. 1.12).

With respect to terrestrial habitats, intermediate-sized terrestrial SSHs (MSS-like habitats) are likely to be patchily distributed on broad geographical scales. For example, it would be difficult to even know where to look for these habitats in areas of little relief, even flat-lying karst areas like the Interior Low Plateau of the USA. At the other extreme, the MSS is nearly ubiquitous on some volcanic terrains, such as La Gomera in the Canary Islands. In mountainous terrains, it seems to be patchily distributed. V. Růžička (pers. comm.) reports that nine per cent of protected mountain areas in the Czech Republic have scree slopes.

Lava tubes perhaps have the most restricted geographical distribution of any SSH, limited to narrow bands of volcanic activity on a global scale, especially around the Pacific Rim (Fig. 8.1). As with other SSHs, on a local scale it can be a very common habitat, with densities of caves exceeding 10 km^{-2} and 5 km of passage per km^2 (Table 8.2). Finally, soil is of course nearly ubiquitous.

All in all, most SSH habitats are patchily distributed (with the exceptions of soil and the hyporheic), and are locally common, with habitats numbering in the hundreds if not thousands.

The question of numerical rarity of stygobionts and troglobionts is a very interesting one (see Box 15.1). A considerable number of troglobionts and stygobionts are known from only a handful of specimens, in some cases a single specimen. The obvious rarity of these species may be more apparent than real because it is likely that the primary habitats of many of these species are not the caves where they are often collected but habitats such as epikarst and phreatic water, and that they are accidentals in caves. Other populations of stygobionts and troglobionts are known to be quite large. The best-known example is that of the Baget ecosystem in France (see chapter 6), where the number of individuals of stygobiotic copepod species that washed out of the system was in the thousands (Rouch 1970).

Thus, the majority of stygobionts and troglobionts are geographically rare, many are likely to be numerically rare, and a few are in rare habitats. This rarity has important conservation implications. Geographically rare species are subject to catastrophic losses as the result of relatively minor and frequent environmental insults, if for no other reason than their geographically restricted range. Numerically rare species are more likely to go extinct than common species because of genetic inbreeding, demographic stochasticity (such as the appearance of a single-sexed population in a generation), and environmental stochasticity (minor environmental insults). Taken together, conservation biologists call these phenomena the extinction vortex (Groom et al. 2005).

15.2.2 Other biological risk factors

Because many stygobionts and troglobionts have relatively low reproductive rates, including SSH species (Crouau-Roy 1987, Rouch 1968), their rate of population growth following an environmental insult will be low, resulting in a smaller population

---

Box 15.1 Estimating population size

There are two general methods available to estimate population size that bring some clarity to the range of observations about population size. One of these is mark–recapture, based on recapturing in a second sample, individuals that were marked in the first sample as well as unmarked individuals. Population size (X) and its standard error can be estimated as follows (Begon 1979):

$$X = an/r$$
$$S.E.(X) = [a^2n(n-r)/r^3 + 1]^{0.5}$$

where a is the number marked in the first sample, n is the number of individuals in the second sample, and r is the number of marked individuals in the second sample. Such studies with subterranean animals are technically difficult and standard errors are often very large because of relatively small sample size but nonetheless very informative. Knapp and Fong (1999) estimated that the epikarst amphipod *Stygobromus emarginatus* had a population of approximately 3500 individuals in drip pools along 300 m of passage in Organ Cave, West Virginia, USA.

Population geneticists use a different measure of population size, N$_e$, 'effective population size', basically the size of

a randomly mating population that would result in the same levels of heterozygote frequency and same levels of genetic variation as the observed population. A population with low genetic variation or low heterozygosity has a smaller effective population size. Empirically, effective population size is estimated using 'coalescence size', which is the ratio of nucleotide diversity (which under neutral theory should be 4μN$_e$) to mutation rate (Lynch and Conery 2003).

Except for a few fish populations, there is little evidence of reduced genetic variability in stygobionts and troglobionts (Sbordoni et al. 2000, 2012). Buhay and Crandall (2005), using mitochondrial DNA sequences, calculated an effective population size for several stygobiotic species of crayfish in eastern USA of between 20,000 and 80,000! Sbordoni et al. (2012) attributed the large estimates of effective population sizes of subterranean species in general to selection for heterozygotes rather than population size per se. In general it appears that subterranean populations, like surface-dwelling populations, can be either large or small and some unknown fraction of stygobionts and troglobionts are numerically rare.

for a longer period of time relative to surface-dwelling populations. This results in increased extinction risk because they are in the extinction vortex longer.

Stygobionts and troglobionts may also be especially sensitive to some kinds of environmental fluctuations. For example, troglobionts in lava tubes are often especially sensitive to changes in relative humidity as a result of exoskeleton thinning (Howarth 1980, 1983). Stygobionts may be more or less sensitive than surface relatives to heavy metals (Notenboom et al. 1994), but like their surface counterparts, stygobionts, especially interstitial species, frequently must cope with heavy metal contamination. This is especially the case for stygobionts in close contact with sediments. Because of the relative insolubility of most heavy metals, the metals remain in sediments (Vesper 2012). Inhabitants of the hypotelminorheic are especially likely to encounter more sediment, especially in periods where they burrow into the clay. Stygobionts and troglobionts appear to be especially sensitive to non-subterranean competitors and predators that can occur in subterranean sites as a result of pollution events, especially organic pollution of streams (Sket 1977).

On the other hand, relative to cave stygobionts and troglobionts, SSH stygobionts and troglobionts are likely to encounter more environmental variability. Examples include:

- greater temperature variability in MSS relative to caves (Jama v Kovačiji, Fig. 4.11);
- occurrence of freezing temperatures in MSS (Figs. 4.11, 4.12);
- periodic drying of the habitat in the hypotelminorheic (Fig. 2.5);
- organic carbon in the hypotelminorheic (Fig. 11.3);
- periods and areas of reduced oxygen in hyporheic habitats (Figs. 6.5, 6.8, and Table 6.3); and
- potentially, epikarst discharge rates relative to cave stream discharge (Fig. 1.10).

Among SSHs, lava tubes (Figs. 1.8, 8.3), calcrete aquifers, and the soil are the most stable, and rival deep caves in this respect. The biotic factors that increase the vulnerability of the SSH fauna are summarized in Table 15.1.

Table 15.1 Biotic factors of the subterranean fauna that increase extinction risk.

Biotic factor	Applicability
Geographical rarity	All SSHs except soil and hyporheic
Numerical rarity	Probably predators and larger-bodied species
Habitat rarity	Hypotelminorheic, calcrete aquifers
Low reproductive rate	Most stygobionts and troglobionts
Sensitivity to environmental fluctuations	More stable SSHs, lava tubes, soil, and calcrete aquifers
Sensitivity to surface-dwelling competitors and predators	Most stygobionts and troglobionts

15.3 Physical threats to the SSH fauna

Threats to the SSH fauna are about as diverse as threats to surface-dwelling species, especially since most environmental disasters in surface habitats are environmental disasters for subsurface habitats as well. Jones et al. (2003) divided threats into three overarching categories:

- Alteration of the physical habitat.
- Water quality and quantity.
- Direct changes to the SSH fauna.

Threats to SSH faunas are present throughout the world, but there are regional differences. Threats in developed and developing countries may be different.

15.3.1 Alteration of the physical habitat

Quarrying of limestone, especially for the making of cement, is the ultimate kind of threat because it completely removes habitats in karst, including epikarst. Worldwide, the area of fastest growth in limestone quarrying is southeast Asia. Annually, 1.75×10^7 tonnes (metric tons) of limestone are quarried from southeast Asia and it is growing at a rate of six per cent per year, higher than any other area of the world (Clements et al. 2006). The limestone karst regions of Indonesia, Thailand, and Vietnam cover an area of 400,000 km² and are 'arks of biodiversity' (Clements et al. 2006), both because they are biodiversity hotspots and because they are relatively untouched by agricultural and forestry practices because of the rugged terrain, such as tower karst and karst pinnacles (Ford and Williams 2007). Both the surface and subsurface biota are very incompletely known, and this is one of those situations where species disappear before they are even discovered (Deharveng et al. 2009a). Because of the island-like nature of these karst outcrops (see chapter 7), endemism of not only the subsurface species, but surface-dwelling species as well, is high. Clements et al. (2006) also pointed out that the subsurface and surface of karst areas in southeast Asia is under-studied, even relative to other habitats in the region. Quarrying is not the result of entirely local factors. Much of the cement produced is exported, and funding for the development of cement plants in southeast Asia is international.

Nearly as destructive as quarrying is the practice of gravel excavation along rivers, which destroys (by removal) the hyporheic zone. Dam construction has nearly as devastating an impact as well. Any new construction, but especially road construction, destroys SSHs, especially epikarst, MSS habitats, and of course soil. The details of the location of highways through karst regions can make a big difference in terms of environmental impacts.

Another type of SSH alteration is development of a lava tube for tourist visitation. The most visited lava tube in the world is Thurston Lava Tube in Hawaii Volcanoes National Park, with upwards of one million visitors a year (Halliday 2004). Little remains of any fauna in the cave, and it has become what is sometimes called a sacrifice cave, in the sense it concentrates damage from human visitation in one cave. Commercialization of a lava tube requires physical alteration of natural passages and installation of lights (although not in the case of Thurston Lava Tube), with the concomitant development of a lampenflora, the growth of plants associated with electric lighting (Mulec 2012). Speleobiologists have frequently warned against excessive commercialization of caves because of these changes (Elliott 2000, Hamilton-Smith and Eberhard 2000).

Buss et al. (2009) pointed out that even many legitimate activities in streams and rivers can have deleterious consequences for the hyporheic zone, including:

- dredging, which removes habitat;
- weirs and impoundments, which increase deposition of fine sediments and clogging of the habitat;
- alteration of the riparian zone with a concomitant increase in sedimentation;
- flood protection, especially channelization of rivers, which reduces river–floodplain connectivity; and
- some river restoration with an overemphasis on aesthetic improvement, which may destroy floodplains, etc.

Hypotelminorheic habitats are extremely easy to destroy, even by such minor activities as trail construction (Culver and Šereg 2004) or livestock grazing, which disrupted and destroyed hypotelminorheic habitats on Nanos Mountain, Slovenia. Deforestation has a dual negative impact by changing organic inputs and altering the local hydrological cycle of the hypotelminorheic. In general, deforestation has a negative impact on all SSHs.

Calcrete aquifers, because they are so shallow, can be destroyed by quarrying or excavation for access for stock watering (Fig. 1.4B).

15.3.2 Water quality and quantity

The hyporheic is subject to most of the same environmental threats as rivers and streams. There are few places left in the world where human activity has not had a negative impact on water quality. In surface habitats, species living in rivers are typically more at risk of extinction than species in other habitats, based on data for the USA fauna (Master et al. 1998). Several environmental drivers are of critical importance. One is agriculture and the nearly universal over-application of fertilizers and pesticides that accompanies it. For example, nitrate levels in groundwater throughout Europe continue to rise, and due to overuse of fertilizers and pesticides, they are frequently found in shallow groundwater (Notenboom 2001). Another important driver is water extraction for agriculture, industry, and urban activities, especially the extensive use of irrigation in agriculture. Groundwater levels have fallen in many areas, more than 30 m in some cases (Danielopol et al. 2003). This results in changed and reduced connections between surface and subsurface with the concomitant reduction in organic carbon and nutrient exchange (see chapter 11).

Even relatively mild alterations to aquifer recharge affect the fauna. In a shallow groundwater aquifer in the city of Lyon, France, Datry et al. (2005) found that artificial recharge of some sites with storm water changed the species composition at those sites, although this was probably the result of the increase in organic carbon rather than in recharge per se.

All SSHs are at risk from spills of toxic materials, typically from trucks on roads, but also from railroads and leaking storage tanks, especially associated with petrol stations. The number of underground storage tanks is staggering. In the mid-1980s, at the start of the USA Leaking Underground Storage Tank (LUST) Trust Fund, there were more than two million storage tanks in the USA, and by 2013, there had been over 500,000 toxic releases (US EPA 2013). These spills either quickly enter SSHs or are buried in SSHs to begin with. Most storage materials are non-aqueous phase liquids (NAPLs), which includes fuels, solvents, and insulators (Loop 2012). NAPLs have, by definition, limited solubility in water and hence are difficult to remove from the subsurface especially because groundwater extraction and treatment ('pump and treat') is very inefficient (Herman et al. 2001). Even the limited solubility of many NAPLs results in groundwater contamination that exceeds the maximum concentration level set by the US Environmental Protection Agency (Loop 2012). NAPLs include hydrocarbons, vinyl chloride, PCE, PCBs, and many insecticides and herbicides such as atrazine. NAPLs in general come to reside in the shallow subsurface, and are moved out of this zone by precipitation, but are often held for extensive periods of time in the soil, epikarst, and regolith, and continue to contaminate the aquifer below. NAPLs can be retained in pores (epikarst and MSS), and in sediments (soil and hypotelminorheic). Natural degradation (attenuation) is faster for hydrocarbons than chlorinated compounds such as chlorinated solvents and PCBs, but for a large spill may take many decades.

15.3.3 Direct changes to the SSH fauna

Overcollecting of stygobionts and troglobionts is a problem in some cases (Elliott 2000). Most speleobiologists have had the experience of collecting

from a site, and returning to find very few animals. Prudence dictates that collecting be kept to the minimum necessary. In Europe, the problem is exacerbated by a collectors' market for some cave animals, such as beetles. A potentially specially damaging collecting technique is pitfall trapping, used in lava tubes and the MSS. Pitfall traps with killing solutions at the bottom, and baited with cheese or rotting meat can attract and kill hundreds of troglobionts in a day or two.

A final human activity may, in the coming years, dwarf all other impacts. This is global warming. Since subterranean habitats in general are less variable than surface habitats, organisms in subterranean habitats will experience less in the way of temperature extremes. However, the rise in average temperature will increase average temperatures of subterranean habitats since their temperature approximates the mean annual temperature. Since any organism, surface or subterranean, is rarely adapted to temperatures it never encounters, rising temperatures may result in lethal conditions for some stygobionts and troglobionts. Temperature changes will have a profound impact on organic carbon storage in the soil, in part by altering decomposition rates (Davidson and Janssens 2006). Since stygobionts and troglobionts have very limited dispersal capability, if temperatures are rising relatively rapidly, species may go extinct. With longer droughts and more intense precipitation events, local and regional water balances (precipitation and evapotranspiration) change. Habitats such as the hypotelminorheic and the hyporheic of small streams may dry up. After prolonged drought, epikarst habitats may dry up as well. Relative humidities in lava tubes and MSS may change, creating profound physiological problems for subterranean terrestrial species (Howarth 1980).

15.4 A landscape approach to conservation and protection of SSHs

A hallmark of the cave fauna, which is much better documented than the SSH fauna (e.g. Culver et al. 2006a, Zagmajster et al. 2008), is the high level of endemism, with many species known from a single cave (Christman et al. 2005). Available information suggests that for many SSHs, high levels of endemism are also present. For example, Pipan et al. (2007a) found that the linear extent of most epikarst copepod populations studied in Slovenia and West Virginia was approximately 100 m, indicating high endemism at the population level. In spite of intensive collecting efforts, the linear extent of the range of the hypotelminorheic amphipods *Stygobromus hayi* and *S. kenki*, which occur in the Washington, DC area, was less than 5 km (Culver et al. 2012a). Even more extreme is the case of the fauna of calcrete aquifers, where nearly all stygobionts are restricted to a single aquifer (Guzik et al. 2010, chapter 5).

We think that it is a mistake to conclude from these patterns that protection of single sites is sufficient to protect the SSH fauna. There are three reasons for this caution. The first is that, if taken literally, the areas of protection become absurdly small. The amphipod *Stygobromus nanus* is known only from drip pools all within 5 m of each other in Piddling Pit, West Virginia (Holsinger 1978), but to protect only this tiny area would be absurd. The epikarst amphipod *Niphargobates orophobata* is known from a single drip in the Postojna Planina Cave system in Slovenia (Sket 1981). As a final example, the planarian *Sphalloplana hypogea* is only known from two tile drain outlets (an artificial hypotelminorheic habitat—see chapter 2) in coastal Virginia, USA (Kenk 1984). It is even more absurd to protect only the tile drain of a cultivated field.

The second reason to take a landscape (regional) view is that even if populations are naturally isolated and highly localized, threats aren't. Of the threats enumerated in section 15.3, the one with the strongest geospatial aspect is that of toxic spills, a threat to both terrestrial and aquatic SSHs. In any toxic spill, there will be some lateral movement of the NAPL, at a minimum because of the mounding effect of the spill. Lateral channels in epikarst and to a lesser extent in the MSS and regolith, will also promote lateral movement. So, protection must take into account threats which are not just local but regional, especially upstream of the populations being protected.

The third reason to take a landscape view is that not all SSH species are limited to small areas, with little or no dispersal. Perhaps it is the case that a stygobiont in a calcrete aquifer is both at an evolutionary

dead end and without possibility of dispersal, but this is not the case for most SSH species. Prendini et al. (2009) argued that troglomorphic scorpions in caves could give rise to troglomorphic scorpions in leaf litter, and that at least in an evolutionary sense, there could be dispersal out of subterranean habitats. Some hyporheic species have ranges much in excess of the 200 km limit for stygobionts suggested by Trontelj et al. (2009). Eme et al. (2013) reported that some *Proasellus* isopod species were dispersed much greater distances along the hyporheic of rivers, which Ward and Palmer (1994) called the interstitial highway. Culver et al. (2009) showed that there were consistent patterns of frequency of occupancy of epikarst copepods that were consistent with differing levels of dispersal among the species (see chapter 11). In chapter 11, we also demonstrated that size of hypotelminorheic amphipods was correlated with range, also suggesting dispersal. Crouau-Roy (1987) showed that *Speonomus* beetles in the MSS in the French Pyrenees had to disperse vertically during the winter. In fact, the initial interest in the MSS was not as a habitat that sustained subterranean populations, but as a dispersal corridor for species occupying disjunct karst regions (Juberthie et al. 1980b).

15.5 Examples of conservation and protection of the SSH fauna

15.5.1 Hypotelminorheic

In the USA, the Endangered Species Act is the primary legal tool used for species protection. Its focus is on species rather than habitats, unlike the Habitats Directive of the European Union, with a stronger focus on habitat. In highly urbanized Washington, DC, there is only one species[1] on the federal endangered species list—*Stygobromus hayi*, a stygobiotic amphipod only known from a total six seepage springs and one hyporheic site (Culver et al. 2012a) in Rock Creek Park (Fig. 15.2), a large urban park that, because it is in the District of Columbia, is administered by the US National Park Service (NPS). One specimen was found in the hyporheic of Rock

[1] The cougar (*Puma concolor*) is also listed but is of course not currently present anywhere near Washington, DC.

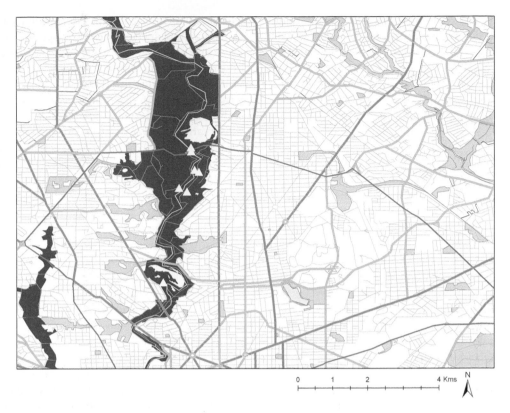

Figure 15.2 Map of the entire distribution (triangles) of the endangered amphipod species *Stygobromus hayi*, found only in Rock Creek Park and the National Zoological Park in Washington, DC. Dark grey areas are part of the US National Park system, and are tributary streams of the Potomac River. Light grey areas are small local parks, and the rest of the area is covered with roads and buildings.

Creek, out of more than 100 Bou–Rouch samples, each the filtrate of 10 L of hyporheic water (see Box 6.1). It seems likely that no permanent population of the species was present in the hyporheic. There is an even rarer species, *Stygobromus kenki*, known from a total of five seepage springs, four in Rock Creek Park and one in a regional park in an adjoining part of Maryland 10 km distant. The sites themselves are rather inconspicuous, except for one on the grounds of the National Zoo, technically part of the lands of the NPS. The zoo site is fenced with a sign indicating the presence of an endangered species. Aside from this site, the other seepage springs are unmarked and there is little information available in Rock Creek Park about the species. This is largely a deliberate policy of not attracting people to seepage springs because of the potential damage both from trampling and collecting. Articles about

the species have appeared in local papers including *The Washington Post*, as well as *The Endangered Species Bulletin* (Pavek 2002).

Major threats to the species are trampling by hikers, sediment clogging of the seep and its runoff, heavy metal contamination, and drying out as a result of climate change. Trails are kept away from seepage springs throughout the park, and stormwater discharge into the park from the city of Washington is being gradually eliminated. Sediments in seeps near the park boundary have elevated levels of heavy metals (especially lead and selenium), the result of runoff from city streets (Culver and Šereg 2004). This remains a problem with no easy solution. Finally, several seepage springs have remained dry for more than five years, when previously they had water from late winter to early summer (W. Yeaman, pers. comm.), suggesting that warming and drying may have an impact on *S. hayi*. Continued protection of the species (and *S. kenki* as well, which has a similar distribution) requires a continuing commitment on the part of the National Park Service to make protection of seeps a priority, even in the face of demands for increased trails, picnic areas, etc.

15.5.2 Epikarst

The most species-rich epikarst sites known are drips in Škocjanske jame and Županova jama. In Škocjanske jame, Pipan and Culver (2007b) reported nine stygobiotic copepod species observed, and another six stygobiotic species expected according to Chao2 estimates (see Box 2.2) if all species were sampled. In Županova jama, they reported 14 stygobiotic copepods observed, and another two expected according to Chao2 estimates. These caves are hotspots of epikarst fauna, and the next most diverse cave is another Slovenian cave, Pivka jama, with a total of 9 stygobiotic copepods. Epikarst sites in Romanian and USA caves were even lower (Meleg et al. 2011b, Pipan and Culver 2005).

Županova jama is a show cave, with four main galleries with a total length of 682 m (Pipan 2003). The drips are in the midst of the highly decorated sections of the cave but not adjoining tourist trails. It has been a show cave since its discovery in 1926. The land above the cave is forested. It is important to recognize that the primary habitat for these epikarst species is the epikarst itself, not pools in the cave that collect epikarst water. In Županova jama, Pipan et al. (2010) found that only seven of the 14 stygobionts collected in drips were found in pools. In addition, one stygophilic species, *Bryocamptus typhlops* was one of the most common species in pools in the cave. As Pipan et al. (2010) pointed out, protection must focus on the epikarst and the area it drains. In Slovenia, caves cannot be privately owned, but show caves are typically managed by local caving clubs, and this is the case for Županova jama. The land above the cave is privately owned, and the likelihood of logging above the cave is uncertain, but logging in this area is typically selective and sustainable, with only a few trees taken at any one time. The cave itself is well protected since all Slovenian caves are protected by national

legislation, including the 2004 decree on protection of wild animal species, including species living in caves and subterranean waters, and a 2004 act on cave protection, including restrictions on number of visitors and on amount of collecting. Based on international treaties, resolutions, recommendations, and state and local community regulations, a nature conservation act was also passed in 2004. This includes Articles 31–33 protecting species and habitat types in ecologically important areas and in sites in the Habitats Directive Natura 2000 network of the European Union.

Škocjanske jame consists of several caves, including the large show cave more than two km in length, as well as the much smaller (<250 m) Tominčeva jama (Pipan 2003). Škocjanske jame has additional protection as a UNESCO Natural Heritage site and as a Ramsar wetland of international importance (Pipan and Culver 2012c). The cave is a major tourist attraction in Slovenia, with one hundred thousand visitors per year. Slightly more than 4 km^2 of largely wooded land, but including several tiny villages, is protected as part of Park Škocjanske jame, a regional park. Tominčeva jama is relatively little visited, with a low fence at the entrance to the cave. Epikarst sites were in both caves, and were separated by greater distances than those of Županova jama. As was the case for Županova jama, epikarst drip pools in Škocjanske jame contained only a fraction of the epikarst fauna. Based on ratios of immature to mature copepods, Pipan et al. (2010) concluded that reproduction in pools was not as frequent. The one exception was a drip pool in Škocjanske jame that had been artificially enlarged to permanently hold water. It had higher frequencies of immatures, but also different copepod species. As with Županova jama, protection should not focus on pools.

Another kind of protection is that for Piddling Pit in West Virginia, USA. It has three species of amphipods in drip pools less than 200 m inside the cave—*Stygobromus emarginatus*, *S. nanus*, and *S. parvus* (Fong et al. 2007). This is a remarkable assemblage of epikarst amphipods, and no other epikarst site in the state has this many species of *Stygobromus* amphipods. Because of its unique biological characteristics, The Nature Conservancy, a non-governmental conservation organization, purchased the cave entrance in 1993, and fortunately purchased the land above the cave as well. Access to both the largely forested land and the cave are highly restricted.

All three of the epikarst sites listed above are in rural areas, with little or no immediate threat of changes in land use. Aside from the risk of logging (not possible at least in principle at Piddling Pit or Škocjanske jame), a toxic spill is conceivably possible at Škocjanske jame, the only site with a road of any kind over the cave and its epikarst.

15.5.3 Intermediate-sized terrestrial shallow subterranean habitats

As far as we can determine, no MSS sites or other intermediate-sized terrestrial shallow subterranean habitats are protected in their own right. However, many scree and talus slopes in the Czech Republic are protected somewhat inadvertently. V. Růžička (pers. comm.) reported that 59 scree slopes were included in the 627 protected areas in the Czech Republic. Protected areas include (National) Nature Reserves, (National) Nature Monuments, and larger-scale areas—National Parks and Protected Landscape Areas. Talus and scree slopes, at least in central Europe, have additional conservation value. Isolated populations of more northern arachnid species occur on scree (see Fig. 4.19) and rare plant communities are present as well (Kubešova and Chytry 2005, Růžička et al. 2012). Nowhere is the conservation importance of scree slopes more recognized than in the Czech Republic. Threats to scree and talus slopes include logging of adjoining forests, which would significantly reduce the resource base of the scree (Růžička et al. 2012), and global warming, since these habitats act as cold climate refuges. On high elevation scree slopes in western North America the pika (*Ochotona princeps*) is frequently present, but is declining in lower elevation sites because of global warming (US Fish and Wildlife Service 2010).

MSS habitats in lava terrains are often protected within National Parks in the Canary Islands, the one volcanic area where the lava MSS has been the subject of extensive biological sampling. For example, on La Gomera in the Canary Islands, the National Park of Garajonay, which comprises 40 km², includes extensive MSS. It is also a UNESCO World Heritage site. La Gomera itself has no lava tubes,

lava tubes having presumably eroded away (Oromí and Izquierdo in press). The central volcano on Tenerife, together with many lava tubes and MSS sites, is part of Teide National Park, which encompasses nearly 190 km² of the island.

15.5.4 Calcrete aquifers

In the Yilgarn and elsewhere in Western Australia, the two main threats to calcrete aquifers and their fauna are the direct mining of the calcretes for use in road beds and elsewhere, and the drawdown of water in the calcrete aquifers. It is this water abstraction that is the major threat to the calcrete aquifer communities (W. Humphreys pers. comm.); water is used for livestock grazing and more importantly for mining operations in the area (Johnson et al. 1999).

Nearly all the efforts at protection of calcrete aquifers are aimed at government agencies, especially the Department of Environment and Conservation of Western Australia. In contrast with the US Endangered Species Act and the European Union Habitats Directive, environmental protection in Western Australia utilizes biological communities as the key protection elements. According to the Department of Environment and Conservation (2013),

Because ecosystems and the links between their community members are so complex, it is impossible to maintain their components on a species by species basis. While it is important to manage individual threatened species of animals and plants, we cannot give the same individual attention as we do to vertebrates and vascular plants to the many thousands of species of invertebrates, non-vascular plants and micro-organisms. To conserve these components of biological diversity, we need to identify, maintain and manage whole ecosystems, their processes and communities.

The first steps towards protection are the identification of priority ecological communities for assessment of the species present. Calcrete aquifers figure prominently in the priority ecological communities list of March 2013 (Department of Environment and Conservation of Western Australia 2013): 32 of 60 ecological communities in the Goldfields (a region that includes the Yilgarn) are calcrete aquifers. Most of these calcrete aquifers are listed as priority 1 ecological communities (ones that are known from very few occurrences with a very restricted distribution [generally ≤ 5 occurrences or a total area of ≤ 100ha]). Following threat assessment, an ecological community can be placed on the list of threatened ecological communities, if it is endorsed by the Minister of the Environment. This current list has 69 communities, only two of which are calcrete aquifers (Ethel Gorge in the Pilbara and Depot Springs in the Yilgarn). Inclusion of calcrete aquifers on the list is difficult because of their high economic value. However, it is clear that the Western Australia government is increasingly aware of the global biological importance of the fauna of calcrete aquifers.

15.5.5 Hyporheic

One of the most diverse hyporheic sites reported is the Lobau wetlands near Vienna, Austria. Danielopol and Pospisil (2001) reported a total of 35 described and undescribed stygobionts from the Lobau section of the Danube River. Unprotected areas in this region disappear rapidly under the pressure of population growth and development. The Danube Floodplain National Park (Nationalpark Donau-Auen) protects the floodplain (including hyporheic habitat) of the Danube River from within the city limits of Vienna to the Slovakian border, with a total of 93 km² of protected land. It includes the Lobau wetlands. The national park was established to protect the last remaining intact wetlands along the Danube River in central Europe, and protection of the hyporheic is a by-product of protection of the wetland. Danielopol and Pospisil (2001) metaphorically liken the Lobau wetlands to 'a museum in which we preserve valuable documents.'

An important site of hyporheic biodiversity in North America is the Flathead River system in northwestern Montana (USA) and southern British Columbia (Canada), which includes Flathead Lake, the largest natural lake in the USA west of the Mississippi. In sharp contrast with the densely populated region around the Danube Floodplain National Park, the Flathead River is a wilderness river with no permanent settlements. It is home to the some of the largest carnivores of North America, including the brown (grizzly) bear, *Ursus arctos horribilis.*

The fauna of the hyporheic and floodplain of the Flathead River was studied using a series of over two hundred 5 m-deep sampling wells drilled at regular intervals, and is one of the most extensively studied hyporheic habitats in the world (Stanford et al. 1994). Stygobiotic copepods, ostracods, amphipods, isopods, and bathynellans were found in the shallow groundwater of the floodplain and river underflow.

The US part of the river is mostly protected as a Wild and Scenic River, but the Canadian part of the river is under threat from the possibility of coal mining and coal methane extraction. Nongovernmental agencies like Nature Conservancy (Canada) and The Nature Conservancy (USA) have taken an active interest in protection of the river system, including the purchase of some key pieces of land. The Flathead Lake Biological Station of the University of Montana has taken an active role in protecting the river and lake.

15.5.6 Lava tubes

Unlike other SSHs, lava tubes are often the direct object of protection efforts. Lava tubes are of geological and sometimes archeological and cultural interest, and a number of national parks (e.g. Teide National Park in Tenerife, Canary Islands and Hawaii Volcanoes National Park in Hawaii, USA). However, most lava tubes are not on protected land, and efforts have been mounted in the Azores, the Canary Islands, and the Hawaiian Islands to explicitly protect lava tubes for their biological diversity.

In their study of lava tubes on the Azorean Islands, Borges et al. (2012) discussed what should be the first step in most conservation efforts—the establishment of priorities for site protection. They took two basic approaches. The first is to assign a value to a site, and they based this on a number of criteria for each cave:

- number of troglobionts;
- number of troglobionts endemic to the island;
- length of the cave;
- geology (particularly the presence of uncommon structures like gas bubbles);
- exploration ease;
- integrity (extent to which the cave has been destroyed and/or vandalized);
- anthropogenic threat; and
- accessibility (including ease of locating the entrance).

Using a generalized linear model that eliminated collinearity, they then assigned a score for each of the 42 lava tubes and lava pits on the Azores for which there are biological data. One may question their scoring scheme (e.g. a cave under threat has a lower score than a cave without threat), but the technique produces an objective priority list of caves. The ten highest ranking caves were on four islands—Pico, São Miguel, São Jorge, and Terceira, with Furna dos Montanheiros being the highest ranked site. These scores, however, do not take into account the possibility of a network of protected sites where all troglobiotic species are protected.

The minimum number of sites required for this network can be determined by complimentarity (Michel et al. 2009), where sites are chosen by their contribution of new species to the already existing network. Borges et al. (2012) determined that 10 sites were needed to include one population of all the known Azorean troglobionts, approximately 25 per cent of the biologically inventoried lava tubes. As more lava tubes are inventoried and new species found, more sites will have to be added to the network. Michel et al. (2009) likewise found that subterranean biodiversity was concentrated in a relatively small percentage of sites. They found that nearly 80 per cent of stygobionts in Europe were included in 10 per cent of the quadrats ($0.2° \times 0.2°$). Borges et al. (2012) found that six caves (on Faial, Terceira, São Jorge, and São Miguel) were irreplaceable, because they harboured single cave endemics. Furna dos Montanheiros on Pico, one of the highest ranked sites, was not irreplaceable, and was only the eleventh most important cave in terms of building a network of protected sites.

Martín Esquivel and Izquierdo Zamora (1999) addressed a different question—what are the elements of a protection plan for a particular lava tube, in this case Cueva del Viento, a more than 20 km-long lava tube that has the most troglobionts of any lava tube in the world (Culver and Pipan 2013). The Canary Islands government developed a management plan, including the declaration of the cave as a Special Natural Reserve, which affords protection to the cave at category IV (Habitat/Species Management Area) of the International Union for the Conservation of Nature. Most of the cave is overlain by forested land, but there is some crop land,

and a developed area with a number of buildings that overlies the Piquetes branch. Among the recommendations were no new building, a new sewage system for existing building, the development of a small visitor centre (not yet built), purchase of land at the entrance, development of a short educational trail in the lava tube, and restriction of the number of visitors to the cave to a maximum of 30 per day. The target of the management/protection plan is the forested area and to a lesser extent the area overlain by crop land. The degraded section, Piquetes, is not addressed in the management plan. Martín Esquivel and Izquierdo Zamora (1999) listed waste water contamination from surface buildings and the impact of visitors in the lava tubes (calcium carbide residues, trash, and even elevation of CO_2 levels) as important threats. The plan emphasized the importance of surface–subsurface connections. At present, educational tours are offered and most of the land above the cave is protected from further development.

The management plan for Kipuka Kanohina in Hawaii also addressed the management plan of an individual lava tube (Cave Conservancy of Hawaii 2003). Kipuka Kanohina is one of the longest lava tubes in the word, with a length of 25 km, covering an area of 100 hectares. No published faunal inventory exists, but it has many sections with *Metrosideros* roots and associated fauna (see Figs 8.4 and 8.5). Entirely on private land, it is in the Ocean View area of Hawaii, an area of single unit family dwellings, each with an individual septic system and a rainwater catchment system for water supply. While densities are currently low, population growth is occurring. Environmental threats to the system include:

- vandalism of formations;
- surface bulldozing which sometimes breaks into the lava tube;
- leaking septic tanks; and
- dumping at entrances.

The goal of the Kipuka Kanohina Cave Preserve is to minimize impact of surface development, control and maintain access, liability protection, and to offer educational seminars. Ideally, the Cave Conservancy of Hawaii would simply purchase the entire area over the cave but that is neither financially nor practically possible. Instead they use a combination of tools including conservation easements and underground leases, as well as direct purchase of property. Conservation easements (similar to the purchase of development rights), are widespread in the United States (Milder and Clark 2011), but their long-term effectiveness in general has yet to be evaluated.

An important aspect of the management plan for Kipuka Kanohina is to ensure access to the cave by scientists and by members of the Cave Conservancy of Hawaii. Particularly in the USA, many caves located on private land are closed to visitors, often due to liability concerns of landowners. The dozens of cave conservancies that have come into being in the last several decades are in part a response to this problem, as well as a concern for species protection and conservation. The access issue is also one faced by rock climbers and other groups of outdoor recreationists. In most circumstances, open access and conservation are compatible, but not always, as is clear from studies of the impact of climbers on the flora of cliff faces (e.g. Larson 2006), and part of the continuing debate in the recreational caving community is about conditions under which access to caves should be restricted due to conservation concerns.

15.6 Summary

Because of their proximity to the surface, SSHs are especially vulnerable to toxic spills. Some SSHs provide significant ecosystem services, especially soil and the hyporheic. All SSH types are sites of rare species found nowhere else.

Most SSHs share the unifying feature of connectivity. They are connected in the sense of threats, especially toxic spills which move laterally, but they are also connected because of the likely dispersal of SSH species laterally. Protection of isolated sites is not an adequate conservation strategy, and a landscape approach is needed.

As is the case with cave fauna, the SSH fauna is for the most part geographically rare, but not neces-sarily numerically rare, or rare because the SSH is a rare habitat. Among SSHs, lava tubes and calcrete aquifers are probably the rarest globally, but both can be quite common locally. Many SSH species are also likely to have low reproductive rates, which, combined with rarity, makes them especially vulnerable to extinction.

Threats to the SSH fauna are diverse and can be grouped into (1) alteration of the physical habitat, (2) changes in water quality or quantity, and (3) direct changes. Examples of physical alteration are the extensive limestone quarrying in southeast Asia, which obliterates epikarst, and impoundments or alteration of the riparian border of streams, which affects the hyporheic. Overuse of nitrogen fertilizers has resulted in a nearly global contamination of shallow subsurface waters. Large numbers of storage tanks, especially for petroleum products (NAPLs), are a major source of SSH contamination. Global warming is likely to have an increasing and major direct effect on SSHs, especially as hydrological regimes and cycles change.

Except for lava tubes, most SSHs are protected as part of a protected landscape or a collateral benefit of protection of some other habitat. Some hypotelminorheic habitats are protected both by being in parks and because the species are rare. Epikarst is usually protected as a collateral result of protection of caves. MSS habitats are typically protected as a part of either a mountain landscape or a volcanic landscape. Among these terrestrial SSHs, perhaps only scree slopes are recognized as important habitats for the protection of biodiversity. Calcrete aquifers in Western Australia are protected by government regulations concerning rare communities. One hyporheic site, the Lobau wetlands near Vienna, Austria, is protected as part of the National-park Donau-Auen, and the hyporheic is of concern to many conservationists especially because of its importance in maintaining water quality and habitat in rivers. Lava tubes are protected in their own right, both in national parks and as the result of local efforts by government and non-government entities.

CHAPTER 16

Epilogue and prospects

16.1 What unites SSHs?

Our working definition of shallow subterranean habitats (SSHs), introduced in chapter 1, was very simple—aphotic habitats less than 10 m from the surface. The 10 m limit is arbitrary, but was designed to include those habitats above caves, in areas where caves are present, as well as subterranean habitats in areas where there are no caves. SSHs we discussed included the hypotelminorheic and associated seepage springs (chapter 2), epikarst (chapter 3), intermediate-sized terrestrial habitats (chapter 4), calcrete aquifers (chapter 5), interstitial habitats along rivers and streams (chapter 6), soil (chapter 7), and lava tubes (chapter 8).

All of these habitats share another feature in addition to the absence of light—the presence of species with reduced or absent eyes and reduced or absent pigment. These characteristics, which are likely moulded by natural selection in an aphotic environment (chapter 9, Culver and Pipan 2009), or perhaps are the result of selectively neutral genetic processes (Wilkens 1988, 2010), indicate absence of light is an important feature of SSHs.

All of these habitats are likely to be aphotic, although there have been very few direct measurements of light in these habitats. The absence of light is not necessarily an obvious feature of all of these habitats, although for some cases it is, such as even a few cm deep in the soil. Our measurements of that most superficial of SSHs, the hypotelminorheic, indicated that even at five cm in depth this was an aphotic habitat. There are other habitats that are little-studied that may also be aphotic. These include the spaces around tree roots (Fig. 4.3), ant nests (Fig. 7.1), mosses (Fig. 7.1), and vertebrate burrows and nests (Fig. 7.1) in the terrestrial

realm, and interstitial habitats in shallow lakes in the aquatic realm.

As a group, SSHs, with the possible exception of calcrete aquifers (chapter 5), share not only the absence of light, but also intimate contact with the surface. We used data from temperature data-loggers as a proxy for surface connectivity. In the absence of close connections, temperature should vary little and cycle even less. With the exception of some lava tubes, SSHs showed annual and often daily variation (Table 10.1). This variation has several important implications, including cyclicity itself which can be important in setting circadian clocks (Fig. 9.2) and input of organic carbon and nutrients (Table 11.1). These characteristics (environmental cyclicity and increased organic carbon inputs) are in contrast with deeper subterranean environments, including caves and deep phreatic water.

A feature that narrow sense SSHs (hypotelminorheic, epikarst, interstitial habitats along rivers and streams, and calcrete aquifers) also share is habitat dimensions intermediate between the large open spaces of caves and the tiny spaces between soil particles or sand grains (Fig. 1.14). When all SSHs are considered, not just narrow sense ones, a wide range of habitat sizes occur. It is also a repeated observation that habitat size (pore size) is a major determinant of body size, especially for the small pore habitats such as soil. Indeed, the only clear morphological pattern to emerge, aside from eye and pigment loss, is the correlation of body size and habitat size (e.g. Figs 3.21 and 12.3, Table 3.14). Differences in body size may serve as an important constraint on the colonization of SSHs by species living in different SSHs. For example, the morphological differences between soil and cave Collembola are extensive (see Fig. 7.13), and cast doubt about the

Shallow Subterranean Habitats. David C. Culver and Tanja Pipan.
© David C. Culver and Tanja Pipan 2014. Published 2014 by Oxford University Press.

likelihood of soil Collembola being ancestral to cave Collembola.

There are several other habitats that we have only mentioned in passing, ones that could be called twilight habitats. The two main such habitats are leaf litter and the benthos of streams (Fig. 6.2). At least occasionally, eyeless and depigmented species are found in these habitats. For example, the depigmented (but eyed) planarian, *Phagocata morgana*, is widespread in cold-water streams and springs in the mid-Atlantic region of the USA (Culver et al. 2012a, Kenk 1972). Symphyla, both eyeless and depigmented, are common denizens of leaf litter (Scheller 1986). Vignoli and Prendini (2009), on the basis of their phylogeny of the scorpion family Typhlochactidae, proposed that troglomorphic cave scorpions gave rise to troglomorphic leaf litter species (Fig. 14.7). Along a similar vein, Heads (2010), in his study of spider crickets, suggested that troglomorphic characteristics (see Table 12.1) in spider crickets evolved in the leaf litter.

One key piece of data that is missing from an analysis of these twilight habitats is the distribution of light, both spatially and temporally. As a starting point, both a rocky stream bottom and leaf litter may have a mosaic of light intensities, with overall intensity declining with depth. From the organism's point of view, it may be possible to avoid light entirely, or, conversely, to avoid darkness. Thus, there may be aphotic habitats within the leaf litter and the stream benthos, and this may explain the presence of species with some troglomorphic features, and the presence of photic habitats within the leaf litter and the stream benthos, would explain the presence of species without any troglomorphic features.

A final feature that unites all SSHs (and deep subterranean habitats for that matter), is the presence of some obligate species that have not lost their eyes and pigment. There are a few rare cases where only stygobionts and troglobionts are known from a subterranean habitat (the cave Vjetrenica in Bosnia & Hercegovina [Sket 2003] and epikarst drips in Županova jama [Pipan 2006]), but these are the exceptions rather than the rule. Most SSHs and most caves have non-troglomorphic species that are not stygobionts or troglobionts (Fig. 9.1, Pipan and Culver 2012a).

16.2 What divides SSHs?

Each SSH type has a characteristic distribution of habitat (pore) sizes (Fig. 1.14), and these are potential divergent selective pressures on different SSHs (Table 12.4). In addition, SSHs, both within a particular type and among types, will differ in the organic carbon flux and in cyclicity of the environment. Whether these are important selective factors remains to be demonstrated in most cases. The documentation of adaptation to low organic carbon and nutrients is mostly with species at the top of the cave food web, especially fish (Hüppop 2000), and the extent of its importance for species in lower trophic levels remains largely untested.

The ground-breaking work of Friedrich (2013) on the connection between microphthalmy and circadian clocks implies that species must in some way 'sample' light in order to reset the circadian clock. Thus, distance to light may be an important feature of SSHs. In the absence of other information, depth below the surface is at least an estimate of this distance.

SSHs also differ in the amount of interspecific competition, and Fišer et al. (2012) demonstrated that the morphology of species in hyporheic communities was more variable than expected by chance (Fig. 6.19), indicating divergence among species as a result of competition. The demonstration by Vergnon et al. (2013) that dytiscid beetles in calcrete aquifers have a limiting similarity (Fig. 5.9) also highlights the importance of interspecific competition.

16.3 Are SSHs staging areas for the colonization of deep subterranean habitats?

For many decades, a prevailing theory of colonization of caves has been that the cave fauna was pre-adapted or exapted (see chapter 13), and at least for the terrestrial cave fauna, was derived from the fauna of leaf litter, which we have called a twilight habitat. It is easy to make the transition from this exaptation hypothesis to the hypothesis that the ancestors of the deep subterranean fauna are the fauna of SSHs. This idea has several appealing features. Firstly, it makes the colonization of what is arguably the extreme environment of caves possible in a series of steps, since SSHs are less extreme environmentally than caves. Secondly, there is a proximity argument. In the vicinity of caves, there are often SSHs, including the epikarst, which can be both an aquatic or terrestrial environment (see Tables 3.11 and 3.13). The diagram in Fig. 16.1, which has been repeated in several similar versions (see Chiesi et al. 2002, Juberthie 2000), suggests that colonization of deep subterranean habitats proceeds in steps, with the species colonizing progressively deeper subterranean habitats. Ortuño and Gilgado (2010) proposed that the uppermost subterranean habitat was the soil (edaphic habitat), which they associated with the A and B horizon of the soil (Fig. 7.2), with the MSS underneath (a hypogean habitat in their terminology), and with caves (labelled MSP [deep subterranean habitat]) below the MSS. By this scenario, the soil is the site of the first colonization of the subterranean environment, followed by the MSS and epikarst, and finally followed by caves.

Attractive as this scenario is, there are several major problems with it. Firstly, the arrangement of the different SSHs and caves in Fig. 16.1 is highly idealized. Soil and MSS are often mixed (Gers 1992, 1998), and there may be an epikarst layer more or less distinct from the MSS layer. More commonly,

one or more of the habitats is not present at all. Secondly, there is not a shred of evidence from phylogenetic studies that this is the case, although the number of such studies is very small (chapter 14). Thirdly, the transition from surface to soil to MSS to caves is a transition from large body size to small body size to large body size, and a transition from normal appendages to shortened appendages to lengthened appendages. This may have occurred, but the morphological changes would not be a straightforward linear progression. Fourthly, it is not clear what factors would force a population deeper and deeper into the subsurface. In most cases, most environmental extremes disappear within a few cm of the surface (Table 10.1). There are certainly cases where climate change forces species deeper. The aridification of western Australia extended to the disappearance of some SSHs, especially surface streams and their associate hyporheic. What is likely is that increased competition and predation pressure could be an effective factor in pushing animals downwards, but this topic needs more careful consideration.

16.4 SSH terminology

In a number of cases, it was difficult to categorize a particular site in terms of the name of the SSH.

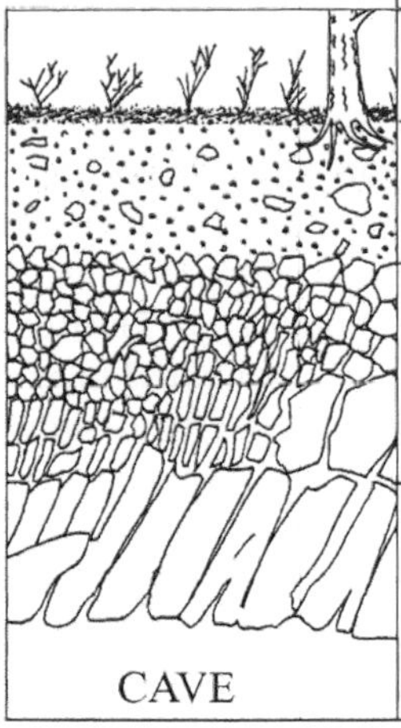

Layer	Environment	Fauna
A	Soil	Humicolous / Edaphic
B	Soil	Edaphic / Endogeous
C1	MSS	Hypogean
C	MSP	Hypogean

CAVE

Figure 16.1 Soil and subsoil layers and an ecological classification of the fauna. From Ortuño and Gilgado (2010).

The following is by no means an exhaustive list of examples:

- Are clinker sites without a soil or moss cover, such as are widespread in the Canary Islands, MSS sites even though initially Juberthie et al. (1980a, 1980b) reserved the name MSS for moss- and soil-covered sites?
- What is the difference, if any, between an MSS site above a cave and an epikarst site that is not water-filled?
- In MSS sites with large amounts of soil present, where does the MSS start and end?
- What is the difference between a hyporheic site isolated from groundwater (Fig. 6.2E) and a hypotelminorheic site (Fig. 2.1)?
- Where does a calcrete aquifer end and where does an alleviated aquifer begin in the Pilbara region of Australia, where the two intermix?

To add to the confusion, different authors give different names to the same habitat. This is nowhere more apparent than with intermediate-sized terrestrial SSHs. For example, a classic MSS habitat, according to Juberthie et al.'s (1980a, 1980b) original definition is a covered (typically moss) talus slope or rock that has eroded in place. Other authors have called this covered scree (e.g. Růžička 2002, Růžička and Zacharda 2010), an epigean site (Barr 1967), and perhaps even a leaf-litter site (Prendini et al. 2009). Near Jama v Kovačiji (Slovenia), there was an MSS site that we termed epikarst (Pipan et al. 2011a, Fig. 4.7) because of it direct connection to the cave. Ortuño et al. (2013) propose yet another type of MSS habitat, one in dry stream beds, which they call 'alluvial mesovoid shallow substratum.' Whether or not the term proves useful, it signifies the overlap between the hyporheic and the MSS, further blurring the distinction among SSHs. In the case of the hypotelminorheic and seepage springs, we suspect that stygobionts have been collected from these habitats, but given different names. The habitat in early North American records was described as a seep. Denton and Scott (2013) describe a hypotelminorheic-like habitat that is only accessible in deeply cut streams, and the only habitat for the amphipod *Stygobromus phreaticus*, as a macropore spring. It differs from the hypotelminorheic in being bound on the top as well as the bottom by clay.

We have resisted the temptation to introduce clarifications of terminology, such as MSS, or new terms, because we do not think this would add any clarity to the field. The terminology used for various SSHs may have some parallel with the terms used for 'pseudokarst', karst-like phenomena (e.g. sinkholes) resulting from non-solutional processes. Eberhard and Sharples (2013) suggested that the terminology associated with pseudokarst is unnecessary and should be replaced with a more standard geomorphological terminology, which is more process oriented. Useful as specialized terms may be to experts, based on the history of terminology in subterranean biology, there is never broad consensus, rather a proliferation of terms (e.g. Sket 2008). We suggest the continued use of general terms such as hyporheic, epikarst, hypotelminorheic, and MSS, even though there are conflicting definitions of each. Thus, the hyporheic, certainly the best-studied SSH, with the exception of soil, is not that clearly delineated from either the stream bed above or the groundwater below (see Table 6.1), and the connection of the hyporheic to deeper subterranean habitats ranges from intimate to non-existent (Fig. 6.2).

Rather than putting forward more detailed definitions, a more appropriate focus is on what characteristics of these habitats impact the fauna, either in terms of the morphology of its inhabitants, or more general issues of faunal composition. We suggest that special attention be paid to the following fluxes and parameters, which are likely to be important environmental factors. The first is light, not only whether any measurable light is present, but how far away light is. Recall Friedrich's (2013) review of many microphthalmic stygobionts and troglobionts, all of which retain circadian clocks. Are these species of aphotic environments, but ones that 'sample' light periodically? The second is habitat dimension, frequently referred to as pore size. This is the one physical factor that has a demonstrable impact on morphology (Table 12.4). The third is vertical depth. In general, depth is a good predictor of environmental variability and cyclicity (Table 10.1) as well as connectivity with the surface. The fourth is the quantity of organic carbon and nutrients. This has an obvious impact on faunal composition (Figs 11.1 and 11.2). Dissolved organic carbon is a key parameter that can now be measured in many environmental science laboratories. The final characteristic of demonstrable importance is the amount of competition and predation in the habitat (Fišer et al. 2012, Trontelj et al. 2012), differences in which promote divergence rather than convergence. Hence species richness is an important parameter.

16.5 What about troglomorphy?

The concept of troglomorphy, now more than 50 years old (Christiansen 1962), has been enormously useful in bringing the study of subterranean biology firmly into the neo-Darwinian fold, and in setting a research agenda for the study of natural selection in subterranean habitats, especially caves. The idea

of caves as evolutionary laboratories (Poulson and White 1969) was a powerful metaphor to help focus this research agenda, especially among North American speleobiologists. However, as with the terms for various shallow subterranean habitats, the term 'troglomorphy' has come to have a variety of meanings. In Christiansen's recent summary (2012), it is nearly synonymous with any character, behaviour, or life history trait that has been selected for in an aphotic environment, even if it is very narrowly restricted in application (see Table 12.1). We suggest a return to the more general evolutionary terminology of parallelism, convergence, and divergence will firmly place the study of SSH fauna and other subterranean fauna in a more general and appropriate evolutionary context.

Literature Cited

Abrams, K.M., Guzik, M.T., Cooper, S.J.B., Humphreys, W.F., King, R.D., Cho, J.L., and Austin, A.D. (2012). What lies beneath: molecular phylogenetics and ancestral state reconstruction of the ancient subterranean Australian Parabathynellidae (Syncarida, Crustacea). *Molecular Phylogenetics and Evolution*, **64**, 130–44.

Ahearn, G.A. and Howarth, F.G. (1982). Physiology of cave arthropods in Hawaii. *Journal of Experimental Zoology*, **222**, 227–38.

Al-fares, W., Bakalowicz, M., Guerin, R.T., and Dukhan, M. (2002). Analysis of the karst aquifer structure of the Lamalou area (Herault, France) with ground penetrating radar. *Journal of Applied Geophysics*, **51**, 97–106.

Allford, A., Cooper, S.J.B., Humphreys, W.F., and Austin. A.D. (2008). Diversity and distribution of groundwater fauna in a calcrete aquifer: does sampling method influence the story? *Invertebrate Systematics*, **22**, 127–38.

Allred, K. (2012). Kazumura Cave, Hawaii. In WB White and DC Culver, eds. *Encyclopedia of caves, second edition*, pp. 438–42. Elsevier/Academic Press, Amsterdam, The Netherlands.

American Public Health Association. (1999). *Standard methods for the treatment of water and wastewater, 20th edition*. American Public Health Association, Washington, DC.

Arechavaleta, M., Oromí, P., Sala, L., and Martin, C. (1996). Distribution of carbon dioxide concentration in Cueva del Viento (Tenerife, Canary Islands). In P Oromí, ed. *Seventh International Symposium on Vulcanospeleology, Santa Cruz de La Palma, Canary Islands, Nov. 1994.*, pp. 11–4. Los Libros de la Frontera, San Cugat del Vallés.

Arechavaleta, M., Sala, L.L., and Oromí, P. (1999). La fauna invertebrada de la Cueva de Felipe Reventón (Icod de los Vinos, Tenerife, Islas Canarias). *Viraea*, **27**, 229–44.

Arnedo, M.A., Oromí, P., Múrria, C., Macías-Hernández, N., and Ribera, C. (2007). The dark side of an island radiation: systematics and evolution of troglobitic spiders of the genus *Dysdera* (Araneae, Dysderidae) in the Canary Islands. *Invertebrate Systematics*, **21**, 623–60.

Ashmole, N.P. and Ashmole, M.J. (2000). Fallout of dispersing arthropods supporting invertebrate communities in barren volcanic habitats. In H Wilkens, DC Culver, and WF Humphreys, eds. *Subterranean ecosystems*, pp. 269–86. Elsevier Press, Amsterdam, The Netherlands.

Ashmole, N.P., Oromí, P., Ashmole, M.J., and Martín, J.L. (1992). Primary faunal succession in volcanic terrain: lava and cave studies in the Canary Islands. *Biological Journal of the Linnean Society*, **46**, 207–34.

Ashmole, N.P., Oromí, P., Ashmole, M.J., and Martín, J.L. (1996). The invertebrate fauna of early successional volcanic habitats in the Azores. *Boletim do Museu Municipal do Funchal*, **48**, 5–39.

Bakalowicz, M. (1995). La zone d'infiltration des aquifers karstiques. Méthodes d'étude. Structure et fonctionnement. *Hydrogéologie*, **4**, 3–21.

Bakalowicz, M. (2004). The epikarst, the skin of karst. In WK Jones, DC Culver, and JS Herman, eds. *Epikarst. Proceedings of the symposium held October 1 through 4, 2003 Shepherdstown, West Virginia*, pp. 16–22. Karst Waters Institute, Charles Town, West Virginia.

Bakalowicz, M. (2012). Epikarst. In WB White and DC Culver, eds. *Encyclopedia of caves, second edition*, pp. 284–8. Elsevier/Academic Press, Amsterdam, The Netherlands.

Ban, R., Pan, G., Zhu, J., Cai, B., and Tan, M. (2008). Temporal and spatial variations in the discharge and dissolved organic carbon of drip waters in Beijing Shihua Cave, China. *Hydrological Processes*, **22**, 3749–58.

Bardgett, R. (2005). *The biology of soil. A community and ecosystem approach*. Oxford University Press, Oxford, UK.

Bardgett, R. and Griffiths, B. (1997). Ecology and biology of soil protozoa, nematodes, and microarthropods. In JD van Elasas, E Wellington, and JT Trevors, eds. *Modern soil microbiology*, pp. 129–63. Marcel Dekker, New York.

Bareth, C. (1983). Diplures campodéidés due milieu souterrain superficiel de la region ariégoise. *Mémoires de Biospéologie*, **10**, 67–72.

Barr Jr., T.C. (1967). A new *Pseudanophthalmus* from an epigean habitat in West Virginia. *Psyche*, **74**, 166–72.

Barr Jr., T.C. (1968). Cave ecology and the evolution of troglobites. *Evolutionary Biology*, **2**, 35–102.

Barr Jr., T.C. and Holsinger, J.R. (1985). Speciation in cave faunas. *Annual Review of Ecology and Systematics*, **16**, 313–37.

Bedos, A., Prié, V., and Deharveng, L. (2011). Focus on soils. In T Bouchet, H Le Guyader, and O. Pascal, eds. *The natural history of Santo*, pp. 288–95. Museum National d'Histoire Naturelle, Paris.

Begon, M. (1979). *Investigating animal abundance: capture-recapture for biologists*. University Park Press, Baltimore, Maryland.

Belles, X. (1991). Survival, opportunism and convenience in the processes of cave colonization by terrestrial faunas. *Oecologia Aquatica*, **10**, 325–35.

Berlese, A. (1905). Apparecchio per raccogliere presto ed in gran numero piccoli Artropodi. *Redia*, **2**, 85–90.

Bernard, E.C. (2008). Reappraisal of the *Stachia–Stachiomella–Pseudostachia* complex (Collembola: Odontellidae), redescriptions of *Stachia minuta* and *Stachiomella oxfordia*, and *Stachia tasgola*, new species. *Proceedings of the Biological Society of Washington*, **21**, 106–16.

Bichuette, M.E. and Trajano, E. (2004). Three new subterranean species of *Ituglanis* from central Brazil (Siluriformes: Trichomycteridae). *Ichthyological Explorations of Freshwaters*, **15**, 243–56.

Bilandžija, H., Ćetković, H., and Jeffery, W.R. (2012). Evolution of albinism in planthoppers by a convergent defect in the first step of melanin biosynthesis. *Evolution and Development*, **14**, 196–203.

Boag, B. and Yeates, G.W. (1998). Soil nematode biodiversity in terrestrial ecosystems. *Biodiversity and Conservation*, **7**, 617–30.

Borges, P.A.V. (1993). First records for the mesocavernous shallow stratum (MSS) from the Azores. *Mémoires de Biospéologie*, **20**, 49–53.

Borges, P.A.V., Cardoso, P., Amorim, I.R., Pereira, F., Constância, J.P., Nunes, J.C., Barcelos, P., Costa, P., Gabriel, R., and Dapkevicius, M.L. (2012). Volcanic caves: priorities for conserving the Azorean endemic troglobiont species. *International Journal of Speleology*, **41**, 101–12.

Borges, P.A.V. and Hortal, J. (2009). Time, area and isolation: factors driving the diversification of Azorean arthropods. *Journal of Biogeography*, **36**, 178–91.

Borgonie, G., Garcia-Moyano, A., Litthauer, D., Bert, W., Bester, A., van Heerden, E., Möller, C., Erasmus, M., and Onstott, T.C. (2011). Nematoda from the terrestrial deep subsurface of South Africa. *Nature*, **474**, 79–82.

Botero-Trujillo, R. and Francke, O.F. (2009). A new species of troglomorphic leaf litter scorpion from Colombia belonging to the genus *Troglotayosicus* (Scorpiones: Troglotayosicidae). *Texas Memorial Museum Speleological Monographs*, **7**, 1–10.

Botosaneanu, L. [ed.] (1986). *Stygofauna mundi*. E.J. Brill, Leiden, The Netherlands.

Bottrell, S.H. and Atkinson, T.C. (1992). Tracer study and storage in the unsaturated zone of a karstic limestone aquifer. In H Hotzl and A Werner, eds. *Tracer hydrology*, pp. 207–11. Balkema, Rotterdam, The Netherlands.

Bou C. (1974). Recherches sur les eaux souterraines -25- Les méthodes de récolte dans les eaux souterraines interstitielles. *Annales de Spéléologie*, **29**, 611–9.

Bou, C. and Rouch, R. (1967). Un nouveau champ de recherches sur la faune aquatique souterraine. *Compte Rendus de l'Académie des Sciences de Paris*, **265**, 369–70.

Boulton, A.J., Dole-Olivier, M.J., and Marmonier, P. (2003). Optimizing a sampling strategy for assessing hyporheic invertebrate biodiversity using a Bou—Rouch method: within site replication and sample volume. *Archiv für Hydrobiologie*, **159**, 327–55.

Boulton, A.J. and Hancock, P.J. (2006). Rivers as groundwater dependent ecosystems: a review of degrees of dependency, riverine processes, and management implications. *Australian Journal of Botany*, **54**, 133–44.

Boulton, A. J., Valett, H. M., and Fisher, S. G. (1992). Spatial distribution and taxonomic composition of the hyporheos of several Sonoran Desert streams. *Archiv für Hydrobiologie*, **125**, 37–61.

Bradford, T., Adams, M., Humphreys, W.F., Austin, A.D., and Cooper, S.J.B. (2010). DNA barcoding of stygofauna uncovers cryptic amphipod diversity in a calcrete aquifer in Western Australia's arid zone. *Molecular Ecology Resources*, **10**, 41–50.

Bradford, T., Humphreys, W.F., Austin, A.D., and Cooper, S.J.B. (2013) Identification of trophic niches of subterranean diving beetles in a calcrete aquifer by DNA and stable isotope analysis. *Marine and Freshwater Research*. doi:10.1071/MF12356

Brancelj, A. (2002). Microdistribution and high diversity of Copepoda (Crustacea) in a small cave in central Slovenia. *Hydrobiologia*, **477**, 59–72.

Brancelj, A. (2007). The epikarst habitat in Slovenia and the description of a new species. *Journal of Natural History*, **40**, 403–13.

Brancelj, A. (2009). Fauna of an unsaturated karstic zone in central Slovenia: two new species of Harpacticoidea (Crustacea: Copepoda), *Elaphoidella millennii* n.sp. and *E. tarmani* n.sp., their ecology and morphological adaptations. *Hydrobiologia*, **621**, 85–104.

Brancelj, A. and Mori, N. (2005). Živalstvo v podtalnici. In I Rejec Brancelj, A Smrekar, and D. Kladnik, eds. *Podtalnica Ljubljanskega polja*, pp. 73–85. Založba ZRC, Ljubljana, Slovenia.

Brancelj, A., Watirayam, S., and Samoamuang, L.-O. (2010). The first record of cave dwelling Copepoda from Thailand and description of a new species—*Elaphoidella namnaoensis* n.sp. (Copepoda, Harpacticoida). *Crustaceana*, **83**, 779–93.

Brucker, R.W. (2012). Mammoth Cave system, Kentucky. In White WB and Culver DC, eds. *Encyclopedia of caves, second edition*, pp. 469–74. Elsevier/Academic Press, Amsterdam, The Netherlands.

Buhay, J.E. and Crandall, K.E. (2005). Subterranean phylogeography of freshwater crayfishes shows extensive gene flow and surprisingly large population sizes. *Molecular Ecology*, **14**, 4259–73.

Bullock, J.A. (1965). The ecology of Malaysian caves (and a note on the faunistic list from Batu Caves). *Malayan Nature Journal*, **19**, 57–64.

Buss, S., Cai, Z., Cardenas B., Fleckenstein, J., Hannah, D. Heppell, K., Hulme, P., Ibrahim, T. Kaeser, D., Krause, S., Lawler, D., Lerner, D., Mant, J., Malcolm, I., Old, G., Parkin, G., Pickup, R., Pinay, G., Porter, J., Rhodes, G., Ritchie, H., Riley, J., Robertson, A., Sear, D., Shields, B., Smith, J., Tellam, J. and Wood, P. (2009). *The hyporheic handbook*. Environmental Agency, Bristol, UK.

Camacho, A.I., Valdecasas, A.G., Rodríguez, J., Cuezva, S., Lario, J., and Sánchez-Moral, S. (2006). Habitat constraints in epikarstic waters of an Iberian Peninsula cave system. *Annales de Limnologie/International Journal of Limnology*, **42**, 127–40.

Camp, C.D. and Jensen, J.B. (2007). Use of twilight zones of caves by plethodontid salamanders. *Copeia*, **2007**, 594–604.

Carlini, D.B., Manning, J., Sullivan, P.J., and Fong, D.W. (2009). Molecular genetic variation and population structure in morphologically differentiated cave and surface populations of the freshwater amphipod *Gammarus minus*. Molecular Ecology, **18**, 19332–45.

Cave Conservancy of Hawai'i. (2003). *Kipuka Kanohina cave preserve management plan*. www.hawaiicaves.org/kanohina.htm.

Chao, A. (2005). Species richness estimation. In N Balakrishnan, CB Read, and B Vidakovic, eds. *Encyclopedia of statistical sciences*, pp. 7909–16. Wiley and Sons, New York.

Chapman, P.R.J. (1985). Are the cavernicoles found in Hawaiian lava tubes just visiting? *Proceedings of the Bristol University Speleological Society*, **17**, 178–92.

Chapman, P.R.J. (1993). *Caves and cave life*. Harper Collins, London.

Chappuis, P.A. (1942). Eine neue Methode zur Untersuchung der Grundwasserfauna. *Acta Scientiarum Mathematica Natura Kolozsvar*, **6**, 1–7.

Chestnut, T.J. and W.H. McDowell. 2000. C and N dynamics in the riparian and hyporheic zones of a tropical stream, Luquillo Mountains, Puerto Rico. *Journal of the North American Benthological Society*, **19**, 199–214.

Chiesi, M., Lapini, L., Latella, L., Muscio, G., Solari, M., and Stoch, F. (2002). *Caves and karstic phenomena. Life in the subterranean world*. Museo Friulano di Stroia Naturale, Udine, Italy.

Cho, J.L., Park, J.G., and Reddy, Y.R. (2006). *Brevisomabathynella* gen. nov. with new species from Western Australia (Bathynellacea, Syncarida): the first definitive evidence of predation in Parabathynellidae. *Zootaxa*, **1247**, 25–42.

Christiansen, K.A. (1961). Convergence and parallelism in cave Entomobryinae. *Evolution*, **15**, 288–301.

Christiansen, K.A. (1962). Proposition pour la classification des animaux cavernicoles. *Spelunca*, **2**, 75–8.

Christiansen, K.A. (1965). Behavior and form in the evolution of cave Collembola. *Evolution*, **19**, 529–37.

Christiansen, K.A. (2012). Morphological adaptations. In WB White and DC Culver, eds. *Encyclopedia of caves, second edition*, pp. 517–28. Elsevier/Academic Press, Amsterdam, The Netherlands.

Christman, M.C. and Culver, D.C. (2001). The relationship between cave biodiversity and available habitat. *Journal of Biogeography*, **28**, 367–80.

Christman, M.C., Culver, D.C., Madden, M., and White, D. (2005). Patterns of endemism of the eastern North American cave fauna. *Journal of Biogeography*, **32**, 1441–52.

Christman, M.C. and Zagmajster, M. (2012). Mapping subterranean biodiversity. In WB White and DC Culver, eds. *Encyclopedia of caves, second edition*, pp. 474–81. Elsevier/Academic Press, Amsterdam, The Netherlands.

Ciani, A., Goss, K.U., and Scharzenbach, R.P. (2005). Light penetration in soil and particulate minerals. *European Journal of Soil Science*, **50**, 561–74.

Cigna, A.A. (2002). Modern trend[s] in cave monitoring. *Acta Carsologica*, **31**, 35–54.

Clemens, T., Hückinghaus, D., Liedl, R., and Sauter, M. (1999). Simulation of the development of karst aquifers: role of the epikarst. *International Journal of Earth Sciences*, **88**, 157–62.

Clements, R., Sodhi, N. S., Schilthuizen, M., and Ng, P. K. (2006). Limestone karsts of Southeast Asia: imperiled arks of biodiversity. *Bioscience*, **56**, 733–42.

Coiffait, H. (1958). Les coléoptères du sol. *Vie et Milieu Supplement*, **7**, 1–204.

Coineau, N. (2000). Adaptations to interstitial groundwater life. In H Wilkens, DC Culver, and WF Humphreys, eds. *Subterranean ecosystems*, pp. 189–210. Elsevier Press, Amsterdam, The Netherlands.

Coineau, N. and Boutin, C. (1992). Biological processes in space and time: colonization, evolution and speciation in interstitial stygobionts. In AI Camacho, ed. *The natural history of biospeleology*, pp. 423–51. Museo Nactional de Ciencias Naturales Monografias, Madrid, Spain.

Coineau, Y., Fize, A., and Delamare Deboutteville, C. (1967). Découverte en France des Acariens Nematalycidae Strenzke à l'occasion des travaux d'aménégment due Languidoc-Rousillon. *Compte Rendus de l'Académie des Sciences de Paris*, **265**, 685–8.

Colwell, R.K. (2009). *EstimateS: Statistical estimation of species richness and shared species from samples. Version 8.2.* User's Guide and application published at http://purl.oclc.org/estimates.

Colwell, R.K., Mao, C.X., and Chang, J. (2004). Interpolating, extrapolating, and comparing incidence-based species accumulation curves. *Ecology*, **85**, 2717–27.

Cooney, T.J. and Simon, K.S. (2009). Influence of dissolved organic matter and invertebrates on the function of microbial films in groundwater. *Microbial Ecology*, **58**, 599–610.

Cooper, S.J.B, Bradbury, J.H., Saint, K.M., Leys, R., Austin, A.D. and Humphreys, W.F. (2007). Subterranean archipelago in the Australian arid zone: mitochondrial DNA phylogeography of amphipods from Western Australia. *Molecular Ecology*, **16**, 1533–44.

Cooper, S.J.B., Saint, M., Taiti, S., Austin, A.D. and Humphreys, W.F. (2008). Subterranean archipelago: mitochondrial DNA phylogeography of stygobitic isopods (Oniscidea: *Haloniscus*) from the Yilgarn region of Western Australia. *Invertebrate Systematics*, **22**, 195–203.

Cornwell, W.K., Shwilk, D.W., and Ackerly, D.D. (2006). A trait-based test for habitat filtering: convex hull volume. *Ecology*, **87**, 1465–71.

Cottarelli, V., Bruno, M.C., Spena, M.T., and Grasso, R. (2012). Studies on subterranean copepods from Italy, with descriptions of two new epikarstic species from a cave in Sicily. *Zoological Studies*, **51**, 556–82.

Covington, M.D., Luhmann, A.J., Wicks, C.M., and Saar, M. (2012). Process length scales and longitudinal damping in karst conduits. *Journal of Geophysical Research—Earth Surface*, **117**, F01025, doi: 10.1029/2011JF002212.

Crouau-Roy, B. (1987). Spéciation and structure génétique des populations chez les coléoptères *Speonomus*. *Mémoires de Biospéologie*, **14**, 1–312.

Crouau-Roy, B. (1988). Genetic structure of cave-dwelling beetles populations: significant deficiencies of heterozygotes. *Heredity*, **60**, 321–7.

Crouau-Roy, B., Crouau, Y., and Ferre, C. (1992). Dynamic and temporal structure of the troglobite beetle *Speonomus hydrophilus* (Coleoptera: Bathysciinae). *Ecography*, **15**, 12–8.

Culver, D.C. (1982). *Cave life*. Harvard University Press, Cambridge, Massachusetts.

Culver, D.C. (2012). Species interactions. In WB White and DC Culver, eds. *Encyclopedia of caves, second edition*, pp. 743–8. Elsevier/Academic Press, Amsterdam, The Netherlands.

Culver, D.C., Brancelj, A., and Pipan, T. (2012b). Epikarst communities. In WB White and DC Culver, eds. *Encyclopedia of caves, second edition*, pp. 288–95. Elsevier/Academic Press, Amsterdam, The Netherlands.

Culver, D.C., Christman, M.C., Elliott, W.R., Hobbs III, H.H., and J.R. Reddell. (2003). The North American obligate cave fauna: regional patterns. *Biodiversity and Conservation*, **12**, 441–68.

Culver, D.C., Deharveng, L., Bedos, A., Lewis, J.J., Madden, M., Reddell, J.R., Sket. B., Trontelj, P., and White, D. (2006a). The mid-latitude biodiversity ridge in terrestrial cave fauna. *Ecography*, **29**, 120–8.

Culver, D.C., Fong, D.W., and Jernigan, R.W. (1991). Species interactions in cave stream communities: experimental results and microdistribution effects. *American Midland Naturalist*, **126**, 364–79.

Culver, D.C., Holsinger, J.R., and Feller, D.J. (2012a). The fauna of seepage springs and other shallow subterranean habitats in the mid-Atlantic Piedmont and Coastal Plain. *Northeastern Naturalist*, **19** (Monograph 9), 1–42.

Culver D.C., Holsinger, J.R., Christman, M.C., and Pipan, T. (2010). Morphological differences among eyeless amphipods in the genus *Stygobromus* dwelling in different subterranean habitats. *Journal of Crustacean Biology*, **30**, 68–74.

Culver, D.C., Kane, T.C., and Fong, D.W. (1995). *Adaptation and natural selection in caves. The case of* Gammarus minus. Harvard University Press, Cambridge, Massachusetts.

Culver, D.C. and Pipan, T. (2008). Superficial subterranean habitats—gateway to the subterranean realm? *Cave and Karst Science*, **35**, 5–12.

Culver, D.C. and Pipan, T. (2009). *The biology of caves and other subterranean habitats*. Oxford University Press, Oxford, UK.

Culver, D.C. and Pipan, T. (2010). Climate, abiotic factors, and the evolution of subterranean life. *Acta Carsologica*, **39**, 577–86.

Culver, D.C. and Pipan, T. (2011). Redefining the extent of the aquatic subterranean biotope—shallow subterranean habitats. *Ecohydrology*, **4**, 721–30.

Culver, D.C. and Pipan, T. (2013). Subterranean ecosystems. In SA Levin, ed., *Encyclopedia of Biodiversity, Second Edition*, pp. 49–62. Academic Press, Waltham, Massachusetts.

Culver, D.C., Pipan, T., and Gottstein, S. (2006b). Hypotelminorheic—a unique freshwater habitat. *Subterranean Biology*, **4**, 1–8.

Culver, D.C., Pipan, T., and Schneider, K. (2009). Vicariance, dispersal, and scale in the aquatic subterranean fauna of karst regions. *Freshwater Biology*, **54**, 918–29.

Culver, D.C. and Poulson, T.L. (1971). Oxygen consumption and activity in closely related amphipod populations from cave and surface habitats. *American Midland Naturalist*, **85**, 74–84.

Culver, D.C. and Šereg, I. (2004). *Kenk's Amphipod* (Stygobromus kenki) *and other amphipods in Rock Creek Park,*

Washington, D.C. Report to Rock Creek Park, National Park Service, Washington, DC.

Culver, D.C. and Sket, B. (2000). Hotspots of subterranean biodiversity in caves and wells. *Journal of Cave and Karst Studies*, **62**, 11–7.

Culver, D.C., Trontelj, P., Zagmajster, M., and Pipan, T. (2012c). Paving the way for standardized and comparable subterranean biodiversity studies. *Subterranean Biology*, **10**, 43–50.

Dalens, H. (1984). Isopodes terrestres recontrés dans les cavités volcaniques de l'île de Tenerife. *Travaux de Laboratoire Ecobiologie de Arthropdes Edaphiques Toulouse*, **5**, 12–9.

Danielopol, D.L. (1998). Conservation protection of the biota of karst: assimilation of scientific ideas through artistic perception. *Journal of Cave and Karst Studies*, **60**, 67.

Danielopol, D.L., Griebler, C., Gunatilaka, A., and Notenboom, J. (2003). Present state and future prospects for groundwater ecosystems. *Environmental Conservation*, **30**, 104–30.

Danielopol, D.L. and Pospisil, P. (2001). Hidden biodiversity in the groundwater of the Danube Flood Plain National Park, Austria. *Biodiversity and Conservation*, **10**, 1711–21.

Danielopol, D.L. and Pospisil, P. (2005). Why and how to take care of subterranean aquatic microcrustaceans? In J Gibert, ed. *World subterranean biodiversity. Proceedings of an international symposium held on 8–10 December in Villeurbanne, France*, pp. 29–35. Equipe Hydrobiologie et Ecologie Souterraines, Université Claude Bernard I, Villeurbanne, France.

Danielopol, D.L., Pospisil, P., and Dreher, J. (2001). Structure and functioning of groundwater ecosystems in a Danube wetland at Vienna. In C Griebler, DL Danielopol, J Gibert, HP Nachtnebel, and J Notenboom, eds. *Groundwater ecology. A tool for management of water resources*, pp. 121–42. European Communities, Luxembourg.

Danielopol, D.L., Pospisil, P., Dreher, J., Mösslacher, F., Torreiter, P., Geiger-Kaiser, M. and Gunatilaka, A. (2000). A groundwater ecosystem in the Danube wetlands at Wien (Austria). In H Wilkens, DC Culver, and WF Humphreys, eds. *Subterranean ecosystems*, pp. 481–511. Elsevier Press, Amsterdam, The Netherlands.

Danks, H.V. and Williams, D.D. (1991). Arthropods of springs, with particular reference to Canada: synthesis and needs for research. *Memoirs of the Entomological Society of Canada*, **123**, 203–17.

Darwin, C. (1881). *The formation of vegetable mould, through the action of worms, with observations on their habits.* John Murray, London.

Datry, T., Malard, F., and Gibert, J. (2005). Response of invertebrate assemblages to increased groundwater recharge rates in a phreatic aquifer. *Journal of the North American Benthological Society*, **24**, 461–77.

Davidson, E.A. and Jassens, I.A. (2006). Temperature sensitivity of soil carbon decomposition and feedbacks to climate change. *Nature*, **440**, 165–73.

Deboué, E.R. and Borowsky, R.L. (2012). Altered rest-activity patterns evolve via circadian independent mechanisms in cave adapted balitorid loaches. *PLoS One*, **7**, e30868. doi:10.1371/journal.pone.0030868.

Deharveng, L. and Bedos, A. (2000). The cave fauna of Southeast Asia. Origin, evolution and ecology. In H Wilkens, DC Culver, and WF Humphreys, eds. *Subterranean ecosystems*, pp. 603–32. Elsevier Press, Amersterdam, The Netherlands.

Deharveng, L. and Bedos, A. (2012). Diversity patterns in the tropics. In WB White and DC Culver, eds. *Encyclopedia of caves, second edition*, pp. 238–50. Elsevier/Academic Press, Amsterdam, The Netherlands.

Deharveng, L., Bedos, A., Le, C.K., Le, C.M., and Truong, Q.T. (2009a). Endemic arthropods of the Hòn Chông hills (Kiên Giang), an unrivaled biodiversity heritage in southeast Asia. In CK Le, QT Truong, and NS Ly, eds. *Beleaguered hills: managing the biodiversity of the remaining karst hills at Kiên Giang, Vietnam*, pp. 31–57. Nhà Xuất Bện Nông Nghiệp, Hô Chi Minh City, Vietnam.

Deharveng, L., Gibert, J., and Culver, D.C. (2012). Biodiversity in Europe. In WB White and DC Culver, eds. *Encyclopedia of caves, second edition*, pp. 219–28. Elsevier/Academic Press, Amsterdam, The Netherlands.

Deharveng, L., Lips, J., and Rahmali, C. (2011). Focus on guano. In T Bouchet, H Le Guyader, and O. Pascal, eds. *The natural history of Santo*, pp. 300–6. Museum National d'Histoire Naturelle, Paris.

Deharveng, L., Stoch, F., Gibert, J., Bedos, A., Galassi, D., Zagmajster, M., Brancelj, A., Camacho, A., Fiers, F., Martin, P., Giani, N., Magniez, G., and Marmonier, P. (2009b). Groundwater biodiversity in Europe. *Freshwater Biology*, **54**, 709–26.

Delay, B. (1968). Donnees sur le peuplement de la zone de percolation temporaire. *Annales de Spéologie*, **23**, 705–33.

Delay, B., Juberthie, C., and Ruffat, G. (1983). Description de *Speonomus colluvii* n.sp. due milieu souterrain superficiel des Pyrénées ariégeoises. *Mémoires de Biospéologie*, **10**, 249–56.

Denton Jr., R.K, and Scott, H. (2013). *Geological survey of the Lower Cretaceous Potomac Formation of Fort Belvoir, Virginia, and its relationship to the habitat of the Northern Virginia Well Amphipod Stygobromus phreaticus.* Proprietary Report, GeoConcepts Engineering, Ashburn, VA, USA.

Department of Environment and Conservation of Western Australia (2013). Western Australia's threatened ecological communities. http://www.dec.wa.gov.au/

management-and-protection/threatened-species/wa-s-threatened-ecological-communities.html

Derkarabetian, S., Steinmann, D.B., and Hedin, M. (2010). Repeated and time-correlated morphological convergence in cave-dwelling harvestmen (Opiliones, Laniatores) from montane western North America. *Plos One*, 5, e10388, doi:10.1371/journal.pone.0010388.

Desutter-Grandcolas, L. (1997). Are troglobitic taxa troglobiomorphic? A test using phylogenetic inference. *International Journal of Speleology*, 26, 1–19.

Desutter-Grandcolas, L. and Grandcolas, P. (1996). The evolution toward troglobitic life: a phylogenetic reappraisal of climatic relict and local habitat shift hypotheses. *Mémoires de Biospéologie*, 23, 57–63.

Desutter-Grandcolas, L., D'Haese, C., and Robillard, T. (2003). The problem of characters susceptible to parallel evolution in phylogenetic analysis: a reply to Marguès and Gnaspini (2001) with emphasis on cave-life phenotypic evolution. *Cladistics*, 19, 131–7.

Didden, W.A.M. (1993). Ecology of terrestrial Enchytraeidae. *Pedobiologia*, 37, 2–29.

Dole, M.J. (1984). Structure biocénotique des niveaux supériurs de la nappe alluviale du Rhône à l'est de Lyon. *Mémoires de Biospéologie*, 11, 17–26.

Dole, M.J. and Mathieu, J. (1984). Etude de la 'pellicule biologique' dans les milieu interstitials de l'Est Lyonnais. *Verhandlungen der Internationalen Vereinigung für Theoretische und Angewandte Limnologie*, 22, 1745–50.

Dole-Olivier, M.J. (2011). The hyporheic refuge hypothesis reconsidered: a review of hydrological aspects. *Marine and Freshwater Research*, 62, 1281–302.

Dole-Olivier, M.J., Castellarini, F., Coineau, N., Galassi, D.M.P., Martin, P., Mori, N., Valdecasas, A., and Gibert, J. (2009). Towards an optimal sampling strategy to assess groundwater biodiversity: comparison across six European regions. *Freshwater Biology*, 54, 777–96.

Dole-Olivier, M.J., Creuzé des Châtelliers, M., and Marmonier, P. (1993). Repeated gradients in subterranean landscape—example of the stygofauna of the alluvial flood plain of the Rhône River (France). *Archiv für Hydrobiologie*, 127, 451–71.

Dole-Olivier, M.J., Galassi, D.M., Marmonier, P., and Creuzé des Châtelliers, M. (2000). The biology and ecology of lotic microcrustaceans. *Freshwater Biology*, 44, 63–91.

Dreybrodt, W. (2011). Comments on processes contributing to the isotope composition of ^{13}C and ^{18}O in calcite deposited in speleothems. *Acta Carsologica*, 40, 233–8.

Dreybrodt, W., Gabrovšek, F., and Romanov, D. (2005). *Processes of speleogenesis. A modeling approach*. ZRC Publishing, Ljubljana, Slovenia.

Ducarme, X., André, H.M., Wauthy, G., and Lebrun, P. (2004a). Comparison of endogeic and cave communities: microarthropod density and mite species richness. *European Journal of Soil Biology*, 40, 129–38.

Ducarme, X., Wauthy, G., André, H.M., and Lebrun, P. (2004b). Survey of mites in caves and deep soil and evolution of mites in these habitats. *Canadian Journal of Zoology*, 82, 841–50.

Dupanloup, I., Schneider, S., and Excoffier, L. (2002). A simulated annealing approach to define the genetic structure of populations. *Molecular Ecology*, 11, 2571–81.

Eamus, D. and Froend, R. (2006). Groundwater-dependent ecosystems: the where, what, and why of GDEs. *Australian Journal of Botany*, 54, 91–6

Eberhard, R. and Sharples, C. (2013). Appropriate terminology for karst-like phenomena: the problem with pseudokarst. *International Journal of Speleology*, 42, 109–13.

Eberhard, S.M., Halse, S.A., Williams, M.R., Scanlon, M.D., Cocking, J., and Barron, H.J. (2009). Exploring the relationship between sampling efficiency and short-range endemism for groundwater fauna of the Pilbara region. *Freshwater Biology*, 54, 885–901.

Eggleton, P., Williams, P.H., and Gaston, K.J. (1994). Explaining global termite diversity: productivity or history. *Biodiversity and Conservation*, 3, 318–30.

Elliott, W.R. (2000). Conservation of the North American cave and karst biota. In H Wilkens, DC Culver, and WF Humphreys, eds. *Subterranean ecosystems*, pp. 665–89. Elsevier Press, Amsterdam, The Netherlands.

Eme, D., Malard, F., Konecny-Dupré, L, Lefébure, T., and Douady, C.J. (2013). Bayesian phylogeographic inferences reveal contrasted colonization dynamics among European groundwater isopods. *Molecular Ecology*, 22, 5685-99. doi: 10.1111/mec 12520.

Engel, A.S. (2012). Chemoautotrophy. In WB White and DC Culver, eds. *Encyclopedia of caves, second edition*, pp. 125–34. Elsevier/Academic Press, Amsterdam, The Netherlands.

Environmental Protection Agency (2013). *Underground storage tanks*. www.epa.gov/oust/.

Español, F. and Ribes, J. (1983). Un nueva especie troglobia de Emesinae (Heter., Reduviidae) de las Islas Canarias. *Speleon*, 26–7, 57–60.

Euliss, N.H., LaBaugh, J.W., Frederickson, L.H., Mushet, D.M., Laubhan, M.K., Swanson, G.A., Winter, T.C., Rosenberry, D.O., and Nelson, R.D. (2004). The wetland continuum: a conceptual framework for interpreting biological studies. *Wetlands*, 24, 448–58.

Excoffier, L., Laval, G., and Schneider, S. (2005). Arlequin Version 3.0: An integrated software package for population genetics data analysis. *Evolutionary Bioinformatics*, 1, 47–50.

Finston, T.L., Johnson, M.S., Humphreys, W.F., Eberhard, S.M., and Halse, S.A. (2007). Cryptic speciation in two widespread subterranean amphipod genera reflects historical drainage patterns in an ancient landscape. *Molecular Ecology*, **16**, 355–65.

Fišer, C., Blejec, A., and Trontelj, P. (2012). Niche-based mechanisms operating within extreme habitats: a case study of subterranean amphipod communities. *Biology Letters*, **8**, 578–81.

Fišer, C., Keber, R., Kereži, V., Moškrič, A., Palandančić, A., Petkovska, H., Potočnik, H., and Sket, B. (2007). Coexistence of species of two amphipod genera: *Niphargus timavi* (Niphargidae) and *Gammarus fossarum* (Gammaridae). *Journal of Natural History*, **41**, 2641–51.

Fišer, C., Konec, M., Kobe, Z., Osanič, M., Gruden, P., and Potočnik, H. (2010). Conservation problems with hypotelminorheic *Niphargus* species (Amphipoda: Niphargidae). *Aquatic Conservation: Marine and Freshwater Ecosystems*, **20**, 602–4.

Fišer, C., Sket, B., and Stoch, F. (2006). Distribution of four narrowly endemic *Niphargus* species (Crustacea: Amphipoda) in the western Dinaric region with description of a new species. *Zoologischer Anzeiger*, **245**, 77–94.

Fong, D.W. (2004). Intermittent pools at headwaters of subterranean drainage basins as sampling sites for epikarst fauna. In WK Jones, DC Culver, and JS Herman, eds. *Epikarst. Proceedings of the symposium held October 1 through 4, 2003, Shepherdstown, West Virginia.* pp. 114–8. Karst Waters Institute, Charles Town, West Virginia.

Fong, D.W., Culver, D.C., Hobbs III, H.H., and Pipan, T. (2007). *The invertebrate cave fauna of West Virginia, second edition.* Bulletin 16, West Virginia Speleological Survey, Barrackville, West Virginia, USA.

Fong, D.W. and Kavanaugh, K.E. (2010). Population dynamics of the stygobiotic amphipod crustacean *Stygobromus tenuis potomacus* and isopod crustacean *Caecidotea kenki* at a single hypotelminorheic habitat over a two-year span. In A Moškrič and P Trontelj, eds. *ICSB 2010 Abstract Book*, pp. 22–3. International Conference on Subterranean Biology, Postojna, Slovenia.

Ford, D. and Williams, P. (2007). *Karst hydrogeology and geomorphology.* John Wiley & Sons, New York.

Foulquier, A., Malard, F., Mermillod-Blondin, F., Montuelle, B., Dolédec, S., Volat, B., and Gibert, J. (2011). Surface water linkages regulate trophic interactions in a groundwater food web. *Ecosystems*, **14**, 1339–53.

Foulquier, A., Simon, L., Gilbert, F., Fourel, F., Malard, F., and Mermillod-Blondin, F. (2010). Relative influences of DOC flux and subterranean fauna on microbial abundance and activity in aquifer sediments: new insights from ^{13}C-tracer tests. *Freshwater Biology*, **55**, 1560–76.

Francke, O.F. (1982). Studies of the scorpion subfamilies Superstitioninae and Typhlochactinae, with description of a new genus (Scorpiones, Chactidae). *Texas Memorial Museum Bulletin*, **28**, 51–61.

Frederickson, J.K., Garland, T.R., Hicks, R.J., Thomas, J.M., Li, S.W., and McFadden, S.M. (1989). Lithotrophic and heterotrophic bacteria in deep subsurface sediments and their relation to sediment properties. *Geomicrobiology Journal*, **7**, 53–66.

Friedrich, M. (2013). Biological clocks and visual systems in cave-adapted animals at the dawn of speleogenomics. *Integrative and Comparative Biology*, **53**, 50–67.

Friedrich, M., Chen, R., Daines, B., Bao, R., Caravas, J., Rai, P.K., Zagmajster, M., and Peck, S.M. (2011). Phototransduction and clock gene expression in the troglobiont beetle *Ptomaphagus hirtus* of Mammoth. *Journal of Experimental Biology*, **214**, 3532–41.

Frisch, J. and Oromí, P. (2006). New species of subterranean *Micranops* Cameron from the Canary Islands (Coleoptera, Staphylinidae, Paederinae), with a redescription of *Micranops bifossicapitatus* (Outerelo & Oromí, 1987). *Deutsche Entomologische Zeitschrift*, **53**, 23–37.

Gabrovšek, F. (2004). Attempts to model the early development of epikarst. In WK Jones, DC Culver, and JS Herman, eds. *Epikarst. Proceedings of the symposium held October 1 through 4, 2003, Shepherdstown, West Virginia*, pp. 50–5. Karst Waters Institute, Charles Town, West Virginia.

Genest, C., and Juberthie, C. (1983a). Description d'*Aphaenops colluvii* (Coléoptères, Trechinae) due milieu souterrain superficiel des Pyrénées ariégeoises. *Mémoires de Biospéologie*, **10**, 295–304.

Genest, C., and Juberthie, C. (1983b). Description de *Paraduvalius rajtchevi* n.sp. (Coléoptères, Trechinae) du milieu souterrain superficiel des Rhodopes centraux (Bulgarie). *Mémoires de Biospéologie*, **10**, 311–4.

Gers, C. (1992). *Ecologie et biologie des Arthropodes terrestres du milieu souterrain superficiel fonctionnement et ecologie evolutive.* Ph.D. Dissertation, Université Paul Sabatier de Toulouse, France.

Gers, C. (1998). Diversity of energy fluxes and interactions between arthropod communities, from soil to cave. *Acta Oecologia*, **19**, 205–13.

Gers, C. and Najt, J. (1983). Note sur les Collemboles (insects, Apterygotes) du milieu souterrain superficiel et description d'une nouvelle espèces d'*Isotoma* (*Desoria*). *Revue de Écologie et Biologie du Sol*, **20**, 427–32.

Giachino, P.M. and Vailati, D. 2010. *The subterranean environment. Hypogean life, concepts and collecting techniques.* World Biodiversity Association Handbook 3, Verona, Italy.

Gibert J. (1986). Ecologie d'un systeme karstique jurassien. Hydrogéologie, dérive animale, transits de matières, dynamique de la population de *Niphargus* (Crustacé Amphipode). *Mémoires de Biospéologie*, **13**, 1–379.

Gibert, J., ed. (2005). *World subterranean biodiversity. Proceedings of an international symposium held on 8–10*

December in Villeurbanne, France. Equipe Hydrobiologie et Ecologie Souterraines, Université Claude Bernard I, Villeurbanne, France.

Gibert, J. and Culver, D.C. (2009). Assessing and conserving groundwater biodiversity: an introduction. *Freshwater Biology*, **54**, 639–48.

Gibert, J. and Deharveng, L. (2002). Subterranean ecosystems: a truncated functional biodiversity. *Bioscience*, **52**, 474–81.

Gibert, J., Dole-Olivier, M.J., Marmonier, P., and Vervier, P. (1990). Surface water/groundwater ecotones. In RJ Naiman and H Décamps, eds. *Ecology and management of aquatic-terrestrial ecotones*, pp. 199–225. Parthenon Publishing, Carnforth, UK.

Gibert, J., Stanford, J.A., Dole-Olivier, M.-J., and Ward, J.V. (1994). Basic attributes of groundwater ecosystems and prospects for research. In J Gibert, DL Danielopol, and JA Stanford, eds. *Groundwater ecology*, pp. 7–40. Academic Press, San Diego, California.

Gilgado, J.D., López, H., Oromí, P., and Ortuño, V.M. (2011). Description of the first larval instar of *Bruscus crassimargo* Wollaston, 1865 (Carabidae: Broscini) and notes about the presence of this species in the MSS of La Gomera (Canary Islands, Spain). *Entomologica Fennica*, **22**, 46–55.

Ginés, A., Knez, M., Slabe, T., and Dreybrodt, W., eds. (2009). *Karst rock features. Karren sculpturing*. Založba ZRC, ZRC Publishing, Postojna-Ljubljana, Slovenia.

Ginet, R. (1985). Redescription du type de l'amphipode hypogé *Niphargus rhenorhodanensis* Schellenberg, 1937. *Crustaceana*, **48**, 225–43.

Ginet, R. and David, J. (1963). Présence de *Niphargus* (Amphipode Gammaridae) dans certaines eaux épigées des forêts de la Dombes (départment de l'Ain, France). *Vie et Milieu*, **14**, 299–310.

Ginet, R. and Decu, V. (1977). *Initiation à la biologie a l'écologie souterraines*. J-P Delarge, Paris.

Glazier, D.S., Horne M.T., and Lehman M.E. (1992). Abundance, body composition and reproductive output of *Gammarus minus* (Crustacea: Amphipoda) in ten cold springs differing in pH and ionic count. *Freshwater Biology*, **28**, 149–63.

Gnaspini, P. (1992). Bat guano ecosystems. A new classification and some considerations, with special references to Neotropical data. *Memoires de Biospeologie*, **19**, 135–8.

Gnaspini, P. (2012). Guano communities. In WB White and DC Culver, eds. *Encyclopedia of caves, second edition*, pp. 357–64. Elsevier/Academic Press, Amsterdam, The Netherlands.

Gnaspini, P. and Trajano, E. (2000). Guano communities in tropical caves. In H Wilkens, DC Culver, and WF Humphreys, eds. *Subterranean ecosystems*, pp. 251–68. Elsevier Press, Amsterdam, The Netherlands.

Goodey, J.B. (1963). *Laboratory methods for work with plant and soil nematodes*. Ministry of Fisheries, Food, and Agriculture. London.

Gordon, N.D., McMahon, T.A., and Finlayson, B.L. (1999). *Stream hydrology: an introduction for ecologists*. Wiley, Chichester, England.

Gottstein, S., Žganec, K., Krnjević, V.C., and Popijač, A. (2010). Life history traits of the epigean populations of *Niphargus dalmatinus* (Crustacea: Amphipoda) along the Cetina River, Croatia. In A Moškrič and P Trontelj, eds. *ICSB 2010 Abstract Book*, pp. 23–4. International Conference on Subterranean Biology, Postojna, Slovenia.

Gould, S.J. (2002). *The structure of evolutionary theory*. Harvard University Press, Cambridge, Massachusetts.

Groom, M.J., Meffe, G.K., and Carroll, C.R. (2005). *Principles of conservation biology, third edition*. Sinauer Associates, Sunderland, Massachusetts.

Gross, J.B., Furterer, A., Carlson, B.M., and Stahl, B.A. (2013). An integrated transcriptome-wide analysis of cave and surface dwelling *Astyanax mexicanus*. *PLoS One*, **8** doi: 10.1371/journal.pone.0055659.

Guzik, M.T., Abrams, K.M., Cooper, S.J.B., Humphreys, W.F., Cho, J.L., and Austin, A.D. (2008). Phylogeography of the ancient Parabathynellidae (Crustacea: Bathynellacea) from the Yilgarn region of western Australia. *Invertebrate Systematics*, **22**, 205–16.

Guzik, M.T., Cooper, S.J.B., Humphreys, W.F., and Austin, A.D. (2009). Fine-scale comparative phylogeography of a sympatric sister species triplet of subterranean diving beetles from a single calcrete aquifer in Western Australia. *Molecular Ecology*, **18**, 3683–98.

Guzik, M.T., Austin, A.D., Cooper, S.J.B., Harvey, M.S., Humphreys, W.F., Bradford, T., Eberhard, S.M., King, R.A., Leys, R., Muirhead, K.A., and Tomlinson, M. (2010). Is the Australian subterranean fauna uniquely diverse? *Invertebrate Systematics*, **24**, 407–18.

Guzik, M.T., Cooper, S.J.B., Humphreys, W.F., Ong, S., Kawakami, T., and Austin, A.D. (2011). Evidence of population fragmentation within a subterranean aquatic habitat in the Western Australian desert. *Heredity*, **107**, 215–30.

Hadley, N.F., Ahearn, G.A., and Howarth, F.G. (1981). Water and metabolic relations of cave adapted and epigean lycosid spiders in Hawaii. *Journal of Arachnology*, **9**, 215–22.

Hahn, H. (2009). A proposal for an extended typology of groundwater habitats. *Hydrology Journal*, **17**, 77–81.

Halliday, W.R., ed. (1976). *Proceedings of international symposium on vulcanospeleology*. Western Speleological Society, Seattle.

Halliday, W.R. (2004). Hawai'i lava tubes, United States. In J. Gunn, ed., *Encyclopedia of caves and karst science*, pp. 415–6. Fitzroy-Dearborn, New York.

Halse, S. and G. Pearson. (2012). Why trap troglofauna: nets provide superior catches. In L Kováč, M Uhrin, A Mock, and P Ľuptáčik, eds. *Proceedings of the 21st International Conference on Subterranean Biology*, pp. 32–3. Košice, Slovakia.

Hamilton-Smith, E. and Eberhard, S. (2000). Conservation of cave communities in Australia. In H. Wilkens, DC Culver, and WF Humphreys, eds. *Subterranean ecosystems*, pp. 647–64. Elsevier Press, Amsterdam, The Netherlands.

Hancock, P.J., Boulton, A.J., and Humphreys, W.F. (2005). Aquifers and hyporheic zones: towards an ecological understanding of groundwater. *Hydrogeology Journal*, **13**, 98–111.

Hathaway, J.J.M., Sinsabaugh, R.L., Deptevicus, M.L.N.E., and Northup, D.E. (2013). Diversity of ammonia oxidation (amoA) and nitrogen fixation (nifH) genes in lava caves of Terceira, Azores, Portugal. *Geomicrobiology Journal* doi:10.1080/01490451.2012.752424

Hawes, R.S. (1939). The flood factor in the ecology of caves. *Journal of Animal Ecology*, **8**, 1–5.

Heads, S.W. (2010). The first fossil spider cricket (Orthoptera: Gryllidae: Phalangopsinae): 20 million years of troglobiomorphosis or exaptation in the dark? *Zoological Journal of the Linnean Society*, **158**, 56–65.

Heath, R.W. (1983). *Basic ground-water hydrology*. United States Geological Survey Water-Supply Paper 2220. Alexandria, Virginia.

Herman, J.S., Culver, D.C., and Salzman, J. (2001). Groundwater ecosystems and the service of water purification. *Stanford Environmental Law Journal*, **20**, 479–95.

Herrando-Perez, S., Baratti, M., and Messana, G. (2008). Subterranean ecological research and multivariate statistics: a review (1945–2006). *Journal of Cave and Karst Studies*, **70**, 120–8.

Hervant, F., and Malard, F. (2012). Responses to low oxygen. In WB White and DC Culver, eds. *Encyclopedia of caves, second edition*, pp. 651–8. Elsevier/Academic Press, Amsterdam, The Netherlands.

Hinaux, H., Poulain, J., Da Silva, C., Noirot, C., Jeffery, W.R., Casane, D., and Rétaux, S. (2013). De novo sequencing of *Astyanax mexicanus* surface and Pachón cavefish transcriptomes reveals enrichment of mutations in cavefish putative eye genes. *PLoS One*, **8**, e53553.

Hobson, C.S. (1997). *A Natural Heritage zoological inventory of U.S. Army Fort Belvoir*. Technical Report 97–5, Division of Natural Heritage, Department of Conservation and Recreation, Commonwealth of Virginia, Richmond, Virginia.

Hoch, H. (2000). Acoustic communication in darkness. In H. Wilkens, DC Culver, and WF Humphreys, eds. *Subterranean ecosystems*, pp. 211–20. Elsevier Press, Amsterdam, The Netherlands.

Hoch, H. (2002). Hidden from the light of day: planthoppers in subterranean habitats (Hemiptera: Auchenorrhyncha: Fulgoromorpha). In W Holzinger and P Gusenleitner, eds. *Zikaden: leafhoppers, planthoppers, and cicadas (Insecta: Auchenorrhyncha: Fulgoromorpha)*, pp. 149–6. Oberösterreichisches Landesmuseum, Linz. Austria.

Hoch, H., Asche, M., Burwell, C., Monteith, G.M., and Wessel, A. (2006). Morphological alteration in response to endogeic habitat and ant association in two new planthopper species from New Caledonia (Hemiptera: Auchenorrhyncha: Fulgoromorpha: Delphacidae). *Journal of Natural History*, **40**, 1867–86.

Hoch, H. and Howarth, F.G. (1999) Multiple cave invasions by species of the planthopper genus *Oliarus* in Hawaii (Homoptera: Fulgoroidea: Cixiidae). *Zoological Journal of the Linnean Society*, **127**, 453–75.

Holsinger, J.R. (1967). Systematics, speciation, and distribution of the subterranean amphipod genus *Stygonectes* (Gammaridae). *Bulletin of the U.S. National Museum*, **259**, 1–176.

Holsinger, J.R. (1969). The systematics of the North American subterranean amphipod genus *Apocrangonyx* (Gammaridae), with remarks on ecology and zoogeography. *The American Midland Naturalist*, **81**, 1–28.

Holsinger, J.R. (1971). Observations on a population of the cavernicolous amphipod crustacean *Crangonyx antennatus* Packard. *Virginia Journal of Science*, **22**, 97.

Holsinger, J.R. (1978). Systematics of the subterranean amphipod genus *Stygobromus* (Crangonyctidae), Part II: species of the eastern United States. *Smithsonian Contributions to Zoology*, **266**, 1–144.

Holsinger, J. R. (2009). Three new species of the subterranean amphipod crustacean genus *Stygobromus* (Crangonyctidae) from the District of Columbia, Maryland, and Virginia. In S. M. Roble and J. C. Mitchell, eds., *A lifetime of contributions to Myriapodology and the natural history of Virginia: A Festschrift in honor of Richard L. Hoffman's 80th Birthday*, pp. 261–276. Virginia Museum of Natural History Special Publication No. 16, Martinsville, VA.

Holsinger, J.R. (2012). Vicariance and dispersalist biogeography. In WB White and DC Culver, eds. *Encyclopedia of caves, second edition*, pp. 849–58. Elsevier/Academic Press, Amsterdam, The Netherlands.

Holsinger, J.R., Mort, J.S., and Recklies, A.D. (1983). The subterranean crustacean fauna of Castleguard Cave, Columbia Icefields, Alberta, Canada, and its zoogeographic significance. *Arctic and Alpine Research*, **(1983)**, 543–49.

Hon, K., Kauahikaua, J., Denlinger, R., and Mackay, K. (1994). Emplacement and inflation of pahoehoe sheet flows: Observations and measurements of active lava flows on Kilauea Volcano, Hawai`i. *Bulletin of the Geological Society of America*, **106**, 351–70.

Howarth, F.G. (1972). Cavernicoles in lava tubes on the island of Hawaii. *Science*, **175**, 325–6.

Howarth, F.G. (1973). The cavernicolous fauna of Hawaiian lava tubes I. Introduction. *Pacific Insects*, **15**, 139–51.

Howarth, F.G. (1980). The zoogeography of specialized cave animals: a bioclimatic model. *Evolution*, **34**, 394–406.

Howarth, F.G. (1983). Ecology of cave arthropods. *Annual Review of Ecology and Systematics*, **28**, 365–89.

Howarth, F.G. (1987a). The evolution of non-relictual tropical troglobites. *International Journal of Speleology*, **16**, 1–16.

Howarth, F.G. (1987b). Evolutionary ecology of aeolian and subterranean habitats in Hawaii. *Trends in Ecology and Evolution*, **2**, 220–3.

Howarth, F..G. and Hoch, H. (2012). Adaptive shifts. In WB White and DC Culver, eds. *Encyclopedia of caves, second edition*, pp. 9–17. Elsevier/Academic Press, Amsterdam, The Netherlands.

Howarth, F.G. and Stone, F.D. (1990). Elevated carbon dioxide levels in Bayliss Cave, Australia: Implications for the evolution of obligate cave species. *Pacific Science*, **44**, 207–18.

Huang, Y.Q., Li, X.K., Zhang, Z.F., He, C.X., Zhao, P., You, Y.M., and Mo, L. (2011). Seasonal changes in *Cyclobalanopsis glauca* transpiration and canopy stomatal conductance and their dependence on subterranean water and climatic factors in rocky karst terrain. *Journal of Hydrology*, **402**, 135–43.

Hubricht, L. and Mackin, J.G. (1940). Description of nine new species of fresh-water amphipod crustaceans with notes and new localities for other species. *American Midland Naturalist*, **23**, 187–218.

Humphreys, W.F. (1999). Physico-chemical profile and energy fixation in Bundera Sinkhole, an anchialine remiped habitat in north-western Australia. *Journal of the Royal Society of Western Australia*, **82**, 89–98.

Humphreys, W.F. (2001). Groundwater calcrete aquifers in the Australian arid zone: the context to an unfolding plethora of stygal biodiversity. *Records of the Western Australian Museum Supplement*, **64**, 63–83.

Humphreys, W.F. (2008). Rising from Down Under: developments in subterranean biodiversity in Australia from a groundwater fauna perspective. *Invertebrate Systematics*, **22**, 85–101.

Humphreys, W.F., Watts, C.H.S., Cooper, S.J.B., and Leijs, R. (2009). Groundwater estuaries of salt lakes: buried pools of endemic biodiversity on the western plateau, Australia. *Hydrobiologia*, **626**, 79–95.

Hüppop, K. (2000). How do cave animals cope with the food scarcity in caves? In H Wilkens, DC Culver, and WF Humphreys, eds. *Subterranean ecosystems*, pp. 159–88. Elsevier Press, Amsterdam, The Netherlands.

Hutchins, B.T., Schwartz, B.F., and Engel, A.S. (2013). Environmental controls on organic matter production and transport across surface–subsurface and geochemical boundaries in the Edwards Aquifer, Texas, USA. *Acta Carsologica*, **42**, 245–59.

Hutchinson, G.E. (1950). Survey of contemporary knowledge of biogeochemistry. 3. The biogeochemistry of vertebrate excretion. *Bulletin of the American Museum of Natural History*, **96**, 1–454.

Ireland, P. (1979). Geomorphological variations of case-hardening in Puerto Rico. *Zeitschrift für Geomorphologie Supplement-Band*, **32**, 9–20.

Izquierdo, I., Martín, J.L., Zurita, N., and Medina, A. (2001). Geo-referenced computer recordings as an instrument for protecting cave-dwelling species of Tenerife (Canary Islands). In DC Culver, L Deharveng, J Gibert, and ID Sasowsky, eds. *Mapping subterranean biodiversity*, pp. 45–8. Karst Waters Institute, Charles Town, West Virginia.

James, J.M. (1977). Carbon dioxide in the cave atmosphere. *Transactions of the British Cave Research Association*, **4**, 417–29.

Jasinska, E.J. and Knott, B. (2000). Root-driven faunas in cave waters. In H Wilkens, DC Culver, and WF Humphreys, eds. *Subterranean ecosystems*, pp. 287–307. Elsevier Press, Amsterdam, The Netherlands.

Jeffery, W.R. (2005a). Adaptive evolution of eye degeneration in the Mexican blind cavefish. *Journal of Heredity*, **96**, 185–96.

Jeffery, W.R. (2005b). Evolution of eye degeneration in cavefish: the return of pleiotropy. *Subterranean Biology*, **3**, 1–11.

Jeffery, W.R. (2009). Regressive evolution in *Astyanax* cavefish. *Annual Review of Genetics*, **43**, 25–47.

Jeffery, W.R. (2010). Pleiotropy and eye degeneration in cavefish. *Heredity*, **105**, 495–6.

Jegla, T.C. and Poulson, T.L. (1968). Evidence of circadian rhythms in a cave crayfish. *Journal of Experimental Biology*, **168**, 273–82.

Jobbágy, E.G. and Jackson, R.B. (2000). The vertical distribution of soil organic carbon and its relation to climate and vegetation. *Ecological Applications*, **10**, 423–36.

Johnson, S.L., Commander, D.P., and O'Boy, C.A. (1999). *Groundwater Resources of the Northern Goldfields, Western Australia*. Water and Rivers Commission, Hydrogeological Record Series, Report HG 2, East Perth, Western Australia.

Jones, D.L. and Willett, V.B. (2006). Experimental evaluation of methods to quantify dissolved organic nitrogen (DON) and dissolved organic carbon (DOC) in soil. *Soil Biology and Biochemistry*, **38**, 991–9.

Jones, J.B. and Holmes, R.M. (1996). Surface–subsurface interactions in stream ecosystems. *Trends in Ecology and Evolution*, **11**, 239–42.

Jones, J.B., Holmes, R.M., Fisher, S.G., Grimm, N.B., and Green, D.M. (1995) Methanogenesis in Arizona dryland streams. *Biogeochemistry*, **31**, 155–73.

Jones, J.B. and Mulholland, P.J. (2000). *Streams and ground water*. Academic Press, San Diego, California.

Jones, W.K., (2013). Physical structure of the epikarst. *Acta Carsologica*, **42**, 311–4.

Jones, W.K., Culver, D.C., and Herman, J.S., eds. (2004). *Epikarst. Proceedings of the symposium held October 1 through 4, 2003, Shepherdstown, West Virginia*. Karst Waters Institute, Charles Town, West Virginia.

Jones, W.K., Hobbs III, H.H., Wicks, C.M., Currie, R.R., Hose, L.D., Kerbo, R.C., Goodbar, J.R., and Trout, J. (2003). *Recommendations and guidelines for managing caves on protected lands*. Karst Waters Institute Special Publication 8, Charles Town, West Virginia.

Jones, W.K. and White, W.B. (2012). Karst. In WB White and DC Culver, eds. *Encyclopedia of caves, second edition*, pp. 430–8. Elsevier/Academic Press, Amsterdam, The Netherlands.

Juberthie, C. (1983). Le milieu souterrain: Etendue et composition. *Mémoires de Biospéologie*, **10**, 17–65.

Juberthie, C. (2000). The diversity of the karstic and pseudokarstic hypogean habitats in the world. In H Wilkens, DC Culver, and WF Humphreys, eds. *Subterranean ecosystems*, pp. 17–40. Elsevier Press, Amsterdam, The Netherlands.

Juberthie, C. and Decu, V. (1994). Structure et diversité du domaine souterrain; particularités des habitats et adaptations des espèces. In C Juberthie and V Decu, eds. *Encyclopaedia biospeologica. Tome I*, pp. 5–22. Société Internationale de Biospéologie, Moulis, France.

Juberthie, C., Delay, B., and Bouillon, M. (1980a). Extension du milieu souterrain en zone non-calcaire: description d'un nouveau milieu et de son peuplement par les coléoptères troglobies. *Mémoires de Biospéologie*, **7**, 19–52.

Juberthie C., Delay, B., and Bouillon, M. (1980b). Sur l'existence d'un milieu souterrain superficiel en zone non calcaire. *Compte Rendus de l'Académie des Sciences de Paris*, **290**, 49–52.

Juberthie-Jupeau, L. (1994). Symphyla. In C Juberthie and V Decu, eds. *Encyclopaedia Biospeologica. Tome I*, pp. 365–6. Société Internationale de Biospéologie, Moulis, France.

Kalbitz, K., Solinger, S., Park, J.-H., Michalzik, B., and Matzner, E. (2000) Controls on the dynamics of dissolved organic matter in soils: A review. *Soil Science*, **165**, 277–304.

Karaman, S. (1935). Die fauna der unterirdischen Gewässer Jugoslawiens. *Verhandlungen der Internationalen Vereinigung für Theoretische und Angewandte Limnologie*, **7**, 46–73.

Karanovic, I. (2007). Candoninae Ostracodes from the Pilbara Region in Western Australia. *Crustaceana Monographs*, **7**, 1–432.

Karanovic, T. and Cooper, S.J.M. (2011). Third genus of parastenocaridid copepods from Australia supported by molecular evidence (Copepoda: Harpacticoidea). In D Defaye, E Guarez-Morales, and JC von Vaupel Klein, eds. *Studies on freshwater Copepoda: a volume in honour of Bernard Dussart*, pp. 293–337. Kominklijke Brill NV, Leiden, The Netherlands.

Kareiva, P., Tallis, H., Ricketts, T.H., and Daily, G.C. (2011). *Natural capital: theory and practice of mapping ecosystem services*. Oxford University Press, Oxford, UK.

Keeley, J.E. and Zedler, P.H. (1998). Characerization and global distribution of vernal pools. In CC Witham, ed. *Vernal pool ecosystems*, pp. 1–14. California Native Plant Society, Sacramento, California.

Kempe, S. (2012). Volcanic rock caves. In WB White and DC Culver, eds. *Encyclopedia of caves, second edition*, pp. 865–73. Elsevier/ Academic Press, Amsterdam, The Netherlands.

Kempe, S., Bauer, I., and Henschel, H.V. (2003). The Pa'auhau Civil Defense Cave on Mauna Kea, Hawai'i, a lava tube modified by water erosion. *Journal and Cave and Karst Studies*, **65**, 76–85.

Kenk, R. (1972). *Freshwater planarians (Turbellaria) of North America*. Biota of freshwater ecosystems identification manual No. 1, Environmental Protection Agency, Washington, DC.

Kenk, R. (1977). Freshwater triclads (Turbellaria) of North America, IX: The genus *Sphalloplana*. *Smithsonian Contributions to Zoology*, **246**, 1–38.

Kenk, R. (1984). Freshwater triclads (Turbellaria) of North America. XV. Two new subterranean species from the Appalachian region. *Proceedings of the Biological Society of Washington*, **97**, 209–16.

King, R.K., Bradford, T., Austin, A., Humphreys, W.F., and Cooper, S.J.B. (2012). Divergent molecular lineages and not-so-cryptic species: the first descriptions of stygobitic chiltoniid amphipods (Talitroidea: Chiltoniidae) from Western Australia. *Journal of Crustacean Biology*, **32**, 465–88.

Klimchouk, A. (2012). Krubera (Voronja) cave. In WB White and DC Culver, eds. *Encyclopedia of caves, second edition*, pp. 443–50. Elsevier/Academic Press, Amsterdam, The Netherlands.

Knapp, S.M. and Fong, D.W. (1999). Estimates of population size of *Stygobromus emarginatus* (Amphipoda: Crangonyctidae) in a headwater stream in Organ Cave, West Virginia. *Journal of Cave and Karst Studies*, **61**, 3–6.

Knez, M. and Slabe, T. (2007). Krasoslovna spremljava gradnje, raziskave ter načrtovanje avtocest prek slovenskega krasa. In M Knez and T Slabe, eds. *Kraški pojavi, razkriti med gradnjo slovenskih avtocest*, pp. 9–22. Založba ZRC, ZRC SAZU, Ljubljana, Slovenia.

Koenemann, S. and Holsinger, J.R. (2001). Systematics of the North American subterranean amphipod genus *Bactrurus* (Crangonyctidae). *Beaufortia*, **51**, 1–56.

Kogovšek, J. (2010). *Characteristics of percolation through the karst vadose zone*. ZRC Publishing, Ljubljana, Slovenia.

Kogovšek, J. and Urbanc, J. (2007). Ocena dinamike premikajoče vode skozi vadozno cono Postojnske jame na osnovi izotopskih značilnosti. *Geologija*, **56**, 477–86.

Kranjc, A. and Opara, B. (2002). Temperature monitoring in Škocjanske jame caves. *Acta Carsologica*, **31**, 85–96.

Krause, S., Hannah, D.M., Fleckenstein, J.H., Heppell, C.M., Kaeser, D., Pickup, R., Pinay, G., Robertson, A.L., and Wood, P.J. (2011). Inter-disciplinary perspectives on processes in the hyporheic zone. *Ecohydrology*, **4**, 481–99.

Kresic, N. (2010). Types and classifications of springs. In N Kresic and Z Stevanovic, eds. *Groundwater hydrology of springs. Engineering, theory, management, and sustainability*, pp. 31–86. Elsevier Press, Amsterdam, The Netherlands.

Kresic, N. (2013). *Water in karst*. McGraw-Hill, New York.

Kroker H. (1983). Catches of Cholevidae with epigeous and hypogeous exposed traps. *Mémoires de Biospéologie*, **10**, 83–4.

Kubešova, S. and Chytry, M. (2005). Diversity of bryophytes on treeless cliffs and talus slopes in a forested central European landscape. *Journal of Bryology*, **27**, 35–46.

Kunz, T.H., Braun de Torrez, E., Bauer, D.M., Lobova, T.A, and Fleming, T.H. (2011). Ecosystem services provided by bats. *Annals of the New York Academy of Sciences*, **1223**, 1–38.

Laiz, L., Groth, I., Gonzalez, I., and Saiz-Jimenez, C. (1999). Microbiological study of the dripping waters in Altamira cave (Santillana del Mar, Spain). *Journal of Microbiological Methods*, **36**, 129–38.

Lamoreux, J. (2004). Stygobites are more wide-ranging than troglobites. *Journal of Cave and Karst Studies*, **66**, 18–9.

Lamprecht, G. and Weber, F. (1992). Spontaneous locomotion behavior in cavernicolous animals: the regression of the endogenous circadian system. In AI Camacho, ed., *The natural history of biospeleology*, pp. 225–62. Museo Nacional de Ciencias Naturales, Madrid, Spain.

Laneyrie, R. (1960). Résumé des connaissances actuelles concernant les Coléoptères hypogés de France. *Annales de la Société Entomologiques de France*, **131**, 89–149.

Larson, C. (2006). Influences of microhabitat constraints and rock-climbing disturbances on cliff-face vegetation communities. *Conservation Biology*, **20**, 821–32.

Laška, V., Mikula, J., and Tuf, I.H. (2008). Jak hluboko žijí půdní bezobratlí? (How deep do soil invertebrates live?). *Živa*, **4**, 169–71.

Laška, V., Kopecký, O., Růžička, V., Mikula, J., Véle, A., Šarapatka, B., and Tuf, I.H. (2011). Vertical distribution of spiders in soil. *Journal of Arachnology*, **39**, 393–8.

Lattinger, R. (1988). *Ekološka diferenciranost faune podzemnih voda Medvednice*. Ph.D. Thesis, University of Zagreb, Croatia.

Lefébure, T., Douady, C.J., Malard, F., and Gibert, J. (2007). Testing dispersal and cryptic diversity in a widely distributed groundwater amphipod (*Niphargus rhenorhodanensis*). *Molecular Phylogenetics and Evolution*, **42**, 676–86.

Leijs, R., Roundew, B., Mitchell, J., and Humphreys, W.F. (2009). A new method for sampling stygofauna from groundwater fed marshlands. *Speleobiology Notes*, **1**, 12–3.

Leijs, R., van Nes, E.H., Watts, C.H., Cooper, S.J.B., Humphreys, W.F., and Hogendoorn, K. (2012). Evolution of blind beetles in isolated aquifers: a test of alternative modes of speciation. *PLoS One*, **7**, e34260.

Levins, R. and Lewontin, R.C. (1985). *A dialectical biologist*. Harvard University Press, Cambridge, Massachusetts.

Lewis, J.J. (2013). *Caecidotea insula*, a new species of subterranean asellid from Lake Erie's South Bass Island, Ohio (Crustacea: Isopoda: Asellidae). *Journal of Cave and Karst Studies*, **75**, 64–7.

Leys, R., Watts, C.H.S., Cooper, S.J.B., and Humphreys, W.F. (2003). Evolution of subterranean diving beetles (Coleoptera: Dytiscidae: Hydroporini: Bidessini) in the arid zone of Australia. *Evolution*, **57**, 2819–34.

Loop, C.M. (2012). Contamination of cave waters by non-aqueous phase liquids. In WB White and DC Culver, eds. *Encyclopedia of caves, second edition*, pp. 166–72. Elsevier/Academic Press, Amsterdam, The Netherlands.

López, H. and Oromí, P. (2010). A pitfall trap for sampling the mesovoid shallow substratum (MSS) fauna. *Speleobiology Notes*, **2**, 7–11.

Luštrik, R., Turjak, M., Kralj-Fišer, S., and Fišer, C. (2011). Coexistence of surface and cave amphipods in an ecotone environment. *Contributions to Zoology*, **80**, 133–41.

Lynch, M. and Conery, J.S. (2003). The origins of genome complexity. *Science*, **302**, 1401–4.

MacDonald, G.A. (1983). *Volcanoes in the sea: the geology of Hawaii*. University of Hawaii Press, Honolulu, Hawaii.

Malard, F., ed. (2003). *Sampling manual for the assessment of regional groundwater biodiversity*. PASCALIS (Protocols for the Assessment and Conservation of Aquatic Life in the Subsurface), Lyon, France. Available at www.pascalis.org.

Malard, F., Boutin, C., Camacho, A.I., Ferreira, D., Michel, G., Sket, B., and Stoch, F. (2009). Diversity patterns of stygobiotic crustaceans across multiple spatial scales in Europe. *Freshwater Biology*, **54**, 756–76.

Malard, F., Ferreira, D., Dolédec, S., and Ward, J.V. (2003a). Influence of groundwater upwelling on the distribution of the hyporheos in a headwater river flood plain. *Archiv für Hydrobiologie*, **157**, 89–116.

Malard, F., Galassi, D., Lafont, M., Dolédec, S., and Ward, J.V. (2003b). Longitudinal patterns of invertebrates in the hyporheic zones of a glacial river. *Freshwater Biology*, **48**, 1709–25.

Malard, F. and Hervant, F. (1999). Oxygen supply and the adaptations of animals in groundwater. *Freshwater Biology*, **41**, 1–30.

Malard, F., Lafont, M., Burgherr, P., and Ward, J.V. (2001). A comparison of longitudinal patterns in hyporheic and benthic oligochaete assemblages in a glacial river. *Arctic, Antarctic and Alpine Research*, **33**, 457–66.

Malard, F., Tockner, K., Dole-Oliver, M.-J., and Ward J.V. (2002). A landscape perspective of surface–subsurface hydrological exchange in river corridors. *Freshwater Biology*, **47**, 621–40.

Malard, F., Ward, J.V., and Robinson, C.T. (2000). An expanded perspective of the hyporheic zone. *Verhandlungen der Internationalen Vereinigung für Theoretische und Angewandte Limnologie*, **27**, 431–7.

Mangin, A. (1973). Sur la dynamique des transferts en aquifer karstique. *Proceedings of the Sixth International Congress of Speleology, Olomouc*, **4**, 157–62.

Mann, A.W. and Horwitz, R.C. (1979). Groundwater calcrete deposits in Australia: some observations from Western Australia. *Journal of the Geological Society of Australia*, **26**, 293–303.

Marmonier, P., Creuzé des Châtelliers, M., Dole-Olivier, M.-J., Plénet, S., and Gibert, J. (2000). Rhône groundwater systems. In H Wilkens, DC Culver, and WF Humphreys, eds. *Subterranean ecosystems*, pp. 513–31. Elsevier Press, Amsterdam, The Netherlands.

Martín, J.L. and Oromí, P. (1987). Tres neuvas especies hipoges de *Loboptera* Brum & W. (Blattaria, Blattellidae) y consideraciones sobre el medio subterráneo de Tenerife (Islas Canarias). *Annales de l'Societe Entomologique de France*, **23**, 313–26.

Martín, J.L. and Oromí, P. (1988). Dos nuevas especies de *Anataelia* Bol. (Dermaptera, Pygidicranidae) de cuevas y lavas recientes del Hierro y La Palma (Islas Canarias). *Mémoires de Biospéologie*, **15**, 49–59.

Martín Esquivel, J.L. and Izquierdo Zamora, I. (1999). The protection of the Viento-Sobrado Cave—a very long volcanic cave in the Canary Islands. In W Burone, R Bonaccurso, and G Lucitra, eds. *Inside Volcanoes: Proceedings of the IXth international symposium on vulcanospeleology of the I.U.S.*, pp. 108–19. Catania, Italy

Mary, N. and Marmonier, P. (2000). First survey of interstitial fauna in New Caledonian rivers: influence of geological and geomorphological characteristics. *Hydrobiologia*, **418**, 199–208.

Master, L.L., Flack, S.R., and Stein, B.A. (1998). *Rivers of life: critical watersheds for protecting freshwater biodiversity*. The Nature Conservancy, Arlington, Virginia.

Mathieu, J., Jennerod, F., Hervant, F., and Kane, T.C. (1997). Genetic differentiation of *Niphargus rhenorhodanensis* (Amphipoda) from interstitial and karst environments. *Aquatic Science*, **59**, 39–47.

Medeiros, M.J., Davis, D., Howarth, F.G., and Gillespie, R. (2009). Evolution of cave living in Hawaiian *Schrankia* (Lepidoptera: Noctuidae) with description of a remarkable new cave species. *Zoological Journal of the Linnean Society*, **156**, 114–39.

Medina, A.L. and Oromí, P. (1990). First data on the superficial underground compartment in La Gomera (Canary Islands). *Mémoires de Biospéologie*, **17**, 87–91.

Medville, D.M. (2009). Hualalai, island of Hawai'i. In AN Palmer and MV Palmer, eds. *Caves and karst of the United States*, pp. 315–7. National Speleological Society, Huntsville, Alabama.

Meinzer, O.E. (1923). *The occurrence of ground water in the United States with a discussion of principles*. U.S. Geological Survey Water-Supply Paper No. 489, Washington, DC.

Meleg, I.N., Fiers, F., and Moldovan, O.T. (2011b). Assessing copepod (Crustacea: Copepoda) species richness at different spatial scales in northwestern Romanian caves. *Subterranean Biology*, **9**, 103–12.

Meleg, I.N., Fiers, F., Robu, M., and Moldovan, O.T. (2011c). Distribution patterns of subsurface copepods and the impact of environmental parameters. *Limnologica*, **42**, 156–64.

Meleg, I.N., Moldovan, O.T., Iepure, S., Fiers, F., and Brad, T. (2011a). Diversity patterns of fauna in dripping water of caves from Transylvania. *Annales de Limnologie/International Journal of Limnology*, **47**, 185–97.

Menge, D.N.L., Hedin, L.O., and Pacala, S.W. (2012). Nitrogen and phosphorus limitation over long-term ecosystem development in terrestrial ecosystems. *PLoS One*, **7**, e42045. doi:10.1371/journal.pone.0042045.

Meštrov, M. (1962). Un nouveau milieu aquatique souterrain: le biotope hypotelminorheique. *Compte Rendus Academie des Sciences, Paris*, **254**, 2677–9.

Meštrov, M. (1964). Différences et relations faunistiques et écologiques entre les milieu souterrains aquatiques. *Spelunca Mémoires*, **4**, 185–7.

Michel, G., Malard, F., Deharveng, L., Di Lorenzo, T., Sket, B., and De Broyer, C. (2009). Reserve selection for conserving groundwater biodiversity in Europe. *Freshwater Biology*, **54**, 861–76.

Milder, J.C. and Clark, S. (2011). Conservation development practices, extent, and land-use effects in the United States. *Conservation Biology*, **25**, 697–707.

Mitchell, R.W. and Peck, S.B. (1977). *Typhlochactas sylvestris*, new eyeless scorpion from montane forest litter in Mexico. *Journal of Arachnology*, **5**, 159–68.

Moe S.J., Stelzer, R.S., Forman, M.R., Harpole, W.S., Daufresne, T., and Yoshida, T. (2005). Recent advances in ecological stoichiometry: insights for population and community ecology. *Oikos*, **109**, 29–39.

Moldovan, O.T., ed. (2006). *Emil George Racovitza. Essay on biospeological problems—French, English, Romanian version.* Casa Cărţi de Ştiinţă, Cluj-Napoca, Romania.

Moldovan O.T., Meleg I.N., and Perşiou A., (2011). Habitat fragmentation and its effects on groundwater populations. *Ecohydrology*, **5**, 445–52.

Moldovan, O.T., Pipan, T., Iepure, S., Mihevc, A., and Mulec, J. (2007). Biodiversity and ecology of fauna in percolating water in selected Slovenian and Romanian caves. *Acta Carsologica*, **36**, 493–501.

Morgan, K.H. (1993). Development, sedimentation and economic potential of palaeoriver systems of the Yilgarn Craton of Western Australia. *Sedimentary Geology*, **85**, 637–56.

Moseley, M. (2008). Estimating diversity and ecological status of cave invertebrates: some lessons and recommendations from Dark Cave (Batu Caves, Malaysia). *Cave and Karst Science*, **35**, 47–52.

Moseley, M. (2009). Are all caves ecotones? *Cave and Karst Science*, **36**, 53–8.

Motas, C. (1958). Freatobiologia, o noura ramura a limnologiei. (Phreatobiology, a new field of limnology). *Natura (Bucharest)*, **10**, 95–105.

Mulec, J. (2012). Lampenflora. In WB White and DC Culver, eds. *Encyclopedia of caves, second edition*, pp. 451–6. Elsevier/Academic Press, Amsterdam, The Netherlands.

Mulec, J., Mihevc, A., and Pipan, T. (2005). Intermittent lakes in the Pivka basin. *Acta Carsologica*, **34**, 543–65.

Musgrove, M. and Banner, J.L. (2004a). Groundwater evolution in two karst aquifers: vadose zone processes, soil influences, and climate connections. In WK Jones, DC Culver, and JS Herman, eds. *Epikarst. Proceedings of the symposium held October 1 through 4, 2003, Shepherdstown, West Virginia*, pp. 62–70. Karst Waters Institute, Charles Town, West Virginia.

Musgrove, M. and Banner, J.L. (2004b). Controls on the spatial and temporal variability of vadose dripwater geochemistry: Edwards Aquifer, central Texas. *Geochimica et Cosmochimica Acta*, **68**, 1007–20.

Nadelhoffer, K.L., Aber, J.D., and Melillo, J.M. (1985). Fine roots, net primary production and soil nitrogen availability. *Ecology*, **66**, 1377–90.

Najt, J. and Weiner, W.M. (2002). A new genus *Pongeia* from France, without mandibles: why does it not belong to Brachystomellidae (Collembola)? *Acta Zoologica Cracoviensia*, **45**, 337–40.

Niemiller, M.L., Near, T.J., and Fitzpatrick, B.M. (2012). Delimiting species using multilocus data: diagnosing cryptic diversity in the southern cavefish, *Typhlichthys subterraneus* (Teleostei: Amblyopsidae). *Evolution*, **66**, 846–66.

Niemiller, M. and Poulson, T.L. (2010). Studies of the Amblyopsidae: past, present, and future. In E Trajano, ME Bichuette, and BG Kapoor, eds. *The biology of subterranean fishes*, pp. 169–280. Science Publishers, Enfield, New Hampshire.

Nitzu, E., Nae, A., Băncilă, R., Popa, I., Giurginca, A., and Plăişu, R. (2014). Scree habitats: ecological function, species conservation and spatial-temporal variation in the arthropod community. *Systematics and Biodiversity*. doi: 10.1080/14772000.2013.878766

Northup, D.E., Connolly, C.A., Trent, A., Peck, V.M., Spilde, M.N., Welbourn, W.C., and Natvig, D.O. (2004). The nature of bacterial communities in Four Windows Cave, El Malpais National Monument, New Mexico, *Association for Mexican Cave Studies Bulletin*, **19**, 119–25.

Notenboom, J. (2001). Managing ecological risks of groundwater pollution. In C. Griebler, D.L. Danielopol, J. Gibert, H.P. Nachtnebel, and J. Gibert, eds. *Groundwater ecology. A tool for management of water resources*, pp. 247–62. Director-General for Research, European Communities, Luxembourg.

Notenboom, J., Plénet, S., and Turquin, M.-J. (1994). Groundwater contamination and its impact on groundwater animals and ecosystems. In J Gibert, DL Danielopol, and JA Stanford, eds. *Groundwater ecology*, pp. 477–504. Academic Press, San Diego.

Novak, T., Perc, M., Lipovšek, S., and Janžekovič, F. (2012). Duality of the terrestrial subterranean fauna. *International Journal of Speleology*, **41**, 181–8.

Novak, T. and Sivec, N. (1977). Biological researches of pegmatite caves in Slovenia (Yugoslavia). In TD Ford, ed. *Proceedings of the 7th international speleological congress Sheffield, England*, pp. 328–9. British Cave Research Association, Somerset, England.

Oda, G., Caldas, I., Piqueira, J., Waterhouse, J., and Marques, M. (2000). Coupled biological oscillators in a cave insect. *Journal of Theoretical Biology*, **206**, 515–24.

Odum, E.P. (1953). *Fundamentals of ecology*. W.B. Saunders, Philadelphia, Pennsylvania.

Orghidan, T. (1959). Ein neuer Lebensraum des unterir-dischen Wassers: Der hyporheische Biotop. *Archiv für Hydrobilogie*, **55**, 392–414.

Oromí, P., ed. (1995). *La Cueva del Viento*. Consejero de Política Territorial, Gobierno de Canarias. La Laguna, Tenerife, Spain.

Oromí, P. (2004). Biospeleology in Macaronesia. *Association for Mexican Cave Studies Bulletin*, **19**, 98–104.

Oromí, P. and Izquierdo, I. (in press) Canary Islands. In C. Juberthie, ed., *Encyclopaedia biospeologica Volume IA*. International Society of Subterranean Biology, Moulis, France.

Oromí, P. and Martín, J.L. (1984). *Apteranopsis canariensis*, un nuevo coleopteran cavernícola de Tenerife (Staphylinidae). *Nouvelle Revue Entomologique (N.S.)*, **1**, 41–8.

Oromí, P. and Martín, J.L. (1992). The Canary Islands. Subterranean fauna: characterization and composition. In AI Camacho, ed., *The natural history of biospeleology*, pp. 527–67. Museo Nacional Ciencias Naturales, Madrid, Spain.

Ortuño, V.M. and Gilgado, J.D. (2010). Update on the Ibero-Balearic hypogean Carabidae (Coleoptera): Faunistics, biology and distribution. *Entomologische Blätter*, **106**, 233–64.

Ortuño, V.M., Gilgado, J.D., Jiménez-Valverde, A., Sendra, A., Pérez-Suárez, G., and Herrero-Borgoñón, J.J. (2013). The 'alluvial mesovoid shallow substratum', a new subterranean habitat. *PloS One*, **8**, e76311.

Owen, J.A. (1995). A pit fall trap for the repetitive sampling of hypogeal arthropod faunas. *Entomologist's Record*, **107**, 222–8.

Ozimec, R. (2012). Ecology, biodiversity and vulnerability of Šipun cave (Cavtat, Dubrovnik, Croatia). *Natura Croatica*, **21**, 86–90.

Palacios-Vargas, J.G. and Gnaspini-Netto, P. (1992). A new Brazilian species of *Acherontides* (Collembola: Hypogastruridae), with notes on its ecology. *Journal of the Kansas Entomological Society*, **65**, 443–7.

Palmer, A.N. (2007). *Cave geology*. Cave Books, Dayton, Ohio.

Palmer, A.N. and Palmer, M.V., eds. (2009). *Caves and karst of the USA*. National Speleological Society, Huntsville, Alabama.

Palmer, M.A., Swan, C.M., Nelson, K., Silver, P., and Alvestad, R. (2000). Streambed landscapes: evidence that stream invertebrates respond to type and spatial arrangement of patches. *Landscape Ecology*, **15**, 563–76.

Papi, F. and Pipan, T. (2011). Ecological studies of an epikarst community in Snežna jama na planini Arto—an ice cave in north central Slovenia. *Acta Carsologica*, **40**, 505–13.

Pavek, D. (2002). Endangered amphipods in our nation's capital. *Endangered Species Bulletin*, **27**, 8–9.

Peck, S.B. (1973). A review of the invertebrate fauna of Volcanic caves in western North America. *Bulletin of the National Speleological Society*, **35**, 99–107.

Peck, S.B. (1975). The life cycle of a Kentucky cave beetle, *Ptomaphagus hirtus* (Coleoptera; Leiodidae; Catopinae). *International Journal of Speleology*, **7**, 7–17.

Peck, S.B. (1980). Climatic change and the evolution of cave invertebrates in the Grand Canyon, Arizona. *Bulletin of the National Speleological Society*, **42**, 53–60.

Peck, S.B. (1986). Evolution of adult morphology and life history in cavernicolous *Ptomaphagus* beetles. *Evolution*, **40**, 1021–30.

Peck, S.B. (1990). Eyeless arthropods of the Galapagos Islands, Ecuador: Composition and origin of the cryptozoic fauna of a young, tropical, oceanic archipelago. *Biotropica*, **22**, 366–81.

Peck, S. B. and Finston, T.L. (1993). Galapagos Islands troglobites: the questions of tropical troglobites, parapatric distributions with the eyed sister-species, and their origin by parapatric speciation. *Mémoires de Biospéologie*, **20**, 19–37.

Pesce, G.L. and Galassi, D.P. (1986). Taxonomic and phylogenetic value of the armature of coxa and antenna in stygobiont cyclopoid copepods. *Bollettino di Zoologia*, **53**, 5.

Petkovski, T.K. (1959). Fauna Copepoda pećine 'Dona Duka' kod rašča—Skopje. *Fragmenta Balcanica*, **2**, 107–23.

Pipan, T. (2003). *Ekologija ceponožnih rakov (Crustacea: Copepoda) v prenikajoči vodi izbranih kraških jam*. Ph.D. Dissertation, University of Ljubljana, Ljubljana, Slovenia.

Pipan, T. (2005a). *Epikarst—a promising habitat*. Založba ZRC, Ljubljana, Slovenia.

Pipan, T. (2005b). Fauna of the Pivka intermittent lakes. *Acta Carsologica*, **34**, 650–9.

Pipan, T. (2006). Ceponožni rakci—predstavniki podzemeljske favne. In D Viršek, ed. *Županova jama, čudežni svet brez sonca*, pp. 53–8. Županova jama—turistično in okoljsko društvo Grosuplje, Slovenia.

Pipan, T., Blejec, A., and Brancelj, A. (2006b). Multivariate analysis of copepod assemblages in epikarstic waters of some Slovenian caves. *Hydrobiologia*, **559**, 213–23.

Pipan, T. and Brancelj, A. (2001). Ratio of copepods (Crustacea: Copepoda) in fauna of percolation water in six karst caves in Slovenia. *Acta Carsologica*, **30**, 257–66.

Pipan, T. and Brancelj, A. (2004). Distribution patterns of copepods (Crustacea: Copepoda) in percolation water of the Postojnska jama cave system (Slovenia). *Zoological Studies*, **43**, 206–10.

Pipan, T., Christman, M.C., and Culver D.C. (2006a). Dynamics of epikarst communities: microgeographic pattern and environmental determinants of epikarst copepods in Organ Cave, West Virginia. *American Midland Naturalist*, **156**, 75–87.

Pipan, T. and Culver, D.C. (2005). Estimating biodiversity in the epikarstic zone of a West Virginia cave. *Journal of Cave and Karst Studies*, **67**, 103–9.

Pipan, T. and Culver, D.C. (2007a). Copepod distribution as an indicator of epikarst system connectivity. *Hydrogeology Journal*, **15**, 817–22.

Pipan, T. and Culver, D.C. (2007b). Regional species richness in an obligate subterranean dwelling fauna—epikarst copepods. *Journal of Biogeography*, **34**, 854–61.

Pipan, T. and Culver, D.C. (2012a). Convergence and divergence in the subterranean realm: a reassessment. *Biological Journal of the Linnean Society*, **107**, 1–14.

Pipan, T. and Culver, D.C. (2012b). Shallow subterranean habitats. In WB White and DC Culver, eds. *Encyclopedia of caves, second edition*, pp. 683–90. Elsevier/Academic Press, Amsterdam, The Netherlands.

Pipan, T. and Culver, D.C. (2012c). Wetlands in cave and karst areas. In WB White and DC Culver, eds. *Encyclopedia of caves, second edition*, pp. 897–904. Elsevier/Academic Press, Amsterdam, The Netherlands.

Pipan, T. and Culver, D.C. (2013). Organic carbon in shallow subterranean habitats. *Acta Carsologica*, **42**, 291–300.

Pipan, T., Holt, N., and Culver, D.C. (2010). How to protect a diverse, poorly known, inaccessible fauna: identification of source and sink habitats in the epikarst. *Aquatic Conservation: Marine and Freshwater Ecosystems*, **20**, 748–55.

Pipan, T., López, H., Oromí, P., Polak, S., and Culver, D.C. (2011a). Temperature variation and the presence of troglobionts in shallow subterranean habitats. *Journal of Natural History*, **45**, 253–73.

Pipan, T., Mulec, J., and Oarga, A. (2011b). Epikarst fauna of selected caves in Yunnan Province. In M Knez, H Liu, and T. Slabe, eds. *South China karst II*, pp. 173–81. Založba ZRC, ZRC-SAZU. Ljubljana, Slovenia.

Pipan, T., Navodnik, V., Janžekovič, F., and Novak, T. (2008). First studies on the fauna of percolation water in Huda Luknja, a cave in the isolated karst in northeast Slovenia. *Acta Carsologica*, **37**, 141–51.

Piscart, C., Navel, S., Maazouzi, C., Montuelle, B., Cornut, J., Mermillod-Blondin, F., Creuze des Chatelliers, M., Simon, L., and Marmonier, P. (2011). Leaf litter recycling in benthic and hyporheic layers in agricultural streams with different types of land use. *Science of the Total Environment*, **409**, 4373–80.

Plénet, S. and Gibert, J. (1995). Comparison of surface water/groundwater interface zones in fluvial and karstic systems. *Compte Rendus de l'Académie des Sciences de Paris, Vie*, **318**, 499–509.

Plénet, S., Gibert, J., and Marmonier, P. (1995). Biotic and abiotic interactions between surface and interstitial systems in rivers. *Ecography*, **18**, 296–309.

Polis, G.A., Anderson, W.B., and Holt, R.D. (1997). Toward an integration of landscape ecology and food web ecology: the dynamics of spatially subsidized food webs. *Annual Review of Ecology and Systematics*, **28**, 289–316.

Poore, G.C.B. and Humphreys, W.F. (1998). First record of Spelaeogriphacea from Australasia: a new genus and species from an aquifer in the arid Pilbara of Western Australia. *Crustaceana*, **71**, 721–42.

Poore, G.C.B. and Humphreys, W.F. (2003). Second species of *Mangkurtu* (Spelaeogriphacea) from north-western Australia *Records of the Western Australian Museum*, **22**, 67–74.

Porter, M.L. and Crandall, K.A. (2003). Lost along the way: the significance of evolution in reverse. *Trends in Ecology and Evolution*, **18**, 541–7.

Pospisil, P. (1994). The groundwater fauna of a Danube aquifer in the wetland Lobau at Vienna, Austria. In H Wilkens, DC Culver, and WF Humphreys, eds. *Subterranean ecosystems*, pp. 347–66. Elsevier Press, Amsterdam, The Netherlands.

Poulson, T.L. (1963). Cave adaptation in amblyopsid fishes. *American Midland Naturalist*, **70**, 257–90.

Poulson, T.L. (1972). Bat guano ecosystems. *Bulletin of the National Speleological Society*, **34**, 55–9.

Poulson, T.L. and Jegla, T.C. (1969). Circadian rhythms in cave animals. *Proceedings of the Fourth International Congress of Speleology, Postojna, Yugoslavia*, **4–5**, 193–5.

Poulson, T.L. and White, W.B. (1969). The cave environment. *Science*, **165**, 971–81.

Prendini, L., Francke, O.F., and Vignoli, V. (2009). Troglomorphism, trichobothriotaxy and typhlochactid phylogeny (Scorpiones, Chactoidea): more evidence that troglobitism is not an evolutionary dead end. *Cladistics*, **25**, 1–24.

Prevorčnik, S., Blejec, A., and Sket, B. (2004). Racial differentiation in *Asellus aquaticus* (L.) (Crustacea: Isopoda: Asellidae). *Archiv für Hydrobiologie*, **160**, 193–214.

Protas, M.E., Hersey, C., Kochanek, D., Zhou, Y., Wilkens, H., Jeffery, W.R., Zon, L.I., Borowsky, R., and Tabin, C.J. (2006). Genetic analysis of cavefish reveals molecular convergence in the evolution of albinism. *Nature Genetics*, **38**, 107–11.

Prous, X., Ferreira, R.L., and Martins, R.D. (2004). Ecotone delimitation: epigean-hypogean transition in cave ecosystems. *Austral Ecology*, **29**, 374–82.

Pulliam, H.R. (1988). Sources, sinks, and population regulation. *American Scientist*, **132**, 652–61.

Rabinowitz, D. (1981). Seven forms of rarity. In H Synge, ed. *Aspects of rare plant conservation*, pp. 205–17. Wiley, New York.

Rabinowitz, D., Cairns, S., and Dillon, T. (1986). Seven forms of rarity and their frequency in the flora of the British Isles. In ME Soulé, ed. *Conservation biology: the science of scarcity and diversity*, pp. 182–204. Sinauer Associates, Sunderland, Massachusetts.

Racoviță, E.G. (1907). Essai sur les problèmes biospéologiques. *Archives de Zoologie Expérimentale et Générale*, 6, 371–488.

Rando, J.C., Sala, L., and Oromí, P. (1993). The hypogean community of Cueva del Llano (Fuertoventura, Canary Islands). *Mémoires de Biospéologie*, 20, 189–93.

Reeves, C.C. (1976). *Caliche: origin, classification, morphology and uses.* Estacado Books, Lubbock, Texas.

Reeves, J.M., De Dekker, T., and Halse, S.A. (2007). Groundwater ostracods from the arid Pilbara region of northwestern Australia: distribution and water chemistry. *Hydrobiologia*, 585, 99–118.

Reid, J.W. (1984). Semiterrestrial meiofauna inhabiting a wet campo in central Brazil, with special reference to the Copepoda (Crustacea). *Hydrobiologia*, 118, 95–111.

Reid, J.W. (2001). A human challenge: discovering and understanding continental copepod habitats. *Hydrobiologia*, 453/454, 201–26.

Remane, R. and Hoch, H. (1988). Cave-dwelling Fulgoroidea (Homoptera: Auchenorrhyncha) from the Canary Islands. *Journal of Natural History*, 22, 403–12.

Rendoš, M., Mock, A., and Jászay, T. (2012). Spatial and temporal dynamics of invertebrates dwelling karst mesovoid shallow substratum of Sivec National Nature Reserve (Slovakia), with emphasis on Coleoptera. *Versita*, 67, 1143–51.

Ribera, C., Ferrández, M.A., and Blasco, A. (1985). Araneidos cavernícolas de Canarias II *Mémoires de Biospéologie*, 12, 51–66.

Rivera, M.A.J., Howarth, F.G., Taiti, S., and Roderick, G.K. (2002). Evolution in Hawaiian cave-adapted isopods: vicariant speciation or adaptive shifts? *Molecular Phylogenetics and Evolution*, 25, 1–9.

Rodrigues, S.G., de Pádua Bueno, A.A., and Ferreira, R.L. (2012). The first hypothelminorheic [sic] Crustacea (Amphipoda, Dogielinotidae, *Hyalella*) from South America. *ZooKeys*, 236, 65–80.

Romero, A. (2009). *Cave biology: life in darkness.* Cambridge University Press, Cambridge, UK.

Romero, A. (2011). The evolution of cave life. *American Scientist*, 99, 144–51.

Romero, A. and Green, S.M. (2005). The end of regressive evolution: examining and interpreting the evidence from cave fishes. *Journal of Fish Biology*, 67, 3–32.

Roncin, E. and Deharveng, L. (2003). *Leptogenys khamm ouanensis* sp. nov. (Hymenoptera: Formicidae). A possible troglobitic species of Laos, with a discussion on cave ants. *Zoological Science*, 20, 919–24.

Rouch, R. (1968). Contribution a la connaissance de Harpacticides hypogés (Crustacés—Copépodes). *Annales de Spéléologie*, 23, 9–167.

Rouch, R. (1970). Recherches sur les eaux souterraines–12–Le système karstique du Baget. I. Le phénomène

d'"hémorragie' au niveau de l'exutoire principal. *Annales de Spéléologie*, 25, 665–709.

Rouch, R. (1988). Sur la répartition spatiale des Crustacés dans le sous-écoulement d'un ruisseau des Pyrénées. *Annales de Limnologie*, 24, 213–34.

Rouch, R. (1991). Structure de peuplement des harpacticides dans le milieu hyporhéique d'un ruisseau des Pyrénées. *Annales de Limnologie*, 27, 227–41.

Rouch, R. (1992). Caractéristiques et conditions hydrodynamiques des écoulements dans les sediments d'un ruisseau des Pyrénées. Implications écologiques. *Stygologia*, 7, 13–25.

Rouch, R., Bakalowicz, M., Mangin, A., and D'Hulst, D. (1989). Sur les caractéristiques de sub-écoulement d'un ruisseau des Pyrénées. *Annales de Limnologie*, 25, 3–16.

Rouch, R. and Danielopol, D.L. (1997). Species richness of Microcrustacea in subterranean freshwater habitats. Comparative analysis and approximate evaluation. *International Revue Gesellshaft für Hydrobiologie*, 82, 121–45.

Růžička, J. (1998). Cave and rock debris dwelling species of the *Choleva agilis* species group from central Europe (Coleoptera, Leiodidae: Cholevinae). In PM Giachino and SB Peck, eds. *Phylogeny and evolution of subterranean and endogean Cholevidae (= Leiodidae Cholevinae). Proceedings of a symposium (30 August 1996, Florence, Italy).* XX International Congress of Entomology, pp. 262–86. Museu Regionaledi Scienze Naturali, Torino, Italy.

Růžička, V. (1982). Modifications to improve the efficiency of pitfall traps. *Newsletter of the British Arachnological Society*, 34, 2–4.

Růžička, V. (1988a). Problems of *Bathyphantes eumenis* and its occurrence in Czechoslovakia (Araneae, Linyphiidae). *Věstník Československé Společnosti Zoologické*, 52, 149–55.

Růžička, V. (1988b). The longtimely exposed rock debris pitfalls. *Věstník Československé Společnosti Zoologické*, 52, 238–40.

Růžička, V. (1990). The spiders of stony debris. *Acta Zoologica Fennica*, 190, 333–7.

Růžička, V. (1998). The subterranean forms of *Lepthyphantes imporbulus, Theonoe minutissima* and *Theridion bellicosum* (Araneae: Linyphiidae, Theridiidae). In PA Selden, ed. *Proceedings of the 17th European Colloquium of Arachnology, Edinburgh 1997,* pp. 101–5. Edinburgh, Scotland.

Růžička, V. (1999a). The first steps in subterranean evolution of spiders (Araneae) in Central Europe. *Journal of Natural History*, 33, 255–65.

Růžička, V. (1999b). The freezing scree slopes and their arachnofauna. *Lebensraum Blackhale. Decheniana-Beihefte*, 37, 141–7.

Růžička, V. (2002). Spatial distribution of spiders (Araneae) on scree slopes in Křivoklátsko and Moravský Kras

Protected Landscape Areas. *Acta Societatis Zoologicae Bohemiae*, **66**, 321–8.

Růžička, V. (2011). Central European habitats inhabited by spiders with disjunctive distributions. *Polish Journal of Ecology*, **59**, 367–80.

Růžička, V., Hajer, J., and Zacharda, M. (1995). Arachnid population patterns in underground cavities of a stony debris field (Araneae, Opiliones, Pseudoscorpionidea, Acari: Prostigmata, Rhagidiidae). *Pedobiologia*, **39**, 42–51.

Růžička, V., Laška, V., Mikula, J., and Tuf, I.H. (2011). Morphological adaptations of *Porrhomma* spiders (Araneae: Linyphiidae) inhabiting soil. *Journal of Arachnology*, **39**, 355–7.

Růžička, V., Mlejnek, R., and Šmilauer, P. (2010). Local diversity *versus* geographical distribution of arthropods occurring in a sandstone rock labyrinth. *Polish Journal of Ecology*, **58**, 533–44.

Růžička, V. and Zacharda, M. (2010). Variation and diversity of spider assemblages along a thermal gradient in scree slopes and adjacent cliffs. *Polish Journal of Ecology*, **58**, 361–9.

Růžička, V., Zacharda, M., Němcova, L., Šmilauer, P., and Nikola, J.C. (2012). Periglacial microclimate in low-altitude scree slopes supports relict biodiversity. *Journal of Natural History*, **46**, 2145–57.

Sbordoni, V., Allegrucci, G., and Cesaroni, D. (2000). Population genetic structure: speciation and evolutionary rates in cave-dwelling organisms. In H Wilkens, DC Culver, and WF Humphreys, eds. *Subterranean ecosystems*, pp. 453–78. Elsevier Press, Amsterdam, The Netherlands.

Sbordoni, V., Allegrucci, G., and Cesaroni, D. (2012). Population structure. In DC Culver and WB White, eds. *Encyclopedia of caves, second edition*, pp. 608–18. Elsevier/Academic Press, Amsterdam, The Netherlands.

Scheller, U. (1986). Symphyla from the United States and Mexico. *Texas Memorial Museum, Speleological Monograph*, **1**, 87–125.

Scheller, U. (2009). Records of Pauropoda (Pauropodidae, Brachypauropodidae, Eurypauropodidae) from Indonesia and the Philippines with descriptions of a new genus and 26 new species. *International Journal of Myriapodology*, **2**, 69–148.

Schneider, K. (2009). *How the availability of nutrients and energy influence the biodiversity of cave ecosystems*. Ph.D. Dissertation, University of Maryland, College Park, Maryland.

Schneider, K. and Culver, D.C. (2004). Estimating subterranean species richness using intensive sampling and rarefaction curves in a high density cave region in West Virginia. *Journal of Cave and Karst Studies*, **66**, 39–45.

Schneider, K., Christman, M.C., and Fagan, W.F. (2011). The influence of resource subsidies on cave invertebrates: results from an ecosystem-level manipulation experiment. *Ecology*, **92**, 765–76.

Schneider, K., Kay, A.D., and Fagan, W.T. (2010). Adaptation to a limiting environment: the phosphorus content of terrestrial cave arthropods. *Ecological Research*, **25**, 565–77.

Schwartz, B.F., Schwinning, S., Gerard, B., Kukowski, B., Stinson, C.L., and Dammeyer, C.L. (2013). Using hydrogeochemical and ecohydrologic responses to understand epikarst processes in semi-arid systems, Edwards Plateau, Texas, USA. *Acta Carsologica*, **42**, 315–25.

Siepel, H. and van de Bund, C.F. (1988). The influence of management practices on the microarthropod community of grassland. *Pedobiologia*, **31**, 339–54.

Simon, K.S. (2013). Organic matter flux in the epikarst of the Dorvan karst, France. *Acta Carsologica*, **42**, 237–44.

Simon, K.S. and Benfield, E.F. (2001). Leaf and wood breakdown in cave streams. *Journal of the North American Benthological Society*, **20**, 550–63.

Simon, K.S. and Benfield, E.F. (2002). Ammonium retention and whole-stream metabolism in cave streams. *Hydrobiologia*, **482**, 31–9.

Simon, K.S., Benfield E.F., and Macko, S.A. (2003). Food web structure and the role of epilithic films in cave streams. *Ecology*, **84**, 2395–406.

Simon, K.S., Pipan, T., and Culver, D.C. (2007). A conceptual model of the flow and distribution of organic carbon in caves. *Journal of Cave and Karst Studies*, **69**, 279–84.

Simon, K.S., Pipan, T., Ohno, T., and Culver, D.C. (2010). Spatial and temporal patterns in abundance and character of dissolved organic matter in two karst aquifers. *Fundamental and Applied Limnology*, **177**, 81–92.

Sket, B. (1977). Gegenseitige Beeinflussung der Wasserpollution und des Hőhlenmilieus. *Proceedings of the 6th International Congress of Speleology, Olomouc, ČSSR*, **4**, 253–62.

Sket, B. (1981). *Niphargobates orophobata* n.g., n.sp. (Amphipoda, Gammaridae s.l.) from cave waters in Slovenia (NW Yugoslavia). *Biološki Vestnik*, **29**, 105–18.

Sket, B. (1986). Ecology of the mixohaline hypogean fauna along the Yugoslav coasts. *Stygologia*, **2**, 317–38.

Sket, B. (1999). The nature of biodiversity in subterranean waters and how it is endangered. *Biodiversity and Conservation*, **8**, 1319–38.

Sket, B. (2003). Životinjski svijet Vjetrenice. In I Lučić, ed. *Vjetrenica: Pogled u dušu Zemlje*, pp. 147–248. Zagreb, Croatia.

Sket, B. (2004). Subterranean habitats. In J Gunn, ed. *Encyclopedia of caves and karst science*, pp. 709–13. Fitzroy-Dearborn, New York.

Sket, B. (2006). An essay about *the essai*.Un hommage a Emil Racoviţă. In OT Moldovan, ed. *Emil George Racovitza. Essay on biospeological problems—French, English,*

Romanian version, pp. 119–25. Casa Cărţi de Ştiinţă, Cluj-Napoca, Romania.

Sket, B. (2008). Can we agree on an ecological classification of subterranean animals? *Journal of Natural History*, **42**, 1549–63.

Sket, B., Trontelj, P., and Žagar, C. (2004). Speleobiological characterization of the epikarst and its hydrological neighborhood: its role in dispersion of biota, its ecology and vulnerability. In WK Jones, DC Culver, and JS Herman, eds. *Epikarst. Proceedings of the symposium held October 1 through 4, 2003, Shepherdstown, West Virginia*, pp. 104–13. Karst Waters Institute, Charles Town, West Virginia.

Sket, B. and Velkovrh, F. (1981). Phreatische Fauna in Ljubljansko polje (Ljubljana—Eben, Jugoslavien)–ihre ökologische Verteilung und zoogeographische Reziehungen. *International Journal of Speleology*, **11**, 105–21.

Smart, P.L. and Friederich, H. (1987). Water movement and storage in the unsaturated zone of a maturely karstified carbonate aquifer, Mendip Hills, England. In *Proceedings of Conference on environmental problems in karst terranes and their solutions*, pp. 59–87. National Water Well Association, Dublin, Ohio.

Smith, I.R. (1975). *Turbulence in lakes and rivers*. Special Publication 29, Fresh Water Biological Association, Far Sawrey, Ambleside, UK.

Soil Science Glossary Terms Committee (2008). *Glossary of soil science terms 2008*. Soil Science Society of America, Madison, Wisconsin.

Springer, A.E. and Stevens, L.E. (2009). Spheres of discharge of springs. *Hydrogeology Journal*, **17**, 83–93.

Stanford, J.A. and Gaufin, A.R. (1974). Hyporheic communities of two Montana rivers. *Science*, **185**, 700–2.

Stanford, J.A., Ward, J.V., and Ellis, B.K. (1994). Ecology of alluvial aquiers of the Flathead River, Montana. In J Gibert, DL Danielopol, and JA Stanford, eds. *Groundwater ecology*, pp. 367–90. Academic Press, San Diego, California.

Stocker, Z.S. and Williams, D.D. (1972). A freezing core method of describing the vertical distribution of sediments in a streambed. *Limnology and Oceanography*, **17**, 136–8.

Stone, F.D., Howarth, F.G., Hoch, H., and Asche, M. (2012). Root communities in lava tubes. In WB White and DC Culver, eds. *Encyclopedia of caves, second edition*, pp. 659–64. Elsevier/Academic Press, Amsterdam, The Netherlands.

Strayer, D.L., May, S.E., Nielsen, P., Wolhein, W., and Ham, S. (1997). Oxygen, organic matter, and sediment granulometry controls on hyporheic animal communities. *Archiv für Hydrobiologie*, **140**, 131–44.

Strenzke, K. 1954. *Nematalycus nematoides* n. gen. n. sp. (Acarina, Trombidiformes) aus dem Grundwasser der Algerischen Küste. *Vie et Milieu*, **4**, 638–47.

Strobl, C., Malley, J., and Tutz, G. (2009). *An Introduction to Recursive Partitioning* Technical Report Number 55, Department of Statistics, University of Munich. Available at http://www.stat.uni-muenchen.de.

Šušteršič, F. (1999). Vertical zonation of the speleogenetic space. *Acta Carsologica*, **28**, 187–201.

Taiti, S. and Humphreys, W.F. (2001). New aquatic Oniscidea (Crustacea, Isopoda) from groundwater calcretes of Western Australia. *Records of the Western Australian Museum Supplement*, **64**, 133–51.

Thibaud, J.M. and Coineau, Y. (1998). Nouvelles stations our le genre *Gordialycus* (Acarien: Nematalycidae). *Biogeographica*, **74**, 91–4.

Tooth, A.F. and Fairchild, I.J. (2003). Soil and karst aquifer hydrologic controls on the geochemical evolution of speleothem-forming drip waters, Crag Cave, southwest Ireland. *Journal of Hydrology*, **273**, 51–68.

Trontelj, P., Blejec, A., and Fišer, C. (2012). Ecomorphological convergence in cave communities. *Evolution*, **66**, 3852–65.

Trontelj, P., Douady, C.J., Fišer, C., Gibert, J., Gorički, Š., Lefébure, C., Sket, B., and Zakšek, V. (2009). A molecular test for cryptic diversity in groundwater: how large are the ranges of macro-stygobionts? *Freshwater Biology*, **54**, 727–44.

Ueno, S.I. (1977). The biospeleological importance of non-calcareous caves. In TD Ford, ed. *Proceedings of the 7th international speleological congress, Sheffield, England*, pp. 407–8. British Cave Research Association, Somerset, England.

United States Fish and Wildlilfe Service (2010). American pika. http://www.fws.gov/mountain-prairie/species/mammals/americanpika/

Vandel, A. (1965). *Biospeleology: the biology of cavernicolous animals*. (BF Freeman, trans.) Pergamon Press, New York.

Vergnon, R., Leijs, R., van Nes, E.G., and Scheffer, M. (2013). Repeated parallel evolution reveals limiting similarity in subterranean diving beetles. *American Naturalist*, **182**, 67–75.

Vervier, P., Gibert, J., Marmonier, P., and Dole-Olivier, M.-J. (1992). A perspective on the permeability of the surface freshwater-groundwater ecotone. *Journal of the North American Benthological Society*, **11**, 93–102.

Vesper, D.J. (2012). Contamination of cave waters by heavy metals. In WB White and DC Culver, eds. *Encyclopedia of caves, second edition*, pp. 160–6. Elsevier/Academic Press, Amsterdam, The Netherlands.

Vignoli, V. and Prendini, L. (2009). Systematic revision of the troglomorphic North American scorpion family Typhlochactidae (Scorpiones: Charctoidea). *Bulletin of the American Museum of Natural History*, **326**, 1–94.

Villacorta, C., Jaume, D., Oromí, P., and Juan, C. (2008). Phylogeography and evolution of the cave-dwelling

Palmorchestia hypogea (Amphipoda: Crustacea) at La Palma (Canary Islands). *BMC Biology*, **6**, 7 doi 10.1186/1741–7007–6–7.

Wall, D.E. and Virginia, R.A. (1999). Controls on soil biodiversity: insights from extreme environments. *Applied Soil Ecology*, **13**, 137–50.

Wang, D. and Holsinger, J.R. (2001). Systematics of the subterranean amphipod genus, *Stygobromus* (Crangonyctidae) in western North America, with emphasis on the *hubbsi* group. *Amphipacifica*, **3**, 39–147.

Ward, J.V. (1977). First records of subterranean amphipods from Colorado with descriptions of three new species of *Stygobromus* (Crangonyctidae). *Transactions of the American Microscopical Society*, **96**, 452–66.

Ward, J.V. and Palmer, M.A. (1994). Distribution patterns of interstitial freshwater meiofauna over a range of spatial scales, with emphasis on alluvial river-aquifer systems. *Hydrobiologia*, **287**, 147–56.

Ward, J.V., Stanford, J.A., and Voelz, N.J. (1994). Spatial distribution patterns of Crustacea in the floodplain aquifer of an alluvial river. *Hydrobiologia*, **287**, 11–7.

Watts, C.H.S. and Humphreys, W.F. (2004). Thirteen new Dytiscidae (Coleoptera) of the genera *Boongurrus* Larson, *Tjirtudessus* Watts and Humphreys and *Nirripirti* Watts and Humphreys, from underground waters in Australia. *Transactions of the Royal Society of South Australia*, **128**, 99–129.

Watts, C.H.S. and Humphreys, W.F. (2006). Twenty-six new Dytiscidae (Coleoptera) of the genera *Limbodessus* Guignot and *Nirripirti* Watts and Humphreys, from underground waters in Australia. *Transactions of the Royal Society of South Australia*, **130**, 123–85.

Watts, C.H.S. and Humphreys, W.F. (2009). Fourteen new Dytiscidae (Coleoptera) of the genera *Limbodessus* Guignot, *Paroster* Sharp, and *Exocelina* Broun, from underground waters in Australia. *Transactions of the Royal Society of South Australia*, **133**, 62–107.

Weishaar, J.L., Aiken, G.R., Bergamaschi, B.A., Fram, M.S., Fujii, R., and Mopper, K. (2003). Evaluation of specific ultraviolet absorbance as an indicator of the chemical composition and reactivity of dissolved organic carbon. *Environmental Science and Technology*, **37**, 4702–8.

Wessel, A., Erbe, P., and Hoch, H. (2007). Pattern and process: Evolution of troglomorphy in the planthoppers of Australia and Hawaii. *Acta Carsologica*, **36**, 199–206.

Wessel, A., Hoch, H., Asche, M., von Rintelen, T., Stelbrink, B., Heck, U., Stone, F.D., and Howarth, F.G. (2013). Founder effects initiated rapid species radiation in Hawaiian cave planthoppers. *Proceedings of the National Academy of Science (USA)*, **110**, 9391–6.

White, W.B. (2004). Contaminant storage and transport in the epikarst. In WK Jones, DC Culver, and JS Herman, eds. *Epikarst. Proceedings of the symposium held October 1 through 4, 2003, Shepherdstown, West Virginia*, pp. 85–91. Karst Waters Institute, Charles Town, West Virginia.

Wilkens, H. (1971). Genetic interpretation of regressive evolutionary processes: studies on hybrid eyes of two *Astyanax* cave populations (Characidae, Pisces). *Evolution*, **25**, 530–44.

Wilkens, H. (1988). Evolution and genetics of epigean and cave *Axtyanax mexicanus* (Characidae, Pisces): support for the neutral mutation theory. *Evolutionary Biology*, **23**, 271–367.

Wilkens, H. (2010). Genes, modules and the evolution of cave fish. *Heredity*, **105**, 413–22.

Williams, P.W. (1983). The role of the subcutaneous zone in karst hydrology. *Journal of Hydrology*, **61**, 45–67.

Williams, P.W. (2008). The role of the epikarst in karst and cave hydrogeology: a review. *International Journal of Speleology*, **37**, 1–10.

Wilson, G.D.F. (2008). Gondwanan groundwater: subterranean connections of Australian phreatoicidean isopoda (Crustacea) to India and New Zealand. *Invertebrate Systematics*, **22**, 301–10.

Wright, S. (1978). *Evolution and genetics of populations. Vol. 4. Variability within and among populations.* University of Chicago Press, Chicago, Illinois.

Yamamoto, Y., Byerly, M.S., Jackman, W.R., and Jeffery, W.R. (2009). Pleiotropic function of embryonic sonic hedgehog expression link jaw and taste bud amplification with eye loss during cavefish evolution. *Developmental Biology*, **330**, 200–11.

Yoshizawa, M., Yamamoto, Y., O'Quin, K.E., and Jeffery, W.R. (2012). Evolution of adaptive behaviour and its sensory receptors promote eye regression in blind cavefish. *BMC Biology*, **10**: 108 doi 10.1186/1741–7007–10–108.

Zacharda, M. (1979). The evaluation of morphological characteristics in Rhagiidae. In JG Rodrigues, ed. *Recent Advances in Acarology II*, pp. 509–14. Academic Press, New York.

Zacharda, M., Gude, M., and Růžička, V. (2007). Thermal regime of three low elevation scree slopes in central Europe. *Permafrost and Periglacial Processes*, **18**, 301–8.

Zagmajster, M., Culver, D.C., and Sket, B. (2008). Species richness patterns of obligate subterranean beetles in a global biodiversity hotspot—effect of scale and sampling intensity. *Diversity and Distributions*, **14**, 95–105.

Zagmajster, M., Culver, D.C., Christman, M.C., and Sket, B. (2010). Evaluating the sampling bias in pattern of subterranean species richness—combining approaches. *Biodiversity and Conservation*, **19**, 3035–48.

Index